AF411322

Old Germplasm for New Needs:
Managing Crop Genetic Resources

Old Germplasm for New Needs: Managing Crop Genetic Resources

Editors

Gregorio Barba-Espín
Jose Ramon Acosta-Motos

MDPI • Basel • Beijing • Wuhan • Barcelona • Belgrade • Manchester • Tokyo • Cluj • Tianjin

Editors

Gregorio Barba-Espín
Departamento de Mejora
Vegetal
CEBAS-CSIC
Murcia
Spain

Jose Ramon Acosta-Motos
Catedra UCAM-Santander de
Emprendimiento en el
Ámbito Agroalimentario
Universidad Católica de
Murcia (UCAM)
Murcia
Spain

Editorial Office
MDPI
St. Alban-Anlage 66
4052 Basel, Switzerland

This is a reprint of articles from the Special Issue published online in the open access journal *Agronomy* (ISSN 2073-4395) (available at: www.mdpi.com/journal/agronomy/special_issues/old_germplasm).

For citation purposes, cite each article independently as indicated on the article page online and as indicated below:

LastName, A.A.; LastName, B.B.; LastName, C.C. Article Title. *Journal Name* **Year**, *Volume Number*, Page Range.

ISBN 978-3-0365-5202-6 (Hbk)
ISBN 978-3-0365-5201-9 (PDF)

Contents

About the Editors

Gregorio Barba-Espín

Crop improvements include several aspects, such as increased stress tolerance, nutritional value, and yield. My research deals with these aspects by providing agronomical and biotechnological solutions, where the importance of the antioxidant metabolism is a key element. In this regard, my main research lines can be synthesized as follows: (1) the role of antioxidant metabolism during germination, plant growth, and stress response; (2) agro-economic improvements and the breeding of black carrots: source of colorants and antioxidants; and (3) in vitro characterization of halophytes for intercropping cultivation and valorization.

My international activity accounts for 6 years in plant departments in Denmark [Technical University of Denmark and University of Copenhagen (KU)], as a Postdoctoral researcher and Associate professor, and in England (University of Leeds) and France (Bayer CropScience, Lyon) as a visiting PhD student.

I have achieved a substantial scientific output of 49 publications (37 SCI articles, 5 book chapters, and 7 non-SCI papers). Out of my 37 SCI articles, 28 (77%) are Q1 and 13 (35%) are D1. I am listed as the first (12 papers) or last author (7 papers) in 19 SCI articles (51% of total), of which 15 (79%) are Q1 and 8 (42%) are D1. I have 1215 citations (Scopus, h-index 19). I am the main inventor of 2 published patent applications, and co-inventor of a published utility model. I have participated in 12 research projects. I speak English, French (B2 Diploma), and Danish (B1 Diploma).

Jose Ramon Acosta-Motos

I obtained my PhD degree on 7 October 2014, working at CEBAS-CSIC. To date, I have completed three Masters Degrees: (1) in Advanced Techniques in Agri-Food Development (TAIDAA) at the Polytechnic University of Cartagena (UPCT-ETSIA); (2) in Current Techniques in Applied Statistics through the University of Distance Education (UNED); and (3) in Bioinformatics and Biostatistics with the Universitat Oberta de Catalunya (UOC) and the Universitat de Barcelona (UB). In relation to my international experience, I enjoyed a one-year postdoctoral stay at the Lancaster Environmental Center (LEC-LU).

I am currently involved in research tasks at CEBAS-CSIC in the framework of the Halofarms project as part of a PRIMA Program. I am also working at the Catholic University of Murcia (UCAM) in a UCAM-Santander Chair Entrepreneurship in the agri-food field, where I am co-directing a PhD entitled: "Extending the shelf life of vegetables and fruits during storage and transport using systems for ethylene removal". Additionally, I am a lecturer in Biotechnology at UCAM, teaching Plant Physiology, and teach the Master's degree in Nutrition and Food Safety teaching statistics. In addition, I collaborate in the Institute of Bioengineering of the Miguel Hernández University (IB-UMH), co-directing a sedcond PhD entitled: "Contribution of the root system to the increase of productivity in elite lines of lettuce during its cultivation in hydroponics".

Regarding my scientific production, I have published 29 SCI papers (from 2009-2022), H-index 10 (WOS, December 9, 2021), with a total number of citations of 974 (data from Scopus). I have provided 28 Congress contributions. I have participated in 11 projects. I serve on th Editorial Board for the {Seeds} journal. I have served as the Guest Editor for 2 Special Issues in *Agronomy*: Old Germplasm for New Needs: Managing Crop Genetic Resources; and Advances in Plant Physiology of Abiotic Stresses.

Preface to "Old Germplasm for New Needs: Managing Crop Genetic Resources"

Dear Colleagues,

The impacts of climate change on crop production are already a reality in Europe and across the rest of the world. In order to mitigate these impacts, access to unexploited genetic crop diversity for the production of new varieties that can thrive in more extreme environmental conditions is of prime importance. Herein, genetic diversity should provide the raw materials for breeding and plant improvements. Despite the vast pool of resources that exist, much of the germplasm richness found in gene banks is poorly documented. To overcome the barriers between germplasm conservation and use, a complete evaluation is necessary to determine the useful diversity they contain.

This Special Issue focuses on "Old Germplasm for New Needs: Managing Crop Genetic Resources". We gathered novel research, reviews, and opinion pieces covering all related topics, including germplasm evaluation, crop genetics and improvements, novel crops, phenotyping, physiological responses of inbred lines, management solutions, modeling, case studies from the field, and policy positions.

Gregorio Barba-Espín and Jose Ramon Acosta-Motos
Editors

Editorial

Crop Genetic Resources: An Overview

Gregorio Barba-Espin [1,*] and José Ramón Acosta-Motos [1,2,*]

1   Group of Fruit Biotechnology, Department of Fruit Breeding, CEBAS-CSIC, 30100 Murcia, Spain
2   Campus de Los Jerónimos, Universidad Católica San Antonio de Murcia, 30107 Guadalupe, Spain
*   Correspondence: gbespin@cebas.csic.es (G.B.-E.); jracosta@ucam.edu (J.R.A.-M.)

Citation: Barba-Espin, G.; Acosta-Motos, J.R. Crop Genetic Resources: An Overview. *Agronomy* **2022**, *12*, 340. https://doi.org/10.3390/agronomy12020340

Received: 20 December 2021
Accepted: 27 January 2022
Published: 29 January 2022

Publisher's Note: MDPI stays neutral with regard to jurisdictional claims in published maps and institutional affiliations.

The impacts of climate change on crop production are already a reality worldwide. Extreme weather events such as droughts, floods and heat waves are increasingly frequent, and this is affecting agriculture due to its high vulnerability. In order to mitigate these impacts, the access to unexploited genetic crop diversity for the production of new varieties which can thrive in more extreme environmental conditions is of prime importance. A multitude of beneficial agronomic traits have been lost over the course of domestication, and, in this sense, crops' wild relatives and landraces contain many genes of potential value for plant breeding. Among these, there are many traits that are relevant for climate change adaptation. Despite the vast pool of resources that exists, much of the crop germplasm richness found in gene banks is underutilized. To overcome the barriers between germplasm conservation and use, evaluation is necessary to discover the useful diversity they contain.

The number of accessions deposited in germplasm banks is continuously growing. Germplasm evaluation refers to the observation, measurement and reporting of heritable plant traits from a collection. The Food and Agriculture Organization of the United Nations stressed, in a recent report, that one of the major constrains affecting the conservation, use, monitoring and reporting of information on plant genetic resources for food and agriculture (PGRFA) concerns data access and communication among researchers, breeders, farmers and governments. In this sense, considering the broad range found in the germplasm collections of cultivated plants, which varies from wild and weedy types to high-yielding varieties, it is of major importance to ensure an accessible standardized format for data compilation and management. Other shared problems rely on developing sampling strategies for representative specimens in natural habitats or the design of technologies for long-term maintenance.

To date, germplasm evaluation has been centered mostly on morphological descriptors, agronomical traits and molecular marker technology. In spite of their reliability, low cost and accessibility, morphological descriptors present some constraints, such as the influence of environment on the genotype, making evaluation and information exchange more complex, and limited polymorphism, which demands the evaluation of more descriptors. On the other hand, evaluation based on agronomic characters is especially used in economically relevant crops, although a complete evaluation in this respect is costly, time-consuming and labor-intensive; as a consequence, nowadays it covers a small fraction of the accessions of interest. However, the final aim of germplasm evaluation and plant breeding is to obtain desirable traits in a highly efficient manner. Thus, traditional evaluation techniques are not sufficient to fulfil demands on food security and sustainability.

Crop performance is determined by complex traits resulting from genetics and epigenetics interactions. Understanding the relationship between genotype and phenotype is important for the sustainable evaluation and conservation of crop germplasm richness. Thus, traditional evaluation techniques are nowadays giving way to germplasm characterization based on molecular methods. In this sense, molecular markers, which reveal DNA sequence polymorphism, constituted a turning point in germplasm characterization. However, approaches based on molecular markers ultimately have the limitation of considering genes as independent functional entities; when traits are the result of a multigenic regulation and have a close interaction with environmental conditions, broader approaches are needed. Nowadays, the rapid expansion of the "-omics" techniques are bringing high-throughput approaches to the frontline of plant breeding, accelerating crop improvement by elucidating the interaction between genotype and phenotype. In this sense, epigenomics, genomics, transcriptomics, proteomics, metabolomics, phenomics and ionomics, together with bioinformatic tools, have made it possible for breeders to design more resilient and/or more productive varieties towards biotic and abiotic stresses. More recently, multi-omic approaches have been proposed for plant evaluation and breeding, integrating datasets from the different -omics together with complex mathematical models.

Through this Special Issue, there are examples of such complexity and diversity of germplasm evaluation on different cash crop species such as carrot, barley, soybean, almond, melon, lettuce, tomato, wheat, cacao and rice [1–11]. Acosta-Motos et al. [1] conducted agromorphological characterization on Eastern carrot landraces; these may be a source of material for carrot breeding programs, particularly black carrot accessions due to their high anthocyanin and flavonoid contents and, concomitantly, antioxidant capacity. In this sense, commercial cultivars are, overall, more productive than landraces; however, landraces are becoming significant sources of genetic variability and a crucial element for agrobiodiversity due to their high variation and adaptability to local environments. Dziurdziak et al. [2] conducted a genome-wide DArTseq analysis of the diversity of 116 spring barley landraces from different countries, preserved in the collection of the Polish gene bank; this revealed remarkable variation among landraces related to the country of origin and the grain type, and supporting the breeding of hulless accessions for spring barley's high quality as a food with beneficial health values.

The interaction between genotype and environment was studied by Abdelghany et al. [3], where the genotypic stability of 135 soybean accessions was evaluated in three Chinese locations. This was achieved by means of mathematical models with a weighted average of absolute scores biplot and a multi-trait stability index. The latter tool proved to be very useful in finding genotypes suitable for both seed performance and stability, by analyzing seven seed composition traits simultaneously. Soybean germplasm diversity was also explored by Jo et al. [4], who analyzed 470 soybean accessions of black seed coat and green cotyledons—rich in chlorophylls, anthocyanin and compounds with anticarcinogenic properties—by 6K single nucleotide polymorphic loci, to determine genetic architecture. As a result, 36 accessions were found to contain 99.5% of the diversity from the total collection analyzed, showing potential for their use in breeding programs.

Twenty-four traditional almond cultivars clearly in decline or close to extinction were analyzed by Pérez- Sánchez et al. [5], from agromorphological and nutritional points of view, using a total of 40 descriptors gathered in international guidelines such as IPGRI and UPOV for flowers, leaves, fruits and vegetative tree habits. As a result, certain cultivars were found to have a high yield and quality of fruit, which constitutes an important step in the conservation of genetic almond resources in the Central-Western Iberian Peninsula.

The application of molecular markers to assess the genetic variability of plant varieties and cultivars is well represented in this Special Issue. Chikh-Rouhou et al. [6] combined 24 phenotypic traits and eight Simple Sequence Repeat (SSR) molecular markers to assess the genetic diversity of a Tunisian melon collection. A considerable phenotypic variability among accessions was measured for several traits of agronomical importance, whereas all of the microsatellites were found to be polymorphic. A precise clustering for landraces and breeding lines was obtained using combined phenotypic–molecular data, which may allow the correct use of these accessions in future breeding programs. Similarly, Caramante et al. [7] also utilized SSR markers to compare 15 traditional landraces with 15 widely used current varieties of tomato, concluding from the data analysis that the landraces conformed to a genetically different population from the commercial varieties, and serving as a milestone for implementing in situ and ex situ conservation programs. Another type of molecular marker, high-density single-nucleotide polymorphism (SNP), has been used by Ganugi et al. [8] to assess the genetic variability of a 265-accession collection of eight tetraploid wheat subspecies. The analysis and data treatment revealed clusters in agreement with the taxonomic classification, shedding light on the wheat's evolutionary history and the phylogenetic relationships among subspecies.

Around 90% of the cacao's production worldwide is located in developing countries, especially in small farming systems. Criollo cacao is highly demanded in Honduras due to is quality attributes. López et al. [9] used 16 SSR molecular markers on 89 samples showing phenotypic traits of Criollo cacao, to assay purity and belonging to that group. As a result, although certain accessions had the genetic traits of Trinitario or other admixtures of cacao types, the genetic purity of Criollo cacao in Honduras was confirmed, providing further evidence of Mesoamerica as a cacao domestication center.

Cultivated lettuce is one of the main leafy vegetables worldwide. Birlanga et al. [10] established, based on a hydroponic system, a root and shoot phenotyping of 12 lettuce genotypes. Tipburn incidence and leaf nutrient content were analyzed, identifying nutrient traits highly correlated with genotype-dependent tipburn, which may lead, in defined nutrient solutions, to select for tipburn-tolerant genotypes that could be adequate for hydroponic cultivation. Chan-in et al. [11] evaluated grain quality traits and allelic variation of the *Badh2* gene—determining fragrance attributes—in 22 Thai rice landraces. Haplotype analysis on the *Badh2* gene revealed a correlation with grain aroma by sensory evaluation. The results indicated that genetic resources could be introduced for fragrant rice breeding programmes to increase the income of highland farmers.

Taken together, the present Special Issue contributes to the efforts on crop germplasm evaluation in order to ensure future food security and commercial profitability. Nevertheless, plant genetic biodiversity is under threat from genetic erosion, a concept introduced by researchers to describe the gradual loss of individual genes and combinations of genes, such as those present in domesticated landraces. The major cause of genetic erosion is the displacement of landraces by contemporary varieties. To overcome these problems, ex situ and in situ conservation approaches must be undertaken. Likewise, the role of indigenous and farming peoples in such efforts should be considered for the preservation of crops' wild relatives and landraces.

Author Contributions: Writing—original draft preparation, G.B.-E. and J.R.A.-M.; writing—review and editing, G.B.-E. and J.R.A.-M. All authors have read and agreed to the published version of the manuscript.

Conflicts of Interest: The authors declare no conflict of interest.

References

1. Acosta-Motos, J.R.; Díaz-Vivancos, P.; Becerra-Gutiérrez, V.; Hernández Cortés, J.A.; Barba-Espín, G. Comparative Characterization of Eastern Carrot Accessions for Some Main Agricultural Traits. *Agronomy* **2021**, *11*, 2460. [CrossRef]
2. Dziurdziak, J.; Gryziak, G.; Groszyk, J.; Podyma, W.; Boczkowska, M. DArTseq Genotypic and Phenotypic Diversity of Barley Landraces Originating from Different Countries. *Agronomy* **2021**, *11*, 2330. [CrossRef]
3. Abdelghany, A.M.; Zhang, S.; Azam, M.; Shaibu, A.S.; Feng, Y.; Qi, J.; Li, J.; Li, Y.; Tian, Y.; Hong, H.; et al. Exploring the Phenotypic Stability of Soybean Seed Compositions Using Multi-Trait Stability Index Approach. *Agronomy* **2021**, *11*, 2200. [CrossRef]
4. Jo, H.; Lee, J.Y.; Cho, H.; Choi, H.J.; Son, C.K.; Bae, J.S.; Bilyeu, K.; Song, J.T.; Lee, J.-D. Genetic Diversity of Soybeans (*Glycine max* (L.) Merr.) with Black Seed Coats and Green Cotyledons in Korean Germplasm. *Agronomy* **2021**, *11*, 581. [CrossRef]
5. Pérez-Sánchez, R.; Morales-Corts, M.R. Agromorphological Characterization and Nutritional Value of Traditional Almond Cultivars Grown in the Central-Western Iberian Peninsula. *Agronomy* **2021**, *11*, 1238. [CrossRef]
6. Chikh-Rouhou, H.; Mezghani, N.; Mnasri, S.; Mezghani, N.; Garcés-Claver, A. Assessing the Genetic Diversity and Population Structure of a Tunisian Melon (*Cucumis melo* L.) Collection Using Phenotypic Traits and SSR Molecular Markers. *Agronomy* **2021**, *11*, 1121. [CrossRef]
7. Caramante, M.; Rouphael, Y.; Corrado, G. The Genetic Diversity and Structure of Tomato Landraces from the Campania Region (Southern Italy) Uncovers a Distinct Population Identity. *Agronomy* **2021**, *11*, 564. [CrossRef]
8. Ganugi, P.; Palchetti, E.; Gori, M.; Calamai, A.; Burridge, A.; Biricolti, S.; Benedettelli, S.; Masoni, A. Molecular Diversity within a Mediterranean and European Panel of Tetraploid Wheat (*T. turgidum* subsp.) Landraces and Modern Germplasm Inferred Using a High-Density SNP Array. *Agronomy* **2021**, *11*, 414. [CrossRef]

9. López, M.; Gori, M.; Bini, L.; Ordoñez, E.; Durán, E.; Gutierrez, O.; Masoni, A.; Giordani, E.; Biricolti, S.; Palchetti, E. Genetic Purity of Cacao Criollo from Honduras Is Revealed by SSR Molecular Markers. *Agronomy* **2021**, *11*, 225. [CrossRef]
10. Birlanga, V.; Acosta-Motos, J.R.; Pérez-Pérez, J.M. Genotype-Dependent Tipburn Severity during Lettuce Hydroponic Culture Is Associated with Altered Nutrient Leaf Content. *Agronomy* **2021**, *11*, 616. [CrossRef]
11. Chan-in, P.; Jamjod, S.; Yimyam, N.; Rerkasem, B.; Pusadee, T. Grain Quality and Allelic Variation of the *Badh2* Gene in Thai Fragrant Rice Landraces. *Agronomy* **2020**, *10*, 779. [CrossRef]

 *agronomy*

Article

Comparative Characterization of Eastern Carrot Accessions for Some Main Agricultural Traits

José R. Acosta-Motos, Pedro Díaz-Vivancos, Verónica Becerra-Gutiérrez, José A. Hernández Cortés and Gregorio Barba-Espín *

Department of Fruit Breeding, CEBAS-CSIC, Campus de Espinardo, 30100 Murcia, Spain; jacosta@cebas.csic.es (J.R.A.-M.); pdv@cebas.csic.es (P.D.-V.); vbecegu@gmail.com (V.B.-G.); jahernan@cebas.csic.es (J.A.H.C.)
* Correspondence: gbespin@cebas.csic.es

Abstract: Background: Unevaluated open-pollinated germplasm represents a promising source of variability to face the problems of worldwide food production under a changing environment. In carrots, this is particularly true for black carrot accessions, which are the most relevant among Eastern carrot germplasm due to their high anthocyanin content and, concomitantly, antioxidant capacity. Methods: In the present work, a comparative characterization was conducted for the first time on 11 Eastern carrot landraces and Night Bird 'F1' as the reference cultivar, grown under glasshouse conditions at temperatures up to 33 °C. Results: Some landraces showed their potential for ulterior evaluation in terms of plant and taproot size, plant compactness, specific leaf area and leaf area ratio, among other traits. The highest anthocyanin and flavonoid contents were found in the reference cultivar, whereas remarkable differences in these variables were observed for the rest of accessions, which in turn may correlate with very distinct coloration patterns. Premature bolting and taproot shape abnormalities were also recorded. Mineral composition analysis showed the nutritional potential of Eastern carrot leaves, which displayed higher concentration than taproot tissue for several macro- and micronutrients. Moreover, several accessions had higher nutrient concentrations than the reference cultivar, which also highlights their profitability. Conclusions: This work contributes to the knowledge on Eastern black carrot germplasm by characterizing some of its main agricultural traits, and opens up the prospect for complementary evaluation on high-yield accessions.

Keywords: anthocyanins; black carrot; bolting; flavonoids; glasshouse cultivation; landraces; leaf; nutrient analysis; taproot

Citation: Acosta-Motos, J.R.; Díaz-Vivancos, P.; Becerra-Gutiérrez, V.; Hernández Cortés, J.A.; Barba-Espín, G. Comparative Characterization of Eastern Carrot Accessions for Some Main Agricultural Traits. *Agronomy* **2021**, *11*, 2460. https://doi.org/10.3390/agronomy11122460

Academic Editor: Leo Sabatino

Received: 20 October 2021
Accepted: 29 November 2021
Published: 2 December 2021

Publisher's Note: MDPI stays neutral with regard to jurisdictional claims in published maps and institutional affiliations.

1. Introduction

Domesticated carrot (*Daucus carota* L. ssp. *sativus*) can be separated into two genetically distinct groups: the Eastern (Asian) and Western (European and American) carrots. The Eastern carrot (*D. carota* L. ssp. *sativus* var. *atroburens*), whose color ranges from yellow to dark purple (black carrot), accumulates anthocyanins as major pigments, whereas Western carrot varieties appear white to orange due to the accumulation of carotenoids [1,2]. Molecular approaches have located Central Asia as the origin for the carrot, with a rapid domestication process that spread carrots into North Africa, Anatolia, Asia and later into Europe by the 14th century [1,3].

Carrot production, mostly based on orange cultivars, has quadrupled during the last 45 years, reaching over 40 million tonnes worldwide, which makes carrot one of the 10 most economically important vegetable crops and the main source of pro-vitamin A worldwide [4]. The main increase in production has been recorded in Asia, which implies that more carrot production is now cultivated in drier and warmer climates than in the past [4]. Therefore, crop improvement should focus efforts in developing cultivars with improved abiotic stress resistance. Commercial cultivars are, overall, more productive than landraces. However, landraces are becoming significant sources of genetic variability in the seeking for genes for tolerance to abiotic and biotic stress factors [5].

The extraordinary antioxidant activity of black carrot, four times higher on average than that of orange carrot [5], results from its very high anthocyanin concentration in the taproot. Easter carrot germplasm displays a wide genetic diversity in terms of anthocyanin content and distribution along different taproot tissues. In this sense, several authors have analyzed anthocyanin pigments in over 30 carrot accessions (accs.) including lines from genbanks, and open-pollinated and hybrid commercial cultivars, reporting a concentration range of 0.5–250 mg/100 g FW [6–9]. In this sense, total monomeric anthocyanin concentration measured by the pH differential method has been frequently utilized, since it provides a robust spectrophotometric method applicable to different anthocyanins [6,8,10]. The interest of the food industry in natural colorants replacing synthetic dyes has increased enormously over the last years, due to both rigorous legal restrictions and consumer concerns [11,12]. Black carrot anthocyanins are an excellent source of natural colorants (labelled E163 in Europe) due to their physiochemical properties (high pH, light, and heat stability), but also for their potential health benefits such as strong dietary antioxidants [6,13]. In addition to anthocyanins, black carrot taproot is a rich source of non-anthocyanin phenolic compounds, such as flavonoids [14], which in turn correlates with the strong antioxidant activity of the taproot extract. Moreover, the onset and accumulation kinetic of anthocyanins and non-anthocyanin phenolic compounds seems to resemble in black carrot taproot, both type of compounds reaching their maximum levels at plant maturity [8]. As non-enzymatic plant antioxidants, phenolic compounds function as scavengers of reactive oxygen species (ROS), participating in the response to environmental stress conditions [15]. In this sense, the accumulation of phenolic compounds increases under a variety of both abiotic and biotic stresses such as heat stress, pathogen attack and UV radiation, among many others [16,17].

Carrot is an outcrossing insect-pollinated crop typically bred for open-pollinated cultivar production until cytoplasmic male sterility was discovered in the 1940s. From that moment on, cultivar development shifted to hybrids, which today represents the majority of large-scale production [18]. Nowadays, Western carrots appear better adapted for commercial production and processing. Eastern carrots often tend to flower early, but may be better adapted to warmer (above 30 °C) and drier climates. In order to improve carrot cultivation traits in the context of a changing environment, the gene pool from open-pollinated Eastern carrot varieties should be incorporated into breeding programs. In the present work, eleven Eastern carrot accs. Originated from India and Middle Eastern countries, and a black carrot commercial cultivar were cultivated under a sub-optimal temperature range (up to 33 °C) and characterized in terms of germination indices, plant performance parameters, mineral nutrient contents and anthocyanin and flavonoid contents of the taproot, measured spectrophotometrically. The results of this work may contribute to expand knowledge on the genetic variation of cultivated carrot and opens up prospects for further evaluation.

2. Materials and Methods

2.1. Plant Material and Germination Indices In Vitro

Seeds of eleven Eastern carrot accs. were obtained from the USDA National Germplasm System and the Warwick Genetic Resources Unit (Table 1). In addition, seeds of the commercial hybrid 'Night Bird' F1, provided by Plant World Seeds (Devon, U.K.), were used as a control.

Table 1. Accession number, source and geographical origin of the carrot accessions used.

Source	Number	Accession	Origin	Life Form	Improvement Status
USDA	2	179689	India	Biennial	Landrace
	3	211024	Afghanistan	n.d.	Landrace
	4	269486	Pakistan	n.d.	Landrace
	5	279776	Egypt	n.d.	Landrace
	6	279777	Egypt	Annual/biennial	Landrace
	7	288242	Egypt	n.d.	Landrace
Warwick	8	006753	India	n.d.	Advanced cultivar
	9	010156	India	Annual	Landrace
	10	10225	India	Annual/biennial	Landrace
	11	10217	Turkey	n.d.	Landrace
	12	13879	n.d.	n.d.	Landrace

Accession n. 1 (not displayed on the list) corresponds to the commercial hybrid 'Night Bird' F1, cultivated as a reference. n.d.: not described.

Twenty-five seeds of each acc. and the commercial cultivar were disinfected in 10% bleach for 1 min, washed with abundant distilled water and placed in 15 cm diameter plastic Petri dishes containing four layers of filter paper moistened with 6 mL water. Petri dishes were then placed in an incubator chamber at 25 °C in the dark for 14 days. The germination percentage, seedling vigour index (mean seedling length × germination percentage/100) and speed of emergence (number of germinated seeds at the starting day of germination/number of germinated seeds at the final day of measurements × 100) were calculated at day 14, following the International Seed Testing Association guidelines [19].

2.2. Glasshouse Cultivation

The experiments were conducted from March to June 2019 over 12 weeks at the glasshouse facilities of the University of Murcia (Espinardo, Region of Murcia, Spain). Two independent repetitions of the experiment were performed with a 1-week interval. The temperature was programmed to daily oscillate between 17 (night) and 33 °C (day) with a mean daily temperature of 24.5, and relative humidity was maintained at 60%. Seeds were sown in 27 cm-long cylindrical pots (18 cm in diameter) with 7 L of a mixture of perlite (particle size 1–5 mm) and garden soil (1/1.5, *v/v*), allowing a proper carrot taproot growth and providing an adequate balance of air and moisture content [20]. The substrate mix was watered until reaching its maximum water holding capacity. Then, four sowing spots per pot were distributed, and three seeds were introduced in each one at 2 cm depths.

The experimental design included 12 pots per acc. and experiment repetition, arranged in three rows of four pots, each row being randomly distributed within the glasshouse. Pots were put on 20 cm diameter plastic trays (Figure 1a). Ten days after sowing, two seedlings per spot were removed, leaving one vigorous plant per place (Figure 1b). An automatic drip irrigation system using one 4 L h^{-1} dripper per pot was programmed to provide water or 1/2 Hoagland solution. Daily water supplied per pot throughout the experiment varied from 100 mL (day 2 to day 15) to 350 mL (day 65 to harvest day). Concerning fertilization, nutrient solution was applied on a weekly basis, from 350 mL (day 14 to day 28) to 500 mL (day 70 to harvest day). At the day of fertilization and the subsequent day, water irrigation was interrupted.

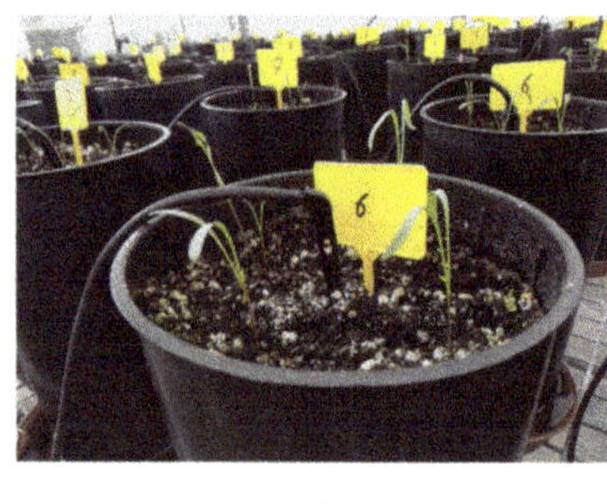

(a)   (b)

Figure 1. Experimental setup of carrot accessions at the glasshouse facilities of the University of Murcia. (**a**): Overview of the setup at day 0. (**b**): Ten-day old seedlings after thinning of less vigorous specimens.

2.3. Harvest, Sample Collection, Growth Measurements and Processing

Twelve weeks after the initiation of the experiment, plants were manually removed from the substrate and taproots washed in abundant tap water followed by immersion in distilled water and air drying. Bolting plants were counted and discarded for further analysis.

The fresh weight (FW) and length of the taproot and top of each plant were registered. Top length corresponded to the length of the longest leaf. Leaf number per plant was recorded. Total leaf area per plant was determined using a LI-3100C area meter (LI-COR Biosciences, Lincoln, NE, USA). For further experiments, biological replicates consisted of: (1) leaf samples, composed of the pool of leaves and stems of the four plants from a pot; and (2) taproot samples, composed of the corresponding four taproots. Then, each combined sample (biological replicate) was processed in liquid nitrogen using a Waring® two-speed commercial blender (VWR-Bie & Berntsen, Herlev, Denmark) and the resulting powder stored at −80 °C.

2.4. Determination of Dry Matter and Nutrient Analysis

An aliquot of 5 g of the ground powder generated above for each taproot and leaf sample was dried in an oven at 60 °C for 2 days. After the samples were dried to a constant weight at 100 °C for 24 h, dry matter (DM) was calculated based on the mass difference between the fresh and dry samples. Another aliquot of 1 g of the ground powder was utilized to measure the different macronutrients and micronutrients, by Inductively Coupled Plasma–Optical Emission Spectrometry (ICP–OES) using a ICAP 6000SERIES spectrometer (Ionomic Services of CEBAS-CSIC; Thermo Scientific, Madrid, Spain) according to standardized protocols.

2.5. Determination of Total Monomeric Anthocyanin Content (TMC) and Total Flavonoid Content (TFC)

A common extraction for the determination of TMC and TFC was conducted. In brief, 5 g of the ground powder was homogenized in a 1.88% sulfuric acid solution (1/4, *w/w*, for a final 1.5% sulfuric acid concentration in the homogenate) using a Waring® two-speed commercial blender (VWR-Bie & Berntsen). Then, the homogenate was centrifuged for 20 min at $3900 \times g$ and the resulting supernatant was collected [8].

TMC measurement followed the pH differential method with minor modifications [8,10,21]. In brief, the supernatant was diluted in a 0.2 M KCl–HCl (pH 1) solution (1/1, *v/v*), and the absorption was determined between 350 and 700 nm using a UV/Vis V-630 Bio spectrophotometer (Jasco, Tokyo, Japan). Finally, the TMC was calculated as cyanidin 3-glucoside equivalents.

TFC determination was based on [22] with minor modifications: in brief, a reaction mix of 200 μL supernatant, 800 μL 50% methanol, 60 μL sodium nitrite (0.5 M) and 60 μL aluminum chloride (0.3 M). After 5 min of incubation, 400 μL sodium hydroxide solution (1 M) was added and the content vigorously mixed. Subsequently, the absorbance at 506 nm was determined against a blank in which the supernatant was replaced by 50% methanol, using a UV-visible spectrophotometer (Thermo Scientific Evolution™ 220, Waltham, MA, USA). A calibration curve for rutin was calculated (from 75 to 750 mg L^{-1}), and TFC was calculated as μg of rutin equivalents (RE) per mL.

2.6. Statistical Analyses

Data from the two independent repetitions were grouped and statistical analysis was carried out jointly, given the homogeneity of the variance ratios and the means tested by an F test and a two sample t-test, respectively. Normality and homoscedasticity of variances for all variables studied were checked by a Shapiro–Wilk and Bartlett tests, respectively. Data related to plant length, weight, leaf area and flowering were taken on individual specimens, which were considered as biological replicates (n = 16 to 67 for the different accs.). The statistical analysis for DM, TMC and TFC determinations were conducted with four biological replicates, each of them consisting of the pool of samples from one pot (four plants). The accs. were compared using a one-way analysis of variance (ANOVA) followed by a Tukey HSD post hoc test ($p \leq 0.05$), using the StatGraphics Centurion XV software (StatPoint Technologies, Warrenton, VA, USA).

Heatmaps were elaborated using the pheatmap package in R [23]. Neighbor joining distance matrixes between accessions were automatically processed from mean values to build the dendrograms and the heatmap representation. Graphs were drawn with GraphPad Prism 9.0.0 for Windows (GraphPad Software, San Diego, CA, USA).

3. Results and Discussion

In the present study, 11 Eastern carrot accs. from different origins were cultivated and evaluated under glasshouse conditions. The geographic origin covered an area between 10 and 40 degrees of latitude, from India to Turkey (Table 1). The commercial carrot cultivar 'Night Bird' F1 was used as a reference. Characterization comprised seed germination indices and leaf and taproot phenotypic traits of interest for ulterior breeding programs such as size, taproot shape, bolting tendency, leaf number, leaf area and those related to taproot antioxidant capacity and use as source of colorants (anthocyanin and flavonoid contents). Firstly, germination indices were determined in in vitro conditions (Table 2). Germination percentage varied 61–96%; this is in the range of that observed for diverse carrot germplasm [24], in which, overall, commercial hybrids displayed higher germination percentages than landraces and wild accessions. Moreover, seedling vigor varied 4.40–11.32, whereas speed of emergence varied between 1.52 and 3.45 (Table 2). In this sense, accs. 2, 3 and 5 showed values above those recorded for the reference cultivar for the three indices measured (Table 2).

Table 2. Germination indices of the carrot accessions. Germination percentage (%), seedling vigor index (mean seedling length × germination percentage/100) and speed of emergence (number of germinated seeds at the starting day of germination/number of germinated seeds at the final day of measurements × 100) were calculated at day 14 after planting.

Accession n.	Germination Percentage [%]	Seedling Vigor Index	Speed of Emergence
1	81	8.67	2.8
2	92	11.32	3.45
3	84	10.00	2.94
4	71	4.40	1.8
5	81	9.48	2.31
6	74	6.96	1.52
7	83	8.96	2.52
8	79	6.95	2.2
9	61	5.25	1.05
10	79	6.64	1.6
11	84	6.30	1.89
12	96	7.10	2.16

Twelve-week old carrot plants showed a variable overall appearance in function of top size (Figure 2a). On the other hand, overall, the taproot appearance consisted of the typical long conical shape with a pointy end, the most frequent shape in both Eastern and Western cultivars [25–27], the taproot shoulder of accs. 2 and 8 being clearly wider than that of the other accs. (Figure 2b). Coloration pattern of periderm and cross taproot sections was highly variable among accs. (Figure 2b,c). Purple coloration was observed for the majority of accs. except for n. 2 and 9, ranging from a purple periderm and non-purple xylem and phloem (for acc. 3, 4, 8 and 10) to a predominant coloration of phloem (for acc. 5 and 7) and a solid purple coloration (for acc. 1 and 12). At the genetic level, the pathway for anthocyanin biosynthesis showed a high overlap among carrot varieties. However, different variants of anthocyanin-related genes result in tissue-specific accumulations of anthocyanins [28], which in turn is manifested in very distinct coloration patterns. Moreover, two MYB transcription factors, DcMYB6 and 7 have been proven to be regulators of anthocyanin pigmentation in purple versus non purple carrot roots [29,30], as well as regulators of anthocyanin glycosylation and acylation [30]. In addition, intra-population genetic diversity of Eastern carrot accs. has been found to be higher than that of the Western carrot, whereas higher allelic richness and variability of landraces has also been observed with respect to F1 hybrids [31], which would explain the color variability within specimens of a same acc. found in this study, especially for accs. 6 and 10. In turn, this suggests that the diversity present in carrot landraces could potentially support carrot breeding efforts in terms of coloration.

Deformed taproot shape and premature bolting are important constrains for carrot cultivation, marketability and carrot breeding [18]. In this work, concerning taproot shape abnormalities, excluding the commercial cultivar as it did not display any forking, the proportion of forked taproots ranged from 3 to 18%, the accs. 7, 9 and 10 showing the highest values (Figure 3a). Again, higher genetic diversity within Eastern carrot accs. may explain the variability on taproot forking found in this study [31], which can rely on differences on optimal plant spacing and nutrient requirements among accs. [32]. As a cool-season vegetable, carrot is normally classified as a biennial species, requiring vernalisation for flowering induction. However, cultivars and landraces adapted to warmer climates—as is the case of the accs. of this study—may need less vernalisation time, and therefore may behave as early flowering or annual [18]. In this work, all accs. except n. 1 (reference cultivar), 3 and 12 showed a certain percentage of bolting specimens (Figure 3b). Accs. 4, 5 and 7 were found to bolt severely, having 80, 47 and 68%, respectively, of bolting plants (Figure 3b). Since premature bolting is an undesirable trait, bolting plants were discarded and not considered for further measurements and analysis. This lead to the reduction of available specimens for accs. 4 and 7 to 16 and 24, respectively, whereas for the rest of accs. the number of plants used for further measurements ranged from 40 to 67. Complementarily, a heatmap dendogram was elaborated to visualize the preponderance of both traits among the accessions (Figure 3c,d); herein, based on similarities, two main groups were distinguished: group 1 for low and group 2 for high forking (Figure 3c) and bolting (Figure 3d).

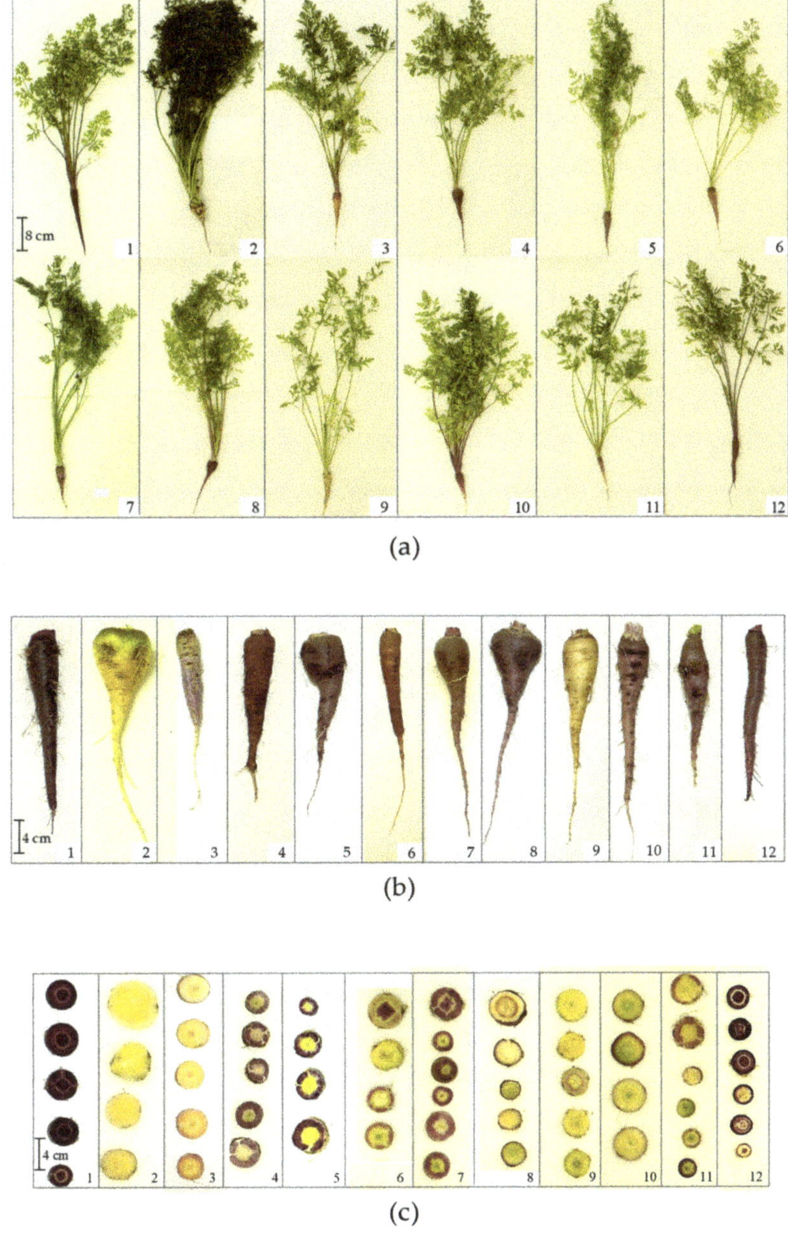

(a)

(b)

(c)

Figure 2. Appearance of whole plant (**a**), taproot (**b**) and transversal taproot sections (**c**) of 90-day-old carrots of twelve accessions. Accession identification (1–12) is show in Table 1.

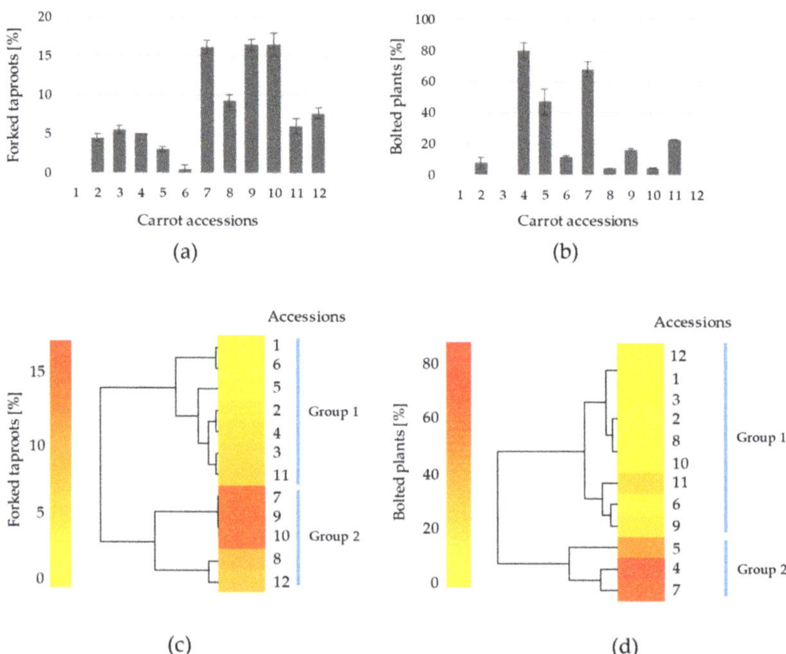

Figure 3. Forked taproots and bolted plants at harvest for the 12 carrot accessions evaluated. (**a**,**b**): Percentages presented as the mean ± standard error of the two experiment repetitions ($n = 2$). (**c**,**d**): Heatmap scoring values. Colored bars indicate the preponderance of both factors in the studied accessions, from low (yellow) to high preponderance (dark orange, red). Accessions were grouped into two groups for both traits. Accession identification (1–12) is shown in Table 1.

Taproot weight is a key factor determining productivity [18]. In the present study, the weight of the reference cultivar (21.9 g) was considerably lower than that of black carrot grown in field experiments [6,8], although similar to that obtained in glasshouse trials under comparable environmental and fertirrigation conditions [9] (Figure 4a). Moreover, temperatures registered during the growing period reached 33 °C and averaged 24.4 °C, whereas ca. 15% of the time, temperature was above 30 °C. These are beyond the suitable temperatures for conventional carrot growth [33,34], which may partially explain the limited taproot growth observed. Compared with the reference cultivar, taproot weight was statistically higher in two accs. (n. 2 and 9) and lower in a single acc. (n. 11), whereas no statistical differences were observed in the rest of accs. (Figure 4a). Remarkably, the weight of acc. 2 was over 160% superior than that of the reference (Figure 4a). Total leaf weight was, to some extent, correlated with taproot weight, although in this case only acc. 2 displayed higher weight values (38%) than the reference cultivar (Figure 4b). Taproot dry matter varied from 9.3% (acc. 6) to 12.6% (reference cultivar) (Figure 4c), which was in the range of the values observed in black carrot [8] and commercial orange carrot cultivars [35]. On the other hand, leaf dry matter levels were higher than those of taproot dry matter, ranging from 10.3% (acc. 6) to 14.5% (acc. 2) (Figure 4d). The root/leaf ratio, measured as the ratio of the dry weight of the root to the one of the top, represents a key plant-adaptive mechanism that reflects biomass allocation [36]. In this study, taproot/leaf ratio varied notably among samples, accs. 2, 3 and 9 showing values higher than that of the reference cultivar. The values registered are in the range of those observed by others authors [37] for Western carrots for the same growth period (12 weeks). The difference among accs. may be tightly linked to the genetic background, although it has also been found that plants accumulate more root biomass under more stressful, low-nutrient and poor climatic conditions [36]. Concerning plant height, acc. 6 displayed the highest mean value (85.6 cm), followed by the reference cultivar (75.3) and accs. 8 (75.8 cm) and 10 (72.5), whereas the rest of accs. displayed values significantly lower than that of the reference cultivar (Figure 4f).

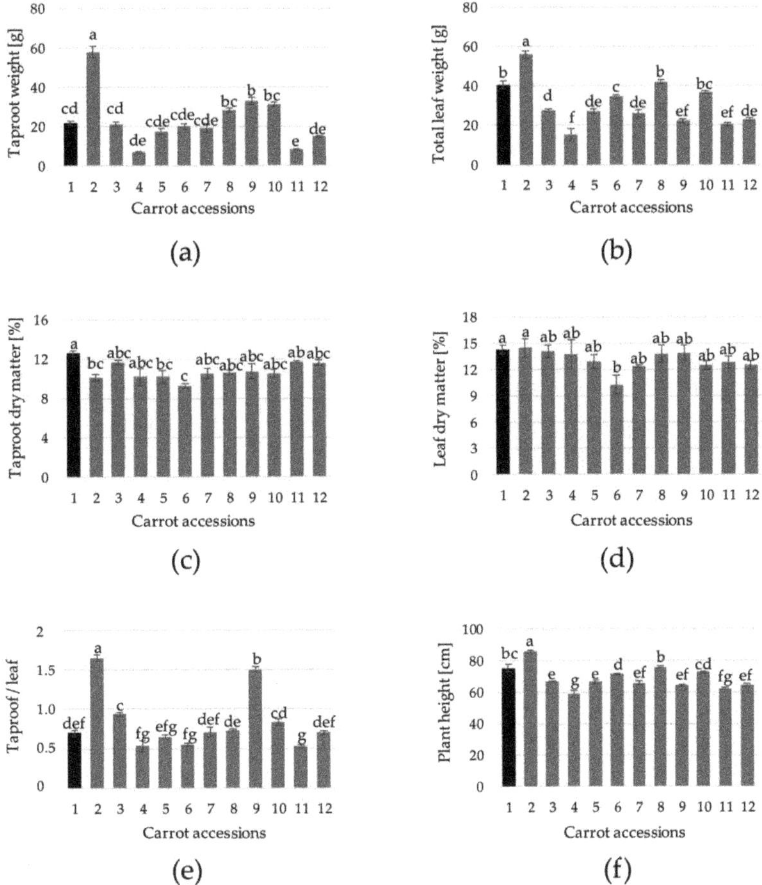

Figure 4. Taproot weight (**a**), total leaf weight (**b**), taproot dry matter (**c**), leaf dry matter (**d**), taproot to leaf ratio (**e**), and plant height (**f**), at harvest for the 12 accessions evaluated. Data are presented as the mean ± standard error, *n* = 20 to 67. Different letters among the columns indicate statistical significance according to Tukey's test ($p \leq 0.05$). Accession identification (1–12) is shown in Table 1.

Leaf number and area, and leaf area-related indices have been reported as some of the traits best indicating potential crop yield [38]. The count of leaf number (6.6 to 11.0 leaves per plant) and total leaf area (154 to 524 cm^2 per plant) provided values in the range of that found in field-grown carrot [39,40] (Figure 5a,b). Specific leaf area (SLA) is defined as the ratio between total leaf area and total leaf dry weight—in other words, the amount of leaf area needed for each unit of biomass produced [41,42]. Carrot SLA varied from 94 cm^2 g^{-1} (acc. 6) to 66 cm^2 g^{-1} (acc. 2) and 68 cm^2 g^{-1} (reference cultivar) (Figure 5c). This may indicate a higher efficiency of accs. 1 and 2 in producing biomass. On the other hand, leaf area ratio (LAR), defined as the ratio of leaf area and total plant weigh showed a great variation (Figure 5d); the accs. with the lowest levels were n. 2, 9, 3 and 1, providing values of 34, 36, 41 and 43 cm^2 g^{-1}, respectively. These results highlight accs. 1 and 2 as the most efficient in producing leaf biomass, while accs. 2 and 9 would produce plant biomass most efficiently (Figure 5d). The variation among accs. of values for plant compactness (Figure 5e), defined as the ratio between the total leaf area and the plant height, resembles the values for total leaf area. Moreover, overall, values for plant compactness, SLA, total leaf area and total leaf weight showed some correlation, which indicates that weight and length differences can be associated with leaf area as the main variable determining productivity [38]. In this sense, a large leaf area may provide better utilization of diminishing growth resources [37].

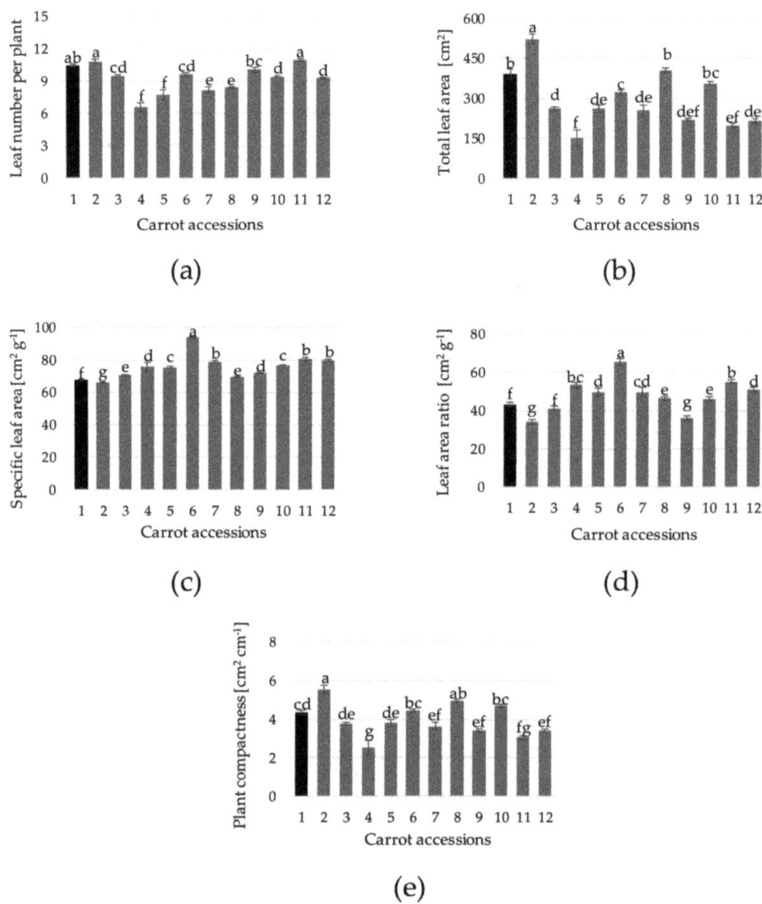

Figure 5. Leaf number per plant (**a**), total leaf area (**b**), specific leaf area (**c**), leaf area ratio (**d**) and plant compactness (**e**) at harvest for the 12 accessions evaluated. Data are presented as the mean ± standard error, n = 16 to 67. Different letters among the columns indicate statistical significance according to Tukey's test ($p \leq 0.05$). Accession identification (1–12) is shown in Table 1.

Anthocyanins and flavonoids are important components of the secondary metabolism of Eastern carrot accs., found to be highly correlated with antioxidant capacity [43]. In this sense, high ROS-scavenging capacity is a common trait of different flavonoids, attributed to the high reactivity of their hydroxyl groups to ROS [44]. Eastern carrot germplasm displays a wide genetic diversity in terms of anthocyanin content and distribution along different taproot tissues [6–9]. Anthocyanin content is directly associated with the color intensity and the extent of taproot tissue covered with purple; in this sense, in the majority of purple genetic accs., anthocyanins are mainly located in the outer root epidermal layer [45]. In this work, from the 12 accs. evaluated, the reference cultivar showed the highest TMC

(2482 µg g^{-1} FW), at values comparable to those reported for commercial varieties [6–9], followed at some distance by acc. 12. (807 µg g^{-1} FW) and then by n. 4, 7, 5, 11, 6, 8, 10 and 3, whereas levels were undetected for accs. 2 and 9 (Figure 6a). This is associated with the coloration observed in the taproot sections (Figure 2). Concerning TFC, a certain correlation with TMC was observed (Figure 6b). However, from the difference between TFC and TMC, it was estimated a higher proportion of non-anthocyanin flavonoids for the accs. displaying lower TMC. Black carrot genotypes display a high proportion of mono-acylated anthocyanins, which confers physio-chemical stability to food color products over their shelf-life [28]. Nevertheless, a signification variation on the ratio of acylated to non-acylated anthocyanins can be found in the carrot germplasm, which is also modulated in function of the environmental conditions [28]. Therefore, further characterization on the accs. of interest in field experiments will be needed.

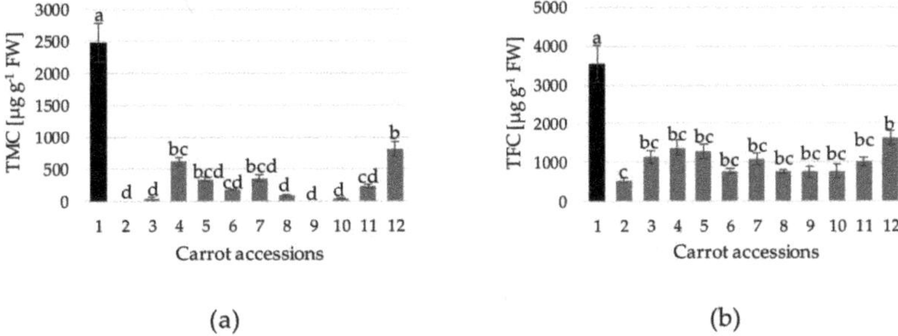

(a) (b)

Figure 6. Total monomeric anthocyanin content [TMC, (**a**)] and total flavonoid content [TFC, (**b**)] at harvest for taproots of the 12 accessions evaluated. Data are presented as the mean ± standard error, *n* = 6. Different letters among the columns indicate statistical significance according to Tukey's test (*p* ≤ 0.05). Accession identification (1–12) is shown in Table 1. FW: fresh weight.

In order to provide a clearer overview of the performance of each acc., a stars and rays graph was elaborated using the main agronomical traits and indices measured so far (Figure 7). In summary, the further the intersection from the polygon center between each axis (trait) and the polygon perimeter, the higher the trait magnitude, which helps visualize the differences among accs. For example, acc. 2 is evidenced as the one fulfilling most of the agronomic requirements, except for TMC and TFC, of which magnitudes are low. On the contrary, the reference cultivar displays high magnitudes for TMC and TFC and intermediate/high values for the rest of traits. On the other hand, accs. such as 4–7 show small magnitudes for most of the traits (Figure 7b), which point the minor potential of these accs. for breeding purposes.

Mineral composition of carrot taproot has been evaluated for Eastern cultivars [46,47], whereas for purple cultivars the literature is scarce [48]. Among the many minerals of the taproot, relatively high amounts of K, Mg, Ca, Na and Fe have been reported, with potassium as the most abundant one [46,47]. On the other hand, the content of Fe, Na and Mg was highly dependent on the carrot variety [47]. In this study, among the macronutrients analyzed, for all the varieties studied K was the most abundant element in the taproot followed by Na and Ca (Table 3). This is in agreement with previous literature [46,47]. Carrot leaves are rich in several minerals such as Na, P, K, Ca, Mg, Mn, Zn and Fe [49]. Herein, the nutrient composition of black carrot leaves was reported for the first time; interestingly, K, Ca and Mg were much abundant in leaf than in taproot tissue (Table 3), highlighting black carrot leaves as a potential nutrient source. Na was the element of which levels varied the most among accs. for both taproot and leaf tissues; this would be in line with previous studies, where Na content in carrot taproot was highly dependent on the fertilization and growing practices [50,51]. Micronutrients are essential to the cell function and as such are extensively involved in primary and secondary metabolisms. Alterations in optimal micronutrients concentrations may, therefore, directly or indirectly impair plant metabolism and increase susceptibility towards environmental stresses [52]. In this study, as for the macronutrients, some variation on the levels of the different micronutrients has been found among accs. (Tables 4 and 5). Remarkably, acc. 8 displayed statistically higher levels than the reference cultivar for most of analyzed micronutrients, whereas acc. 11 showed an opposite trend (Tables 4 and 5). Whether this variation among accs. may imply an improved stress tolerance or not, it is matter requiring further research. Overall, the yield potential of each cultivar may influence nutrient demand, as reported [35], and therefore may be reflected in mineral concentrations of both taproot and leaf tissues. In this sense, broader trials should be conducted to know fertilization needs accurately.

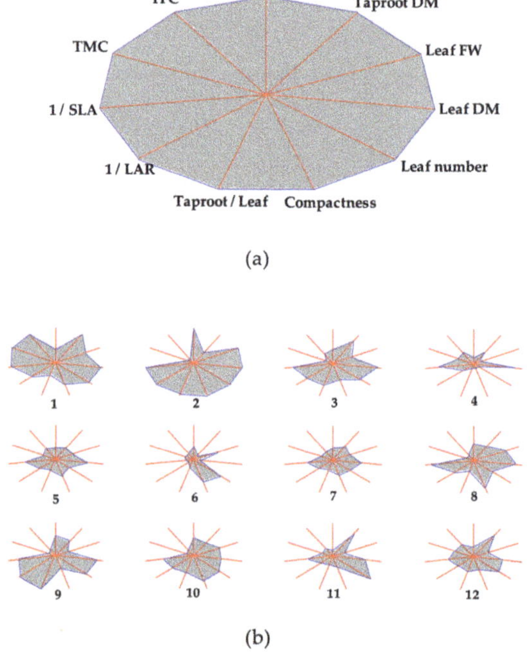

Figure 7. Stars and rays graphs displaying the differences between accessions for the morphological traits analyzed, total monomeric anthocyanin content (TMC) and total flavonoid content (TFC). Each axis represents one variable, and its intersection with a vertex of the polygon indicates the relative magnitude for that variable. (**a**): Reference polygon for variable identification. Leaf area ratio (LAR) and specific leaf area (SLA) are shown as their inverses to visually correlate higher magnitude with a positive trait. (**b**): Individual graphs for each accession. Accession identification (1–12) is shown in Table 1. DM: dry matter (%). FW: fresh weight.

A Pearson's correlation matrix was elaborated in order to associate variations in macronutrient contents, DM, TMC and TFC at the taproot level (Table 6). For each variable, data for all accs. were grouped and treated jointly. As a result, except for the interaction P/S, the rest of macronutrient pairs provided significant and positive correlation coefficients, which highlights the shared underlying physiological mechanisms implied in nutrient uptake and the potential linked mechanisms in maintaining nutrient stoichiometry [53]. Since the increase in accumulation of all these five macronutrients may precede DM accumulation, a correlation between these variables can be of interest as predictors in growth models [54]. Furthermore, our results showed a negative correlation between DM and Ca and Na contents. On the other hand, DM highly correlated with TMC and TFC. This may be due to a time-dependent accumulation of DM [37] and phenolic compounds [8] during carrot taproot growth (Table 6).

Table 3. Macronutrients detected by Inductively Coupled Plasma–Optical Emission Spectrometry at harvest for taproots (T) and leaf (L) samples of the 12 accessions evaluated. Data are presented as the mean, $n = 4$ to 6. Different lowercase letters within the columns indicate statistical significance according to Tukey's test ($p \leq 0.05$). Accession identification (1–12) is shown in Table 1. DW: dry weight.

Acc.	K		Ca		Na		S		P		Mg	
	T	L	T	L	T	L	T	L	T	L	T	L
1	1.51 ab	3.38 abc	0.27 b	2.84 a	0.73 bc	0.84 bcde	0.089 ab	0.85 ab	0.36 b	0.38 abcd	0.15 bc	0.32 abcd
2	1.21 ab	2.39 c	0.36 ab	2.16 abc	1.40 ab	1.37 a	0.086 ab	0.46 bc	0.41 ab	0.30 bcd	0.13 c	0.31 abcd
3	1.79 a	3.45 abc	0.30 ab	2.43 abc	0.66 bc	0.70 de	0.077 b	0.41 c	0.43 ab	0.43 ab	0.16 bc	0.33 abc
4	1.62 a	3.29 abc	0.39 ab	3.03 a	1.25 abc	1.06 abcd	0.077 b	0.86 a	0.42 ab	0.35 abcd	0.24 a	0.41 a
5	1.68 a	3.36 abc	0.45 a	1.91 bc	1.57 a	1.23 abc	0.097 ab	0.52 abc	0.43 ab	0.26 cd	0.17 abc	0.28 bcd
6	1.44 ab	4.28 a	0.34 ab	2.11 abc	1.12 abc	1.32 ab	0.083 b	0.53 abc	0.50 ab	0.38 abc	0.16 bc	0.30 abcd
7	1.19 ab	3.46 abc	0.37 ab	1.80 bc	1.29 abc	1.37 a	0.110 ab	0.53 abc	0.57 a	0.32 abcd	0.20 ab	0.30 abcd
8	0.98 b	2.75 bc	0.37 ab	2.65 abc	1.51 a	1.12 abcd	0.133 a	0.84 ab	0.38 b	0.37 abcd	0.17 bc	0.34 ab
9	1.45 ab	2.71 bc	0.25 b	2.04 c	0.65 c	0.55 e	0.106 ab	0.30 c	0.33 b	0.25 d	0.16 bc	0.24 cd
10	1.30 ab	3.12 bc	0.35 ab	2.35 abc	1.06 abc	1.09 abcd	0.084 b	0.58 abc	0.34 b	0.42 ab	0.14 bc	0.31 abcd
11	1.49 ab	2.75 bc	0.25 b	1.89 bc	0.87 abc	0.79 cde	0.071 b	0.45 bc	0.49 ab	0.46 a	0.11 c	0.21 d
12	1.59 a	3.65 ab	0.29 ab	2.15 abc	0.68 bc	0.73 de	0.093 ab	0.50 abc	0.43 ab	0.37 abcd	0.15 bc	0.32 abcd

Macronutrient [g/100 g DW]

Table 4. Micronutrients (Si, Mn, Sr, Fe, Zn, B, Al, Rb and Li) detected by Inductively Coupled Plasma–Optical Emission Spectrometry at harvest for taproots (T) and leaf (L) samples of the 12 accessions evaluated. Data are presented as the mean, n = 4 to 6. Different lowercase letters within the columns indicate statistical significance according to Tukey's test ($p \leq 0.05$). Accession identification (1–12) is shown in Table 1. DW: dry weight. n.d.: not detected.

	Micronutrients [mg/kg DW]								
Acc.	Si	Mn	Sr	Fe	Zn	B	Al	Rb	Li
Taproot									
1	31.1 [b]	15.0 [ab]	36.8 [ab]	33.4 [a]	11.1 [a]	15.1 [a]	7.5 [b]	5.84	n.d.
2	56.2 [ab]	15.6 [ab]	60.8 [ab]	34.2 [a]	15.2 [a]	15.8 [a]	12.3 [ab]	5.21 [bc]	n.d.
3	59.9 [ab]	13.0 [b]	44.5 [ab]	36.9 [a]	12.3 [a]	15.8 [a]	17.0 [ab]	9.60 [a]	n.d.
4	35.9 [ab]	16.3 [ab]	54.5 [ab]	30.7 [a]	10.5 [a]	14.1 [a]	17.2 [ab]	7.31 [abc]	n.d.
5	38.5 [ab]	16.2 [ab]	58.1 [ab]	34.5 [a]	14.1 [a]	15.1 [a]	10.6 [ab]	8.38 [ab]	n.d.
6	48.0 [ab]	15.4 [ab]	54.4 [ab]	38.9 [a]	11.5 [a]	13.9 [a]	10.6 [ab]	7.14 [abc]	n.d.
7	53.4 [ab]	15.8 [ab]	56.1 [ab]	37.8 [a]	12.8 [a]	14.0 [a]	10.2 [ab]	5.72 [bc]	n.d.
8	106 [a]	21.2 [a]	65.3 [a]	47.2 [a]	14.4 [a]	16.0 [a]	29.5 [a]	3.89 [c]	n.d.
9	52.1 [ab]	13.2 [b]	32.2 [b]	34.2 [a]	14.0 [a]	11.8 [a]	7.7 [b]	5.91 [bc]	n.d.
10	56.8 [ab]	17.4 [ab]	46.0 [ab]	34.6 [a]	11.5 [a]	14.8 [a]	14.9 [ab]	5.17 [bc]	n.d.
11	32.5 [b]	14.2 [ab]	37.4 [ab]	28.4 [a]	12.3 [a]	12.8 [a]	6.3 [b]	5.47 [bc]	n.d.
12	45.0 [ab]	13.4 [ab]	40.2 [ab]	35.8 [a]	12.6 [a]	14.6 [a]	8.5 [b]	6.33 [abc]	n.d.
Leaf									
1	211 [ab]	157 [ab]	165 [ab]	56.2 [ab]	35.8 [b]	41.1 [ab]	27.9 [a]	13.0 [bc]	3.94 [ab]
2	306 [a]	141 [ab]	156 [abc]	53.7 [ab]	38.5 [b]	38.4 [ab]	26.3 [a]	9.2 [c]	5.45 [a]
3	164 [ab]	188 [ab]	169 [ab]	63.6 [ab]	56.5 [ab]	44.5 [a]	17.9 [a]	11.9 [bc]	2.65 [ab]
4	232 [ab]	236 [a]	160 [abc]	103.5 [a]	79.8 [a]	39.7 [ab]	26.5 [a]	9.8 [bc]	3.54 [ab]
5	195 [ab]	192 [ab]	98 [c]	49.9 [b]	49.9 [ab]	38.5 [ab]	16.9 [a]	14.1 [ab]	3.35 [ab]
6	174 [ab]	91 [ab]	139 [abc]	51.1 [ab]	33.0 [b]	32.6 [ab]	21.8 [a]	18.7 [a]	2.39 [b]
7	171 [ab]	222 [a]	122 [bc]	37.7 [b]	59.3 [ab]	39.0 [ab]	13.6 [a]	14.8 [ab]	3.72 [ab]
8	292 [a]	202 [a]	189 [a]	72.0 [ab]	43.1 [ab]	36.1 [ab]	17.5 [a]	10.2 [bc]	4.14 [ab]
9	188 [ab]	141 [ab]	121 [bc]	50.4 [b]	31.4 [b]	29.5 [b]	19.9 [a]	10.2 [bc]	2.66 [ab]
10	195 [ab]	106 [ab]	148 [abc]	59.2 [ab]	29.0 [b]	28.5 [b]	14.4 [a]	13.0 [bc]	2.82 [ab]
11	80 [b]	72 [b]	139 [abc]	37.6 [b]	26.8 [b]	30.6 [ab]	10.6 [a]	10.1 [bc]	1.71 [b]
12	178 [ab]	117 [ab]	132 [bc]	59.8 [ab]	32.0 [b]	30.3 [ab]	16.2 [a]	14.8 [ab]	2.28 [b]

Table 5. Micronutrients (Mo, Cu, Ti, Pb, As, Cr, Ni, Cd and Co) detected by Inductively Coupled Plasma–Optical Emission Spectrometry at harvest for taproots (T) and leaf (L) samples of the 12 accessions evaluated. Data are presented as the mean, *n* = 4 to 6. Different lowercase letters within the columns indicate statistical significance according to Tukey's test (*p* ≤ 0.05). Accession identification (1–12) is shown in Table 1. DW: dry weight. n.d.: not detected.

	Micronutrients [mg/kg DW]								
Acc.	Mo	Cu	Ti	Pb	As	Cr	Ni	Cd	Co
				Taproot					
1	n.d.	1.94 [ab]	0.60 [a]	0.20 [ab]	n.d.	0.29 [a]	0.10 [a]	n.d.	n.d.
2	n.d.	2.19 [ab]	1.74 [a]	0.28 [ab]	n.d.	0.43 [a]	0.37 [a]	n.d.	n.d.
3	n.d.	1.73 [ab]	0.83 [a]	0.18 [b]	n.d.	0.23 [a]	0.08 [a]	n.d.	n.d.
4	n.d.	1.82 [ab]	0.45 [a]	0.24 [ab]	n.d.	0.18 [a]	0.03 [a]	n.d.	n.d.
5	n.d.	1.43 [ab]	0.51 [a]	0.17 [b]	n.d.	0.17 [a]	0.06 [a]	n.d.	n.d.
6	n.d.	1.93 [ab]	1.29 [a]	0.26 [ab]	n.d.	0.16 [a]	0.03 [a]	n.d.	n.d.
7	n.d.	1.30 [b]	0.70 [a]	0.30 [ab]	n.d.	0.42 [a]	0.18 [a]	n.d.	n.d.
8	n.d.	1.96 [a]	1.70 [a]	0.37 [a]	n.d.	0.40 [a]	0.27 [a]	n.d.	n.d.
9	n.d.	1.50 [b]	2.84 [a]	0.13 [b]	n.d.	0.21 [a]	0.16 [a]	n.d.	n.d.
10	n.d.	1.56 [ab]	1.05 [a]	0.14 [b]	n.d.	0.18 [a]	0.04 [a]	n.d.	n.d.
11	n.d.	1.11 [b]	0.92 [a]	0.11 [b]	n.d.	0.27 [a]	0.09 [a]	n.d.	n.d.
12	n.d.	1.55 [ab]	1.06 [a]	0.16 [b]	n.d.	0.23 [a]	0.09 [a]	n.d.	n.d.
				Leaf					
1	4.06 [a]	2.61 [a]	0.83 [b]	0.74 [ab]	0.61 [bcde]	0.28 [a]	0.35 [ab]	0.37 [ab]	0.06 [a]
2	2.01 [bc]	2.07 [a]	1.54 [a]	0.99 [a]	0.84 [ab]	0.50 [a]	0.35 [ab]	0.18 [d]	0.11 [a]
3	4.28 [a]	2.48 [a]	0.71 [b]	0.89 [ab]	0.73 [abcd]	0.60 [a]	0.29 [bc]	0.30 [bc]	0.12 [a]
4	2.91 [abc]	2.46 [a]	1.14 [ab]	0.85 [ab]	1.07 [a]	0.64 [a]	0.49 [a]	0.43 [a]	0.16 [a]
5	2.23 [abc]	1.83 [a]	0.81 [b]	0.37 [b]	0.58 [bcde]	1.52 [a]	0.32 [abc]	0.16 [d]	0.10 [a]
6	2.37 [abc]	2.53 [a]	1.14 [ab]	0.61 [ab]	0.58 [bcde]	0.72 [a]	0.34 [ab]	0.21 [cd]	0.07 [a]
7	2.14 [abc]	1.82 [a]	0.73 [b]	0.75 [ab]	0.49 [cde]	0.53 [a]	0.31 [bc]	0.23 [cd]	0.11 [a]
8	4.04 [a]	2.77 [a]	1.17 [ab]	0.82 [ab]	0.83 [ab]	0.44 [a]	0.33 [ab]	0.35 [ab]	0.13 [a]
9	2.10 [bc]	1.67 [a]	1.12 [ab]	0.50 [ab]	0.46 [de]	0.19 [a]	0.22 [bc]	0.17 [d]	0.13 [a]
10	1.68 [c]	1.90 [a]	0.99 [ab]	0.58 [ab]	0.78 [abc]	0.56 [a]	0.27 [bc]	0.22 [cd]	0.04 [a]
11	2.04 [bc]	1.81 [a]	0.70 [b]	0.34 [b]	0.36 [e]	0.28 [a]	0.17 [c]	0.22 [cd]	0.06 [a]
12	3.32 [ab]	2.21 [a]	0.80 [b]	0.66 [ab]	0.57 [bcde]	0.30 [a]	0.29 [bc]	0.18 [d]	0.09 [a]

Table 6. Pearson's correlation matrix for macronutrients, dry matter (%), total monomeric anthocyanin content (TMC) and total flavonoid content (TPC) in the taproot. Significant interactions are highlighted in bold.

	Dry Matter	TMC	TFC	K	Ca	Na	Mg	P
TMC	**0.457 ****							
TFC	**0.545 *****	**0.953 *****						
K	0.265	0.125	0.218					
Ca	**−0.290 ***	−0.182	−0.154	0.135				
Na	**−0.377 ***	−0.256	−0.276	−0.241	**0.833 *****			
Mg	−0.233	0.041	0.046	0.164	**0.646 *****	**0.498 *****		
P	−0.063	−0.106	−0.115	0.140	**0.490 *****	**0.490 *****	**0.385 ***	
S	−0.056	−0.085	−0.098	−0.158	**0.464 *****	**0.556 *****	**0.382 ***	0.253

(*), (**) and (***) indicate significant interactions at $p \leq 0.05$, $p \leq 0.01$ and $p \leq 0.001$, respectively.

4. Conclusions

In the present work, a comparative characterization was conducted for the first time on 11 Eastern carrot accessions. The reference cultivar, Night Bird 'F1', proved to be by far the highest anthocyanin and flavonoid-accumulating accession. Nevertheless, some landraces showed their potential for ulterior breeding purposes. In this sense, acc. 2 displayed the highest plant and taproot size, leaf area and number, and plant compactness, but the lowest SLA and LAR and a relatively low prevalence of premature bolting and taproot shape abnormalities; however, TMC and TFC were very low in acc. 2. Except for the reference cultivar, acc. 12 displayed the highest TMC and TFC and a plant size statistically comparable to the reference cultivar. Mineral composition analysis showed the nutritional potential of Eastern carrot leaves, which displayed a higher concentration than taproot tissue for several macro- and micronutrients. Several accs. had higher nutrient concentrations than the reference cultivar, which also highlights their potential profitability. At the taproot level, a clear correlation between DM and TMC and TFC was found, whereas macronutrient accumulation was positively correlated. Interestingly, accs. tested proved their ability to grow under a high temperature regime. In summary, this work contributes to the knowledge on Eastern black carrot germplasm, determining agricultural traits of relevance on both taproot and leaf organs. Further evaluation is needed in order to incorporate the accs. of interest to ulterior breeding programs.

Author Contributions: Conceptualization, G.B.-E.; methodology, J.R.A.-M., J.A.H.C. and G.B.-E.; formal analysis, J.R.A.-M., P.D.-V., V.B.-G., J.A.H.C. and G.B.-E.; investigation, J.R.A.-M., V.B.-G. and G.B.-E.; data curation, J.R.A.-M..; writing—original draft preparation, G.B.-E. and J.R.A.-M.; writing—review and editing, G.B.-E., J.A.H.C., J.R.A.-M., and P.D.-V. All authors have read and agreed to the published version of the manuscript.

Funding: This research was funded by the "Fundación Séneca"–Agency of Science and Technology of the Region of Murcia, grant number 20405/SF/17.

Data Availability Statement: The data presented in this study are available on request from the corresponding author.

Acknowledgments: The authors thank Almudena Gutiérrez and José A. Sánchez, from the glasshouse facilities of the University of Murcia, for their support during the development of the experiment.

Conflicts of Interest: The authors declare no conflict of interest.

References

1. Iorizzo, M.; Ellison, S.; Senalik, D.; Zeng, P.; Satapoomin, P.; Huang, J.; Bowman, M.; Iovene, M.; Sanseverino, W.; Cavagnaro, P.; et al. A high-quality carrot genome assembly provides new insights into carotenoid accumulation and asterid genome evolution. *Nat. Genet.* **2016**, *48*, 657–666. [CrossRef] [PubMed]
2. Bolton, A.; Klimek-Chodacka, M.; Martin-Millar, E.; Grzebelus, D.; Simon, P.W. Genome-Assisted Improvement Strategies for Climate-Resilient Carrots. In *Genomic Designing of Climate-Smart Vegetable Crops*; Springer International Publishing: Cham, Switzerland, 2020.
3. Iorizzo, M.; Senalik, D.A.; Ellison, S.L.; Grzebelus, D.; Cavagnaro, P.F.; Allender, C.; Brunet, J.; Spooner, D.M.; Van Deynze, A.; Simon, P.W. Genetic structure and domestication of carrot (*Daucus carota* subsp. *sativus*) (Apiaceae) 1. *Am. J. Bot.* **2013**, *100*, 930–938. [CrossRef]
4. Simon, P.W. Economic and Academic Importance. In *The Carrot Genome. Compendium of Plant Genomes*; Simon, P., Iorizzo, M., Grzebelus, D., Baranski, R., Eds.; Springer: New York, NY, USA, 2019; pp. 1–8.
5. Giusti, M.M.; Wrolstad, R.E. Acylated anthocyanins from edible sources and their applications in food systems. *Biochem. Eng. J.* **2003**, *14*, 217–225. [CrossRef]
6. Montilla, E.C.; Arzaba, M.R.; Hillebrand, S.; Winterhalter, P. Anthocyanin composition of black carrot (*Daucus carota* ssp. *sativus* var. *atrorubens* Alef.) Cultivars antonina, beta sweet, deep purple, and purple haze. *J. Agric. Food Chem.* **2011**, *59*, 3385–3390. [CrossRef] [PubMed]
7. Algarra, M.; Fernandes, A.; Mateus, N.; de Freitas, V.; Esteves da Silva, J.C.G.; Casado, J. Anthocyanin profile and antioxidant capacity of black carrots (*Daucus carota* L. ssp. *sativus* var. *atrorubens* Alef.) from Cuevas Bajas, Spain. *J. Food Compos. Anal.* **2014**, *33*, 71–76. [CrossRef]
8. Barba-Espín, G.; Glied, S.; Crocoll, C.; Dzhanfezova, T.; Joernsgaard, B.; Okkels, F.; Lütken, H.; Müller, R. Foliar-applied ethephon enhances the content of anthocyanin of black carrot roots (*Daucus carota* ssp. *sativus* var. *atrorubens* Alef.). *BMC Plant Biol.* **2017**, *17*, 70–80. [CrossRef]
9. Müller, R.; Acosta-Motos, J.R.; Großkinsky, D.K.; Hernández, J.A.; Lütken, H.; Barba-Espin, G. UV-B exposure of black carrot (*Daucus carota* ssp. *sativus* var. *atrorubens*) plants promotes growth, accumulation of anthocyanin, and phenolic compounds. *Agronomy* **2019**, *9*, 323. [CrossRef]
10. Dzhanfezova, T.; Barba-Espín, G.; Müller, R.; Joernsgaard, B.; Hegelund, J.N.; Madsen, B.; Larsen, D.H.; Martínez Vega, M.; Toldam-Andersen, T.B. Anthocyanin profile, antioxidant activity and total phenolic content of a strawberry (*Fragaria* × *ananassa* Duch) genetic resource collection. *Food Biosci.* **2020**, *36*, 100620. [CrossRef]
11. McCann, D.; Barrett, A.; Cooper, A.; Crumpler, D.; Dalen, L.; Grimshaw, K.; Kitchin, E.; Lok, K.; Porteous, L.; Prince, E.; et al. Food additives and hyperactive behaviour in 3-year-old and 8/9-year-old children in the community: A randomised, double-blinded, placebo-controlled trial. *Lancet* **2007**, *370*, 1560–1567. [CrossRef]
12. Carocho, M.; Barreiro, M.F.; Morales, P.; Ferreira, I.C.F.R. Adding molecules to food, pros and cons: A review on synthetic and natural food additives. *Compr. Rev. Food Sci. Food Saf.* **2014**, *13*, 377–399. [CrossRef]
13. Akhtar, S.; Rauf, A.; Imran, M.; Qamar, M.; Riaz, M.; Mubarak, M.S. Black carrot (*Daucus carota* L.), dietary and health promoting perspectives of its polyphenols: A review. *Trends Food Sci. Technol.* **2017**, *66*, 36–47. [CrossRef]
14. Smeriglio, A.; Denaro, M.; Barreca, D.; D'Angelo, V.; Germanò, M.P.; Trombetta, D. Polyphenolic profile and biological activities of black carrot crude extract (*Daucus carota* L. ssp. *sativus* var. *atrorubens* Alef.). *Fitoterapia* **2018**, *124*, 49–57. [CrossRef]
15. Barba-Espin, G.; Nicolas, E.; Almansa, M.S.; Cantero-Navarro, E.; Albacete, A.; Hernández, J.A.; Díaz-Vivancos, P. Role of thioproline on seed germination: Interaction ROS-ABA and effects on antioxidative metabolism. *Plant Physiol. Biochem.* **2012**, *59*, 30–36. [CrossRef]
16. Sharma, A.; Shahzad, B.; Rehman, A.; Bhardwaj, R.; Landi, M.; Zheng, B. Response of phenylpropanoid pathway and the role of polyphenols in plants under abiotic stress. *Molecules* **2019**, *24*, 2452. [CrossRef] [PubMed]
17. Tak, Y.; Kumar, M. Phenolics: A Key Defence Secondary Metabolite to Counter Biotic Stress. In *Plant Phenolics in Sustainable Agriculture*; Springer: Singapore, 2020.
18. Simon, P.W. Domestication, Historical Development, and Modern Breeding of Carrot. In *Plant Breeding Reviews*; Janick, J., Ed.; Wiley: New Jersey, NJ, USA, 2019; Volume 19, pp. 157–190.
19. Aveling, T.A.S. Global standards in seed health testing. In *Global Perspectives on the Health of Seeds and Plant Propagation Material*; Springer: Dordrecht, The Netherlands, 2014; Volume 6.
20. Asaduzzaman, M.; Kobayashi, Y.; Mondal, M.F.; Ban, T.; Matsubara, H.; Adachi, F.; Asao, T. Growing carrots hydroponically using perlite substrates. *Sci. Hortic.* **2013**, *159*, 113–121. [CrossRef]
21. Barba-Espín, G.; Chen, S.-T.; Agnolet, S.; Hegelund, J.N.; Stanstrup, J.; Christensen, J.H.; Müller, R.; Lütken, H. Ethephon-induced changes in antioxidants and phenolic compounds in anthocyanin-producing black carrot hairy root cultures. *J. Exp. Bot.* **2020**, *71*, 7030–7045. [CrossRef]
22. Ahmed, D.; Khan, M.M.; Saeed, R. Comparative analysis of phenolics, flavonoids, and antioxidant and antibacterial potential of methanolic, hexanic and aqueous extracts from *Adiantum caudatum* leaves. *Antioxidants* **2015**, *4*, 394. [CrossRef]
23. Dessau, R.B.; Pipper, C.B. ["R"–project for statistical computing]. *Ugeskr. Laeger* **2008**, *170*, 328–330. [PubMed]

24. Bolton, A.; Nijabat, A.; Mahmood-Ur-Rehman, M.; Naveed, N.H.; Majharul Mannan, A.T.M.; Ali, A.; Rahim, M.A.; Simon, P. Variation for Heat Tolerance during Seed Germination in Diverse Carrot [*Daucus carota* (L.)] Germplasm. *HortScience* **2019**, *54*, 1470–1476. [CrossRef]

25. Hernández, J.A.; Diaz-Vivancos, P.; Barba-Espín, G.; Clemente-Moreno, M.J. On the role of salicylic acid in plant responses to environmental stresses. In *Salicylic Acid: A Multifaceted Hormone*; Springer: Singapore, 2017.

26. Chen, C.; Ma, J.; Ma, J.; Ma, W.; Yang, J. Analysis of main agronomic traits in different varieties of carrots. *J. Phys. Conf. Ser.* **2020**, *1549*, 032053. [CrossRef]

27. Sharma, R.; Agarwal, A.; Mamatha, H.R. Classification of carrots based on shape analysis using machine learning techniques. In Proceedings of the 3rd International Conference on Intelligent Communication Technologies and Virtual Mobile Networks, ICICV 2021, Tirunelveli, India, 4–6 February 2021.

28. Iorizzo, M.; Curaba, J.; Pottorff, M.; Ferruzzi, M.G.; Simon, P.; Cavagnaro, P.F. Carrot anthocyanins genetics and genomics: Status and perspectives to improve its application for the food colorant industry. *Genes* **2020**, *11*, 906. [CrossRef] [PubMed]

29. Bannoud, F.; Ellison, S.; Paolinelli, M.; Horejsi, T.; Senalik, D.; Fanzone, M.; Iorizzo, M.; Simon, P.W.; Cavagnaro, P.F. Dissecting the genetic control of root and leaf tissue-specific anthocyanin pigmentation in carrot (*Daucus carota* L.). *Theor. Appl. Genet.* **2019**, *132*, 2485–2507. [CrossRef] [PubMed]

30. Xu, Z.S.; Yang, Q.Q.; Feng, K.; Xiong, A.S. Changing carrot color: Insertions in DcMYB7 alter the regulation of anthocyanin biosynthesis and modification. *Plant Physiol.* **2019**, *181*, 195–207. [CrossRef]

31. Maksylewicz, A.; Baranski, R. Intra-population genetic diversity of cultivated carrot (*Daucus carota* L.) assessed by analysis of microsatellite markers. *Acta Biochim. Pol.* **2013**, *60*, 753–760. [CrossRef]

32. Appiah, F.K.; Sarkodie-Addo, J.; Opoku, A. Quality of Carrot (*Daucus Carota* L.) As Influenced by Green Manure and Plant Spacing on Forest Ochrosols in Ghana. *J. Biol. Agric. Healthc.* **2017**, *7*, 36–44.

33. Hussain, S.I.; Hadley, P.; Pearson, S. A validated mechanistic model of carrot (*Daucus carota* L.) growth. *Sci. Hortic.* **2008**, *117*, 26–31. [CrossRef]

34. Colley, M. Organic carrot production. In *Carrots and Related Apiaceae Crops*; Geoffriau, E., Simon, P., Eds.; CABI: Boston, MA, USA, 2020; pp. 197–215.

35. Aquino, R.F.B.A.; Assunção, N.S.; Aquino, L.A.; Aquino, P.M.D.; Oliveira, G.A.D.; Carvalho, A.M.X.D. Nutrient Demand by the Carrot Crop Is Influenced by the Cultivar. *Rev. Bras. Ciência Solo* **2015**, *39*, 541–552. [CrossRef]

36. Qi, Y.; Wei, W.; Chen, C.; Chen, L. Plant root-shoot biomass allocation over diverse biomes: A global synthesis. *Glob. Ecol. Conserv.* **2019**, *18*, e00606. [CrossRef]

37. Suojala, T. Growth of and partitioning between shoot and storage root of carrot in a northern climate. *Agric. Food Sci. Finl.* **2000**, *9*, 49–59. [CrossRef]

38. Forbes, J.C.; Watson, R.D. *Plants in Agriculture*; Cambridge University Press: Cambridge, UK, 1992. [CrossRef]

39. Vega Rojas, T.; Soto, C.M.; Montero, W.R. Análisis del crecimiento de cinco híbridos de zanahoria (*Daucus carota* L.) mediante la metodología del análisis funcional. *Agron. Costarric.* **2012**, *36*, 29–46. [CrossRef]

40. Clautilde, M.; Lucien, T.; Eric, N.; Abba, M.; Hamadou, B. Field Productivity of Carrot (*Daucus carota* L.) in Adamawa Cameroon and Chemical Properties of Roots According To Chicken Manure Pretreatments and Vivianite Powder. *IOSR J. Agric. Vet. Sci.* **2017**, *10*, 16–23. [CrossRef]

41. Blackman, G.E.; Evans, G.C. The Quantitative Analysis of Plant Growth. *J. Ecol.* **1973**, *61*, 617–619. [CrossRef]

42. Kimball, B.A.; Kobayashi, K.; Bindi, M. Responses of agricultural crops to free-air CO2 enrichment. *Adv. Agron.* **2002**, *77*, 293–368. [CrossRef]

43. Singh, B.K.; Koley, T.K.; Maurya, A.; Singh, P.M.; Singh, B. Phytochemical and antioxidative potential of orange, red, yellow, rainbow and black coloured tropical carrots (*Daucus carota* subsp. *sativus* Schubl. & Martens). *Physiol. Mol. Biol. Plants* **2018**, *24*, 899–907. [CrossRef]

44. Dimitrić Marković, J.M.; Milenković, D.; Amić, D.; Popović-Bijelić, A.; Mojović, M.; Pašti, I.A.; Marković, Z.S. Energy requirements of the reactions of kaempferol and selected radical species in different media: Towards the prediction of the possible radical scavenging mechanisms. *Struct. Chem.* **2014**, *25*, 1795–1804. [CrossRef]

45. Cavagnaro, P.F.; Bannoud, F.; Iorizzo, M.; Senalik, D.; Ellison, S.L.; Simon, P.W. Carrot anthocyanins: Nutrition, diversity and genetics. *Acta Hortic.* **2019**, *1264*, 101–106. [CrossRef]

46. Sharma, K.D.; Karki, S.; Thakur, N.S.; Attri, S. Chemical composition, functional properties and processing of carrot—A review. *J. Food Sci. Technol.* **2012**, *49*, 22–32. [CrossRef]

47. Que, F.; Hou, X.L.; Wang, G.L.; Xu, Z.S.; Tan, G.F.; Li, T.; Wang, Y.H.; Khadr, A.; Xiong, A.S. Advances in research on the carrot, an important root vegetable in the Apiaceae family. *Hortic. Res.* **2019**, *6*, 1–15. [CrossRef]

48. Nicolle, C.; Simon, G.; Rock, E.; Amouroux, P.; Rémésy, C. Genetic variability influences carotenoid, vitamin, phenolic, and mineral content in white, yellow, purple, orange, and dark-orange carrot cultivars. *J. Am. Soc. Hortic. Sci.* **2004**, *129*, 523–529. [CrossRef]

49. Goneim, G.; Ibrahim, F.; El-Shehawy, S. Carrot Leaves: Antioxidative and Nutritive Values. *J. Food Dairy Sci.* **2011**, *2*, 1–9. [CrossRef]

50. Warman, P.R.; Havard, K.A. Yield, vitamin and mineral contents of organically and conventionally grown carrots and cabbage. *Agric. Ecosyst. Environ.* **1997**, *61*, 155–162. [CrossRef]

51. Smoleń, S.; Rozek, S.; Strzetelski, P.; Ledwozyw-Smoleń, I. Preliminary evaluation of the influence of soil fertilization and foliar nutrition with iodine on the effectiveness of iodine biofortification and mineral composition of carrot. *J. Elem.* **2011**, *16*, 103–114. [CrossRef]
52. Broadley, M.; Brown, P.; Cakmak, I.; Rengel, Z.; Zhao, F. Function of Nutrients: Micronutrients. In *Marschner's Mineral Nutrition of Higher Plants*, 3rd ed.; Petra Marschner: Adelaide, GA, USA, 2011.
53. Griffiths, M.; Roy, S.; Guo, H.; Seethepalli, A.; Huhman, D.; Ge, Y.; Sharp, R.E.; Fritschi, F.B.; York, L.M. A multiple ion-uptake phenotyping platform reveals shared mechanisms affecting nutrient uptake by roots. *Plant Physiol.* **2021**, *185*, 781–795. [CrossRef] [PubMed]
54. Niedziela, C.E.; Nelson, P.V.; Dickey, D.A. Growth, Development, and Mineral Nutrient Accumulation and Distribution in Tulip from Planting through Postanthesis Shoot Senescence. *Int. J. Agron.* **2015**, *2015*, 1–11. [CrossRef]

Article

DArTseq Genotypic and Phenotypic Diversity of Barley Landraces Originating from Different Countries

Joanna Dziurdziak, Grzegorz Gryziak , Jolanta Groszyk , Wiesław Podyma and Maja Boczkowska *

National Centre for Plant Genetic Resources, Plant Breeding and Acclimatization Institute—National Research Institute, Błonie, 05-870 Radzików, Poland; j.nocen@ihar.edu.pl (J.D.); g.gryziak@ihar.edu.pl (G.G.); j.groszyk@ihar.edu.pl (J.G.); w.podyma@ihar.edu.pl (W.P.)
* Correspondence: m.boczkowska@ihar.edu.pl

Abstract: Landraces are considered a key element of agrobiodiversity because of their high variability and adaptation to local environmental conditions, but at the same time, they represent a breeding potential hidden in gene banks that has not yet been fully appreciated and utilized. Here, we present a genome-wide DArTseq analysis of the diversity of 116 spring barley landraces preserved in the collection of the Polish gene bank. Genetic analysis revealed considerable variation in this collection and several distinct groups related to the landraces' country of origin and the grain type were identified. The genetic distinctness of hulless accessions may provide a basis for pro-quality breeding aimed at functional food production. However, the variable level of accession heterogeneity can be a significant obstacle. A solution to this problem is the establishment of special collections composed of pure lines that are accessible to breeders. Regions lacking genetic diversity have also been identified on 1H and 4H chromosomes. A small region of reduced heterogeneity was also present in the hulless forms in the vicinity of the *nud* gene that determines the hulless grain type. However, the SNPs present in this area may also be important in selection for traits related to grain weight and size because their QTLs were found there. This may support breeding of hulless forms of spring barley which may have applications in the production of high-quality foods with health-promoting values.

Keywords: barley; genetic diversity; germplasm; *Hordeum vulgare*; landrace; DArTseq; population structure

Citation: Dziurdziak, J.; Gryziak, G.; Groszyk, J.; Podyma, W.; Boczkowska, M. DArTseq Genotypic and Phenotypic Diversity of Barley Landraces Originating from Different Countries. *Agronomy* **2021**, *11*, 2330. https://doi.org/10.3390/agronomy11112330

Academic Editors: Gregorio Barba-Espín and Jose Ramon Acosta-Motos

Received: 2 September 2021
Accepted: 16 November 2021
Published: 18 November 2021

Publisher's Note: MDPI stays neutral with regard to jurisdictional claims in published maps and institutional affiliations.

1. Introduction

The progressive genetic erosion of agricultural ecosystems and the associated irreversible loss of this vital part of the global biodiversity essential to sustaining humankind is a fact that no one discusses anymore. Agrobiodiversity is recognized as a finite global resource that is known to be eroded or lost in part because of imprudent, unsustainable human practices [1]. Landraces are considered to be a key element of agrobiodiversity [2]. Several definitions of landrace exist in the scientific literature so far. Among them is the one proposed by V. Negri, which combines several other definitions. "A landrace of a seed-propagated crop can be defined as a variable population, which is identifiable and usually has a local name. It lacks "formal" crop improvement, is characterized by a specific adaptation to the environmental conditions of the area of cultivation (tolerant to the biotic and abiotic stresses of that area) and is closely associated with the traditional uses, knowledge, habits, dialects, and celebrations of the people who developed and continue to grow it" [1]. The list of factors that contribute to the genetic erosion of landraces is long. Changes in agricultural practices and land use are highlighted as the main ones. Mechanization, crop protection chemicals, and irrigation promote the displacement of landraces by modern cultivars. National registration and certification systems that restrict the sale of crop seeds if a cultivar is not included in the national or regional list of registered cultivars are factors as well. Moreover, the depopulation of rural areas and the resulting loss of traditional knowledge and cultivation systems for landraces have a tremendous impact. Changes in consumer habits and food standards that limit the supply of landraces are also important. Unfortunately, one of the factors is the lack of education and consequent awareness of the value of plant genetic resources as a valuable local and national heritage. In addition, a major threat that is becoming more serious every year is global warming and changing weather patterns. This is because landraces are very often cultivated on marginal land under conditions that are close to species limits. An irretrievable loss of landraces also occurs due to armed conflicts and political instability [3].

Loss of diversity is common in modern crops of major species that have been developed by large, focused, "industrial" breeding programs. Overall, the ideotype is well defined by market needs and regional adaptations [4]. In barley, as in most crops, the current elite cultivars are less genetically diverse than their wild relatives or early domesticated forms at the majority of loci [5–7]. Nearly all modern cultivars arose from the reshuffling of selected alleles, leading to a limited number of alleles present within the gene pool [8]. Barley breeders use almost exclusively elite lines with which they are familiar to meet short-term breeding goals [9]. The major emphasis has been put on high yields and tolerance to biotic stresses [10]. Elite cultivars produce high yields under optimal conditions; however, they can fail under harsh environmental conditions [11]. Barley landraces can produce up to 61% higher grain yield under unfavorable conditions compared to improved cultivars [12]. However, it should be kept in mind that under conditions of modern, intensive farming, landraces yield at a much lower level than commercial cultivars and are unable to feed the world [13]. Precisely, these unique morphological, physiological, and genetic traits of landraces that allowed them to survive and to be productive despite the pressure of biotic and abiotic factors should be used to develop new cultivars resistant to climatic conditions and new aggressive races of pathogens [10].

While elite lines have been widely used in breeding programs, introgression of potentially new genes from landraces has received too little attention [14]. Underutilization of landraces in breeding programs is most likely due to the time-consuming and labor-intensive nature of pre-breeding associated with both the necessity to break linkage drag and the potential loss of important gene complexes [10]. Nevertheless, landraces have been used in breeding as a source of resistance to viruses, fungal pathogens, and pests [15–19]. The potential of landraces as a source of valuable traits for breeding has been widely discussed by Dawson, et al. [20], Pietrusińska, et al. [21], Hernandez, et al. [22] and Kumar et al. [10]. Introgression of the landrace's gene pool into ongoing breeding programs is a prerequisite for improving tolerance of biotic and abiotic stresses, grain nutritional value, and ensuring future food security.

Over the years, the diversity of barley germplasm has been the subject of many studies concerning agronomic, morphological, as well as genetic traits. A wide range of biochemical and molecular techniques, i.e., Restriction Fragment Length Polymorphisms (RFLP), Random Amplified Polymorphic DNA (RAPD), Amplified Fragment Length Polymorphism (AFLP), Inter Simple Sequence Repeat (ISSR), or Simple Sequence Repeat (SSR), has been used to characterize barley germplasm [23–37]. The portfolio of techniques used has changed over time and followed the latest trends and availability of cutting-edge tools. A group of tools for SNP-based genotyping has recently been put into the hands of scientists working on barley genetic variation. Among them are those based on the Kompetitive Allele Specific PCR (KASP) technology, microarrays, and next-generation sequencing (NGS) [38–46].

In this study, DArTseq derived SNPs were used (a) for genetic analysis and population structure of barley spring landraces preserved in Polish gene bank; (b) to assess the level of within-accession heterogeneity; (c) to evaluate chromosome level diversity with an emphasis on the region adjacent to the *nud* gene; and (d) to assist in the selection of materials useful for breeding programs. By doing so, it is hoped to increase breeders' interest in barley germplasm. A side goal of the study was a comparison of results obtained with classical molecular markers and a modern NGS-based method to prove that a properly performed old-school analysis can generate valid results. This study analyzed unique material that had never been genotyped by sequencing before.

2. Materials and Methods

2.1. Plant Material

In the study, 116 spring barley accessions were investigated. All of them were assigned as landrace/traditional cultivars and are preserved by the National Center for Plant Genetic Resources (NCPGR), i.e., the Polish gene bank. The accessions originating from Poland make up 60% of the studied accessions and their detailed biological, geographical agro-morphological characteristics were described in detail in Dziurdziak et al. [32]. The remaining 40% of the studied materials come from foreign field expeditions (Table S1, Figure 1). In the 1970s and 1980s, accessions were collected in the USSR (13% of the foreign accessions) and Czechoslovakia (4%). The remaining accessions were acquired in 2004 from Iran (15% of the remaining accessions) and Georgia (7%), and during three field expeditions in the years 2011–2013 from Lithuania (60% of the remaining accessions). Only 12 accessions represent hulless barley, the rest were hulled. All hulless accessions originated from Poland (Table S1). Scans of grains in the active collection of the NCPGR long-term storage facility were made for all accessions examined. Each accession was represented by approximately 400 grains sourced from an active collection. The grains were scattered uniformly on a CanoScan LiDE 700 F flatbed desktop scanner surface. Scans with a resolution of 300 dpi were saved in jpg format and forwarded to the EGISET database [47] (Figure 2).

Figure 1. Map of barley accessions stored at the National Center of Plant Gene Resources that have landrace status and were covered by DArTseq-based SNP analysis, highlighted in blue.

(a) (b) (c) (d)

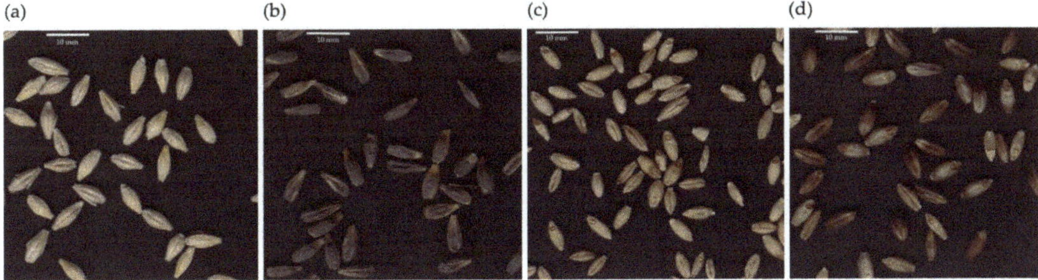

Figure 2. Grain color variation within hulled and hulless spring barley landraces: (**a**) PL 41634; (**b**) PL 502171; (**c**) PL 41867; (**d**) PL 42122.

2.2. Agro-Morphological Features

Data were obtained from EGISET, the NCPGR database [47]. Accessions were evaluated as described by Dziurdziak et al. [32]. Data were collected between 1978 and 2016. The variation coefficient was defined as:

$$Cv = \frac{\sigma}{\mu} \tag{1}$$

where σ is the standard deviation and μ is an arithmetic mean, and was used to measure the dispersion of traits.

2.3. DNA Isolation

Seeds obtained from long-term storage were sown and tissue was harvested from healthy seedlings that were in the second leaf stage and the middle part of the second leaf (about 10 mm long) was taken. Each accession was represented by eight random plantlets that formed a bulk sample. Total genomic DNA was extracted using a modified CTAB protocol [48,49]. The quality and quantity of DNA samples were assessed by spectrophotometric analysis by a NanoDrop ND-1000 spectrophotometer (NanoDrop Technologies, Willmington, DA, USA) followed by agarose gel electrophoresis (1.5% agarose).

2.4. DArTseq Genotyping

Genotyping of 116 spring barley landraces was performed using the genome-wide profiling DArTseq method as previously described [50]. Extracted genomic DNA samples were sent to Diversity Arrays Technology Pty Ltd (http://www.diversityarrays.com accessed on 1 September 2021). The resulting sequences were aligned against the Morex barley genome assembly [51]. Raw data are available on the platform Center for Open Science at https://osf.io/mh4tn/ (accessed on 23 October 2021). (doi 10.17605/OSF.IO/MH4TN).

2.5. ISSR Analysis

Previous data for some accessions studied here, obtained using the ISSR method, were also used. The ISSR analysis and its results are described in detail by Dziurdziak et al. [32]. Results of previous studies were used to compare different methods of genetic diversity analysis.

2.6. Marker Data Analysis

The results of DArTseq genotyping were generated as a table listing single nucleotide polymorphisms (codominant) markers that were detected in the sequenced fragments of genome representations. SNP markers were then transformed into a binary matrix. To preserve their codominant nature, each locus was represented by two consecutive lines. The presence of a SNP relative to the reference sequence was denoted as 1, while the absence of a SNP was denoted as 0. Thus, in the array, homozygotes were denoted as 1/1 or 0/0 and heterozygotes as 1/0. The analyzed markers were filtered by reproducibility (RepAvg $\geq$ 0.95), call rate (CallRate $\geq$ 0.95), and the minor allele frequency (MAF > 0.01).

The percentage of polymorphic fragments and the polymorphic information content ratios (PIC) were calculated. The PIC formula was:

$$PIC = 1 - \sum_{i=1}^{n} p_i^2 \tag{2}$$

where i is the ith allele of the jth marker, n is the number of alleles of the jth marker, and p is an allele frequency.

The expected heterozygosity (uH_e) was estimated using Nei's gene diversity coefficient was calculated as follows:

$$uH_e = \frac{2N}{2N - 1}\left(1 - \sum_{i=1}^{n} p_i^2\right) \tag{3}$$

where p_i is the frequency of the ith allele, n is the number of alleles of the jth marker, and N is the sample size.

The observed heterozygosity for the codominant data (uH_o) was calculated as follows:

$$uH_o = \frac{2N}{2N-1}\left(\frac{No.\_of\_Hets}{N}\right) \qquad (4)$$

where the number of heterozygotic loci was determined by direct counting and N is the sample size. The Fixation Index for codominant data (F) was calculated as follows:

$$F = \frac{uHe - uHo}{uHe} \qquad (5)$$

where uH_e is the expected heterozygous and uH_o is the observed heterozygosity.

Allelic richness for groups was calculated based on rarefaction due to a different number of accessions originating from each country [52].

PIC and Ho values along chromosomes were assessed by a sliding window approach with 500 kb windows at 250 positions along the chromosomes.

The data for observed heterozygosity were divided into groups based on accession origin countries. The means in these groups were compared using analysis of variance (ANOVA) and Tukey's post hoc test. The binary data matrix was used to calculate the Jaccard dissimilarity coefficient. The Principal Coordinate Analysis (PCoA) was performed to determine the relationship between the accessions.

The genetic structure of the germplasm was analyzed by clustering based on the Bayesian model implemented in STRUCTURE v.2.3.4 [53]. The search for the most probable K value was performed in the range from 1 to 10 with ten independent repetitions for each K value. The number of burn-ins and MCMC replicates were 5×10^4 and 1.5×10^5 respectively in each run. Batch runs were carried out on a LINUX cluster hosted by the Interdisciplinary Centre for Mathematical and Computational Modelling at the Warsaw University.

The determination of the number of true clusters was performed based on the posteriori data probability for a given K and ΔK [54]. The best match for replicated cluster analysis results was performed using the full search algorithm. The maximum probability coefficient was used to assign landrace to clusters with a 0.8 probability limit of being assigned to a cluster.

Correlation between the dissimilarity matrix of morphometric and genetic data was performed using the Mantel test (10^4 permutations). A consensus configuration for these two sets of data was obtained by the Generalized Procrustes Analysis (GPA) [55].

All above-mentioned analyses were performed using the Microsoft Excel 2016, XL-STAT Ecology (Addinsoft, Inc., Brooklyn, NY, USA), GenAlEx 6.501 [56], HP-RARE 1.1 [57], STRUCTURE v2.3.4 [53], CLUMPAK [58]. The data analysis was performed within the framework of the Computational Grant (G72-19) of the Interdisciplinary Centre for Mathematical and Computational Modelling, University of Warsaw (ICM UW).

3. Results

3.1. Agro-Morphological Diversity

All agro-morphological data were historical and sourced from field observations between 1978 and 2016. So far, no evaluation has been conducted for five accessions. A summary of all analyzed traits was presented in Table 1.

Table 1. Descriptive statistics of average agro-morphological characters based on EGISET database [47]. The accession number is in parentheses.

	Trait	Unit	% of Evaluated Accessions	Mean	Min.	Max.	Variation (Cv)
Phenological	days to heading	days	95%	69.7	52.0 (PL 42389)	93.0 (2 access.)	7%
	grain filling duration	days	85%	35.4	22.0 (PL 505304)	44.0 (PL 40554)	13%
	days to maturity	days	85%	105.5	81.7 (PL 42768)	117.0 (PL 41532)	10%
Metrical	plant length	cm	96%	81.3	52.0 (PL 502172)	112.0 (PL 36315)	13%
	ear length	cm	76%	8.6	4.6 (PL 502171)	11.8 (PL 501971)	16%
	number of grains per ear	no.	64%	16.8	7.7 (PL 40553)	43.6 (PL 502170)	36%
	thousand seed weight	g	95%	47.7	30.5 (PL 36315)	58.8 (PL 502169)	11%
Bonitations	lodging resistance	scale [1]	85%	7.7	2.0 (PL 41267)	9.0 (42 access.)	19%
	powdery mildew resistance	scale [1]	88%	6.9	1.0 (3 access.)	9.0 (21 access.)	28%
	net blotch resistance	scale [1]	91%	8.5	5.0 (PL 501968.)	9.0 (71 access.)	10%
	stem rust resistance	scale [1]	61%	8.8	7.0 (5 access.)	9.0 (65 access.)	6%
	scald resistance	scale [1]	67%	8.7	6.0 (PL 505304)	9.0 (63 access.)	8%

[1] scale (1—very low, . . . , 9—very good).

3.2. Genetic Analysis

3.2.1. Data Quality Analysis

By using the genome reduction method, 77,817 polymorphic SNP loci were generated. To avoid errors in statistical analysis, loci with low reproducibility (RepAvg $\leq$ 0.95), low call rate (CallRate $\leq$ 0.95), and low minor allele frequency (MAF < 0.01) were removed. Therefore, 66,815 SNP loci were excluded from the analysis, and 11,002 SNP loci met all quality parameters and were classified for further analysis.

Analysis of the distribution of SNP loci on chromosomes before and after filtering the results was performed. Filtering of the results did not disturb their distribution, and they were positioned almost equally across chromosomes, ranging from 10% on chromosome 1H and 6H to a maximum of 15% on chromosome 2H (Figure S1). There was a relatively high proportion (18% raw data, 16% filtered data) of loci with an unknown chromosomal location.

Analysis was also performed to examine loci distribution along each chromosome (Figure 3). For all chromosomes, there was the same pattern of distribution of the studied loci, i.e., their proportion was higher at the ends of the chromosome arms and decreased toward the centromeres. Moreover, on both chromosomes 1H and 4H, a negligible number of loci among the analyzed DArTseq loci were in the centromeric and pericentromeric regions. On 1H, this "empty" region was about 105 Mbp long and contained only 6 analyzed loci and on 4H, only 24 loci were present in a fragment of about 243 Mbp surrounding the centromere. As a result, the average PIC and Ho for these regions were very low. For the other chromosomes, the number of loci analyzed in the centromere region was higher, although still significantly lower than for the regions located closer to the telomeres. PIC and Ho values were evenly distributed along chromosome 3H. In contrast, there were two regions of low heterogeneity on 2HL, with one directly adjacent to the centromere (~104 Mbp) and the other slightly shorter, i.e., ~89 Mbp terminating at the midpoint of the arm length. On 5H, the region of reduced heterogeneity occurred on the short arm in the direct vicinity of the centromere and was ~54 Mbp long.

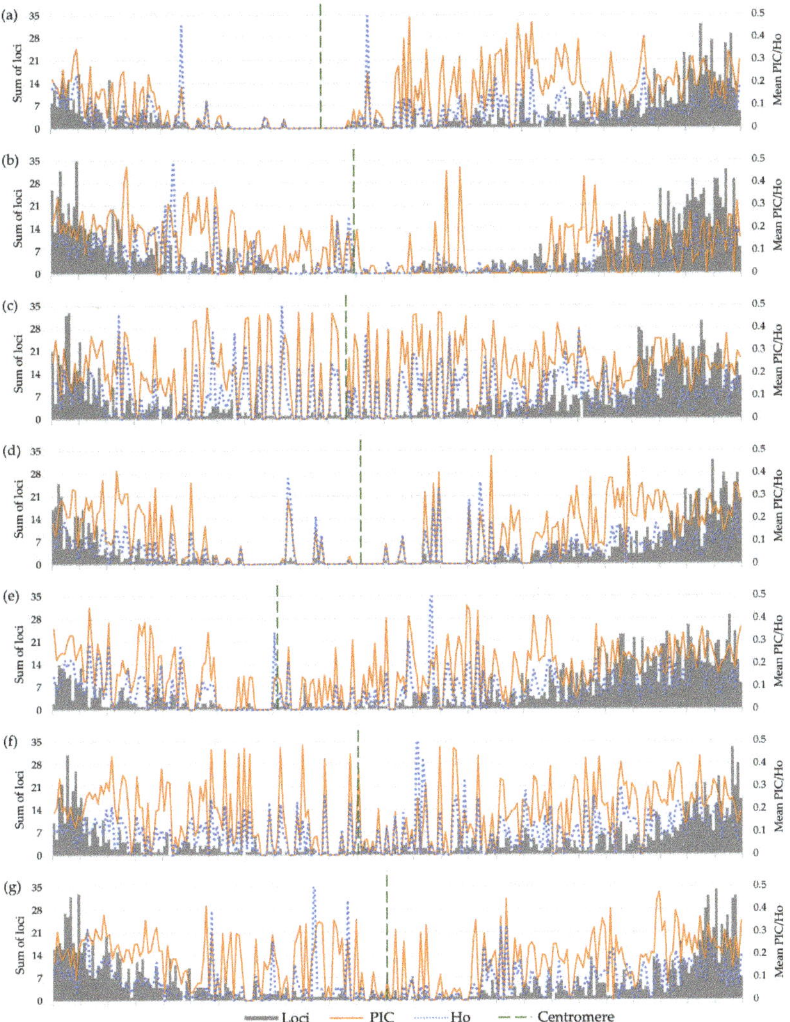

Figure 3. Frequency of analyzed loci and mean PIC and Ho values along chromosomes were assessed by a sliding window approach with 500 kb windows at 250 positions along the full length of barley chromosomes based on the genome assembly: IBSC_v2 [51] all chromosomes have been normalized to a standard length; (**a**–**g**) chromosome 1H—7H, respectively.

3.2.2. Genetic Diversity

For all analyzed loci, the mean polymorphic information content (PIC) value was 0.22 and the median was 0.20. About 30% of loci had PIC below 0.1, while about 18% of loci had PIC above 0.4 thus were highly informative (Figure S2).

The genetic variation parameters, such as observed heterozygosity (uHo), expected heterozygosity (uHe), and fixation index (F) for the entire set of accessions had the values 0.114, 0.232, and 0.519 respectively. Because the analysis was performed on pooled samples, the observed heterozygosity coefficient actually shows the heterogeneity of individuals within the accessions. The lowest observed heterogeneity (uHo) had PL 41430 (0.010) while the highest was found in PL 503844 (0.422) (Table S1). This indicates the presence of significant differences in the level of heterogeneity of the accessions studied, which may have significant consequences for the phenotypic differentiation within the accessions. In groups according to country of origin, accessions originating from Georgia had the highest observed heterogeneity (0.206) and the lowest fixation index (-0.628) while accessions originating from Iran had the lowest uHo (0.072) and the highest F (0.722). This indicates significant internal differentiation among the accessions from Georgia. However, there is little qualitative difference in SNPs between the accessions. The opposite situation occurred in the case of accessions from Iran, where there were significant qualitative differences in SNPs between accessions, while individuals within each accession were genetically very similar. Allele richness was highest in the group of accessions from Georgia and lowest in those from Poland. All three coefficients were also calculated considering their chromosomal location. Observed heterogeneity for individual chromosomes ranged from 0.093 (chr 1H) to 0.124 (chr 6H) and this difference was significant (Figure 4). However, when it was combined with a classification by country, it turned out that for Poland, Iran, and Georgia, the greatest number of heterogenous loci occurred on chromosome 6H, for Lithuania on chr 5H, and the USSR on chr 4H. The greatest homozygosity was observed on chromosome 1H in the accessions from Lithuania and the USSR, on chr 4H in the accessions from Poland, and chr 5H in those from Georgia and Iran (Figure 4). These differences were statistically significant.

The number of unique SNPs for the studied countries was also determined. The maximum number of unique SNPs was found in accessions originating from Iran (775) and the minimum in those from the former Czechoslovakia (13). Their distribution on chromosomes was as follows: the highest number of unique SNPs occurred on chr 2H (326) and the lowest on chr 6H (162) (Figure 4)

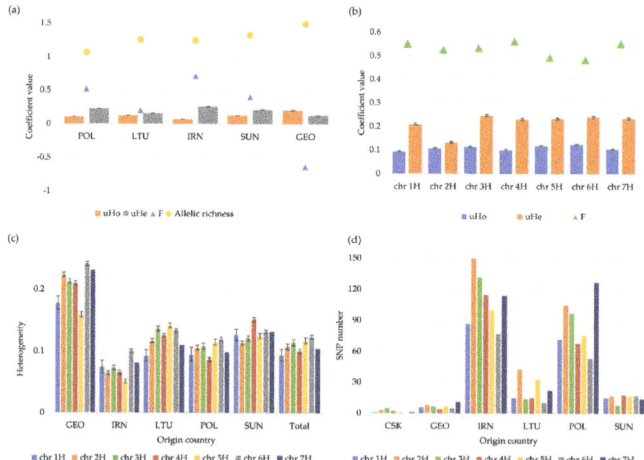

Figure 4. Diversity coefficients (**a**) unbiased coefficient of variation (uHe), unbiased observed heterogeneity (uHo), and Fixation Index (F) for 116 spring barley landraces based on DArTseq data calculated for groups of landrace in accordance with their origin; (**b**) diversity coefficients for 116 spring barley landraces based on DArTseq data, calculated for groups according to the location of the tested loci on barley chromosomes; (**c**) heterogeneity on individual chromosomes according to the country of accessions origin; (**d**) number of unique SNPs on barley chromosomes according to accession country of origin.

3.2.3. Principal Coordinate Analysis

Principal Coordinate Analysis (PCoA) showed that 42.5% of the variation was explained by the first three axes (27.2%, 8.2%, and 7.2% of the variation, respectively). Graphical presentation of the results in a 3D plot of the first three coordinates divided the studied accessions into three groups (Figure 5). The first and the largest group on the left side of the plot comprised 85 accessions, as a conglomeration of landraces collected in Poland, Lithuania, and former Czechoslovakia. Although the accessions from Poland and Lithuania form this large group, one can see a shift of the Lithuanian accessions in relation to the second and third coordinates. This indicates that gene pools from these neighboring countries are similar, but not identical. Outliers from this group were PL 41633, PL 43357, and PL 503844. The second distinct group, on the right side of the 3D chart, was formed by 11 accessions that originated from Poland. All of them were hulless and accession PL 41867 was an outlier from this group. In the space between these two main groups, a low-density cluster of accessions was found. Its upper part consisted of 6-row landraces collected in Poland, while the lower one was composed of accessions originating from Iran, Georgia, and the former Soviet Union.

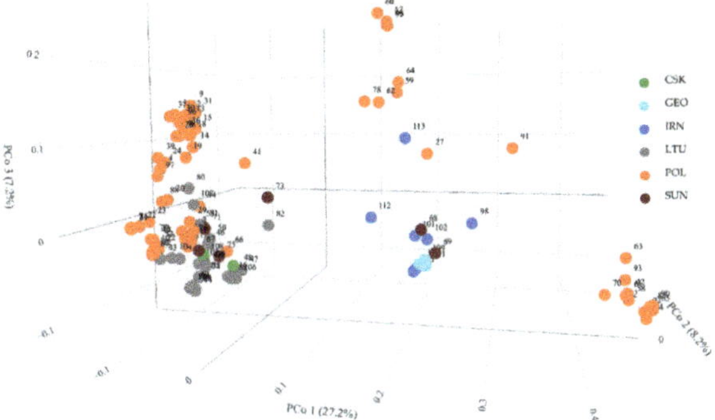

Figure 5. Results of Principal Coordinate Analysis (PCoA) for 116 spring barley landraces with an indication of the country of origin. The accession numbers according to Table S1. RotaTable 3D figures with an indication of the country of origin, grain type, and ear type can be found in the Supplementary Materials (Figures S3–S5).

3.2.4. Population Structure

Analysis of molecular variance AMOVA for barley landraces showed that only 15% of genetic variation could be associated with the country of accession origin. Grain coverage had a far greater influence on population structure. Here, 41% of the variability was related to this trait.

The STRUCTURE analysis [53] showed that population structure is evident in the studied spring barley landraces. According to an ad hoc statistic ΔK based on the second-order rate of change of the likelihood function regarding K, it was assumed that in the investigated landraces, a primary structure existed in which two gene pools took part. The presence of a low-order structure was also detected, indicating three gene pools. Individuals were assigned into groups based on an 80% membership threshold. At the primary level, 82 accessions were classified into the first gene pool and 11 accessions were classified into the second one. The remaining 23 accessions displayed a varying admixture. Here, as in previous analyses, a group of hulless accessions was isolated (Figure 6a). Three gene pools were partitioned at the lower level (Figure 6b). The first comprised 80 accessions. Compared to the primary structure, two accessions were mixed (PL 43357 and PL 41633). The second gene pool corresponded fully to the hulless one separated in the primary structure. In contrast, the third gene pool was formed from 19 accessions originally considered as admixed. Six accessions that were not assigned unambiguously to either gene pool in the PCoA analysis were considered as outliers. The results of the population structure and PCoA analysis were consistent.

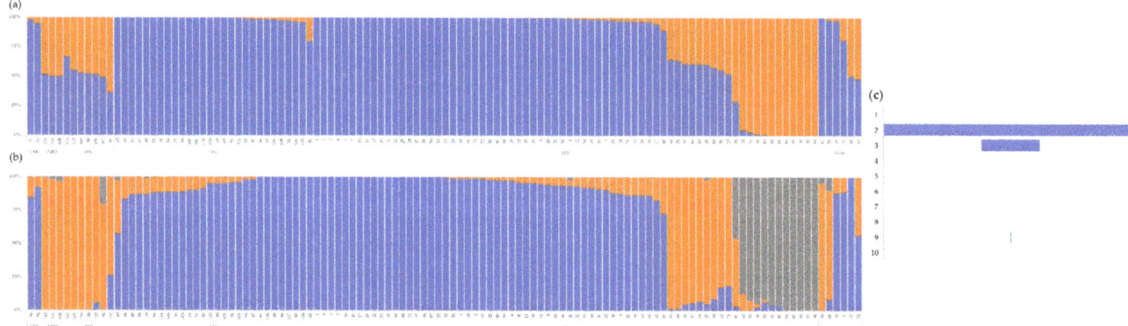

Figure 6. (**a**) Results of 100,000 iterations of the STRUCTURE program [53] for 116 spring barley landraces based on DArTseq data at K = 2, where K is the number of ad hoc groups formed; each vertical bar represents one accession, which is labeled with a sequence number (Table S1). The length of the colored bar indicates the estimated proportion of that sample's membership in each group. (**b**) Results of 100,000 iterations of the STRUCTURE program [53] for 116 spring barley landraces based on DArTseq data at K = 3, where K is the number of ad hoc groups formed; each vertical bar represents one accession, which is labeled with a sequence number (Table S1). The length of the colored bar indicates the estimated proportion of that sample's membership in each group; (**c**) the results of ad hoc measure ΔK [54] generated by CLUMPAK software [58].

3.3. Joint Analysis

Combining the results of PCoA analysis and agro-morphological data enabled identifying accessions exhibiting both high levels of resistance to the studied diseases and genetic distinctiveness. Accessions showing high levels of resistance to three out of four tested diseases were in all three groups (Figure 7). Most of them belonged to the largest group, previously described as group one. These accessions originated from Poland and were hulled. On the other hand, accessions originated from Lithuania showed resistance to lodging. Among the group of hulless accessions, nine were highly resistant to the three diseases studied, but none of them was resistant to lodging. The middle group included seven accessions resistant to three diseases and four ones resistant to lodging. Landraces of Iranian origin, although genetically distinct, showed a high level of resistance only to scald.

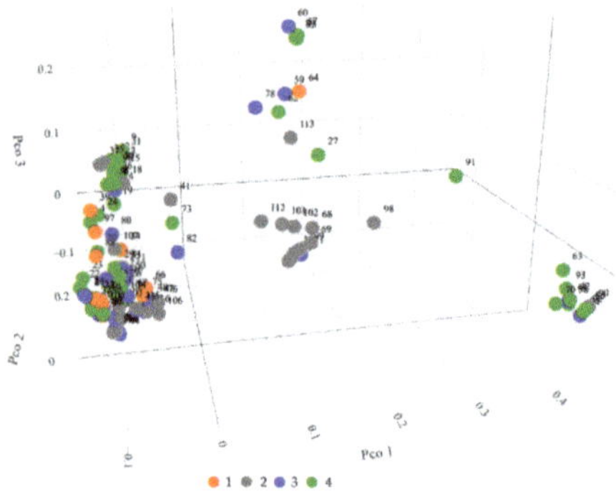

Figure 7. Results of Principal Coordinate Analysis (PCoA) for 116 spring barley landraces with indication of disease resistance. The accession numbers according to Table S1. The RotaTable 3D figure can be found in the Supplementary Materials (Figure S6).

3.4. Diversity of Chromosome 7H

Due to the genetic distinctness of the hulless and hulled forms, diversity analysis was performed within chr 7H on which the *nud* gene for hullessness is located. None of the analyzed SNPs was located adjacent to the *nud* gene. However, there was a group of 25 SNPs in its proximity, most of which were unique to hulless forms. At the same time, there was a very low proportion of heterozygous loci in this region, which indicates a significant homogeneity of accessions in this region. PCoA analysis showed that variation occurring within chromosome 7H differentiated accessions into hulled and hulless groups (Figure 8).

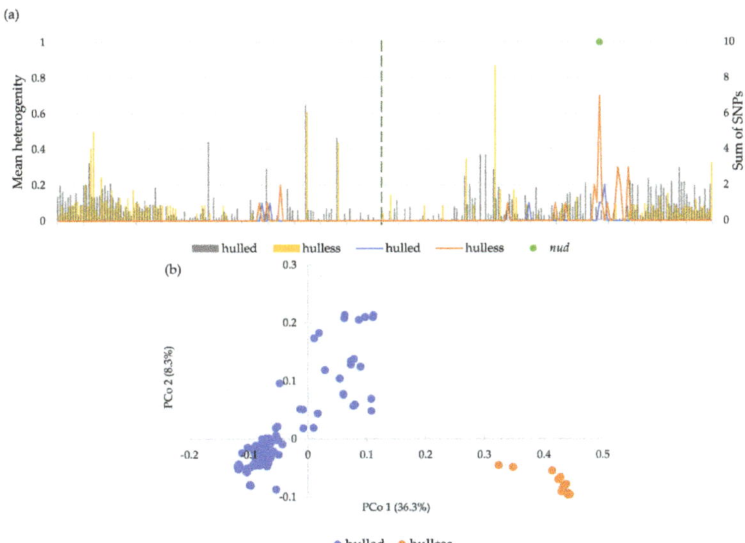

Figure 8. Genetic variation on the 7H chromosome for 116 spring barley landraces; (**a**) level of heterogeneity and unique SNPs assessed by a sliding window approach with 500 kb windows at 250 positions along the entire chromosome; (**b**) results of Principal Coordinate Analysis (PCoA) with indication the grain type.

3.5. Comparison between DArTseq and ISSR Data

In the previous study, 64 landraces from Poland were used and genetic analysis was performed using ISSR markers [32]. From the results obtained here, a subgroup containing the same accessions was extracted. For obtained dissimilarity matrices, the Mantel test was performed. It showed a moderate uphill correlation between results obtained by ISSR and DArTseq method (0.552, $p < 0.0001$). Generalized Procrustes Analysis (GPA) was carried out to obtain a consensus configuration of ISSR and DArTseq data. The most effective transformation was scaling and followed by a rotation. Both types of data matched the consensus configuration at a similar level as pointed out by the equal value of residuals by configuration after the transformation. The results of the consensus test indicated the authenticity of the configuration. The first three factors contributed to 74.72% of the original variability. A scatter plot showed that the studied accessions formed three groups (Figure 9). On the left side of the coordinate system, there were all the accessions with hulless grain type. Out of them, three outliers were identified, i.e., PL 42867, PL 41869, and PL 41871.

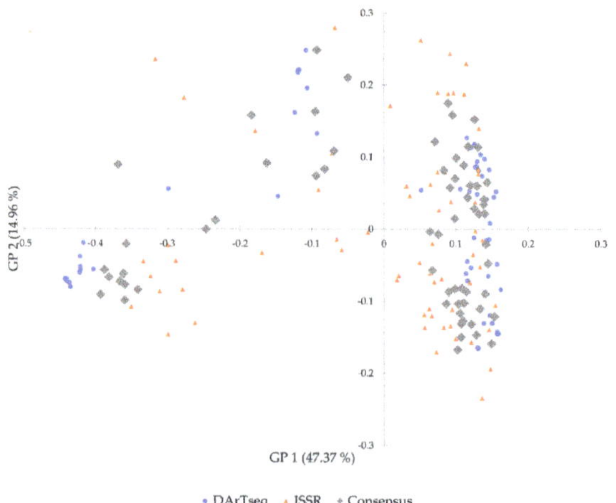

Figure 9. Generalized Procrustes Analysis (GPA) of 116 spring barley cultivars landraces based on ISSR and DArTseq data.

4. Discussion

The analysis presented here is the initial step towards genome-wide high-density genotyping of the barley collection preserved in the Polish gene bank. The results obtained should complement the passport data, characterization, and evaluation of accessions in the collection. Molecular and especially genome-wide studies provide extremely valuable information that will become increasingly important to gene bank managers. Molecular data enable identification of duplicates in a collection, assessment of loss of genetic integrity, and an informed selection of materials for further study and breeding.

Here, 116 spring barley landraces acquired during expeditions between 1977 and 2013 were evaluated. The value of the coefficient of variation (uHe = 0.232) indicates rather average genetic variability of the collection. Detailed analysis showed that this value is significantly influenced by a substantial proportion of loci with low levels of variation. In earlier studies, using ISSR markers, the diversity of landraces from Poland was assessed [32]. The level of variation detected there was 0.185 and was a little lower than that recorded now (0.226). These differences may be derived from the greater uniformity of SNP distribution in the genome, their greater number, and higher resolution compared to the ISSR technique. However, ISSR markers are considered as a group of highly informative, multi-allelic loci that provide highly discriminating information with good reproducibility and are relatively common [59].

Among the accessions studied here, there was a large variation in observed heterozygosity values ranging from 0.01 to 0.42. In the sample, 51% of all investigated accessions had a heterozygosity level above 0.1. The occurrence of heterozygous loci in the barley collection, which is a self-pollinating species, directly results from the way the analysis was performed. In the study presented here, pooled samples of eight individuals representing the accession were used. Thus, the level of heterozygosity of an accession indicates the heterogeneity of the comprising individuals. The use of pooled samples significantly reduced the cost of analysis, but the heterogeneity assessed in this way is affected by higher error than in the analysis of individuals. Furthermore, the frequency of SNPs in the accession cannot be assessed in this way. Heterogeneity of landraces is consistent both with their definition and with previous results of analyses conducted on a group of individuals representing landraces of self-pollinated cereals [7,60,61]. The obtained results provide an important indication for researchers and breeders that the seed sample obtained from the gene bank will be a mixture of different genotypes. Keeping the accessions in heterogeneous form is advantageous from the point of view of maintaining diversity, but it is a significant obstacle for breeders. Using heterogeneous accessions in breeding programs forces an additional workload associated with the selection of pure lines. Heterozygous loci were identified on all chromosomes. Looking at the entire collection, the proportion of heterozygous loci on individual chromosomes was stable at about 10%.

Detailed analysis of loci distribution, their polymorphism, and heterogeneity showed significant variation at the chromosome level. All chromosomes showed a similar pattern of loci occurrence, i.e., a decrease in their frequency from distal parts towards the centromere. This pattern is consistent with previous results of high-throughput SNP analyses for barley, durum wheat, and soybean [51,62,63]. "Empty" regions adjacent to centromeres were present for certain chromosomes. The lack of analyzed loci in these regions was a consequence of polymorphism absence in the loci occurring there. As a result of data filtering, these loci were removed from the analysis. Similar results were obtained by Mascher et al. [51] who analyzed homozygous inbred elite barley lines. Within spring forms, these authors identified segments completely lacking variation in the centromeric and pericentromeric regions of chromosomes 1H, 2H, and 7H, while in winter lines at 5H. In the spring landraces analyzed here, highly homogeneous regions bilaterally surrounded the centromere on 1H and 4H. On chromosome 5H, such a region was present only on the short arm, whereas on 2H on the long arm but not in the direct vicinity of the centromere. For the remaining chromosomes, the level of variation remained relatively high along their entire length. The different pattern of highly homogenous regions, landraces, and elite lines may reflect the different selection pressures applied to elite materials during the breeding process and landraces during their establishment on farms. Considering that the landraces studied here came from different geographic locations, it can be assumed that the regional reduction in variability in these regions may also be the result of domestication, as has been demonstrated for durum wheat [64]. Moreover, both the level of heterozygosity and unique SNPs frequency on individual chromosomes showed differentiation in groups consistent with the country of origin. A different share of heterozygous loci and unique SNPs on individual chromosomes may result from the adaptation of landraces to specific eco-climatic conditions or from a targeted selection by farmers for a specific trait. It should be kept in mind, however, that the number of accessions analyzed from different countries showed considerable variation. Therefore, the results obtained in this study cannot be generalized to the genetic diversity of landraces occurring in the countries concerned, nor to the size of the native gene pools. The results obtained during this analysis present only the diversity of accessions that were collected by the Polish gene bank. Only the accessions from Poland can be considered representing the full preserved genetic diversity of native landraces. This study is even more valuable because Polish landraces have not previously been subjected to high-throughput genotyping using SNPs. In a study by Milner et al. [43], only eight, and in Bustos-Korts et al. [65] only four landraces from Poland were included. Unique results were also obtained for accessions originating from Lithuania. Accessions from this country were not included in previous studies either [43,65]. Overall, the accessions from Lithuania and Poland had a similar average level of heterogeneity. Based on the results of the PCoA analysis, the gene pools from these countries are similar, but not identical. Eight of the twenty-eight Lithuanian accessions were highly homogeneous. Because expeditions took place relatively recently, in 2011–2013, and the collected grain samples were not critically small, it can be assumed that these accessions may not be landraces, but are cultivars, although not from the group of the most contemporary ones. This is also supported by high genetic similarity of two of them, i.e., PL 503854 and PL 501965. Out of 22 K SNPs examined, these accessions differed only by 130, even though they were collected in different locations.

The group of accessions originating from Iran had the lowest level of heterozygous loci. This may result from collecting too few seeds during the expedition or losing some material during reproduction in the gene bank. Nevertheless, this collection is characterized by high diversity among the accessions. The high number of unique alleles is also significant. Although in the studied set only seven accessions originated from Iran, as many as 775 unique alleles were identified for them, whereas for the collection of 70 accessions of Polish landraces, 598 SNPs were unique, including alleles unique for hulless forms. This indicates that landraces from Iran may contain a significant number of alleles absent in the European gene pool. Accessions from the Middle East also showed genetic distinctness from those originated from Europe and the former Soviet Union in earlier studies [43]. Considering the above results and the data of agro-morphological traits evaluation, it should be assumed that the collection of Iranian landraces stored in the Polish gene bank can be a valuable source of genetic variability. These accessions were characterized by high resistance to scald; there were also accessions highly resistant to lodging among them, which may relate to low plant height. Thus, they make up a reservoir of variability which may be valuable for future breeding.

The accessions originating from Georgia were characterized by a high level of internal heterogeneity and low differentiation between accessions. This may indicate that their gene pools overlap significantly and that they may share a common ancestor. This will also influence the strategy for finding useful traits in collections from these countries. In the case of accessions from Georgia, screening should be conducted on a much larger number of individuals representing the accession than in the case of accessions from Iran, where more emphasis should be placed on the number of accessions analyzed. The results of DArTseq analysis also showed some distinctness of the six-row forms from the two-row ones. However, it was not as significant as in studies of other barley germplasm collections [66–68]. In contrast, in our previous study of only Polish landraces using ISSRs, the distinctiveness of these groups was not apparent [32]. Some distinctness of the six-row forms here may be derived from both an increase in the number of accessions studied and a change in the analysis method.

The result that integrates the previous analysis by ISSR and the current one by DArTseq is the distinctness of hulless and hulled accessions [32]. The correlation level of the dissimilarity matrix of both methods was at a moderate level. It was significantly lower compared to previous barley studies using SSRs and SNPs. However, it should be taken into account that ISSRs are dominant markers in contrast to SNPs and SSRs [33,34]. Moreover, in the present and previous analysis, although the same accessions were analyzed, the DNA matrices were not the same, i.e., for ISSR analysis, bulk samples comprising 24 individuals were used, whereas for DArTseq only eight individuals per accession were used. The distinctness of hulless and hulled accessions was also indicated by grain morphometry. The grain of the hulless forms is considerably smaller than hulled ones. It is believed that the domestication of hulless barley followed the hulled type and this was around 6500 BC [69]. The most likely origin of hulless barley is monophyletic. It probably arose from a single mutation of either wild barley (*H. vulgare* subsp. *spontaneum*) or domesticated hulled barley [70,71]. The hulless grain is a very stable trait and environmental conditions have little influence on its expression. It is determined by a single recessive gene *"nud"* on the long arm of chromosome 7H [72]. At this locus, there is a gene encoding a transcription factor of the Ethylene Response Factor (ERF) family belonging to the WIN1/SHN1 (Wax Inducer 1/Shine 1) transcription factor group [73]. The hulless grain trait is associated with either a 17kb deletion or amino acid conversion of T643A at the *NUD* locus [74]. Compared to hulled barley, hulless barley is rich in nutrients such as limiting amino acids (lysine, methionine, threonine, and tryptophan), starch, fiber, and β-glucan [75]. The grain of hulless barley is not only cholesterol-free but also has cholesterol-lowering properties due to its high β-glucan content. In addition, barley fiber is a source of niacin (vitamin B), which decreases platelet aggregation, total cholesterol, lipoproteins, and free radicals [76,77]. It is also a valuable source of thiamine, selenium, iron, magnesium, zinc, phosphorus, and copper. Depending on the genotype, the content of mineral nutrients ranges from 2 to 3% [78]. All these properties make hulless barley a part of a healthy diet. All except one of the hulless accessions analyzed here originated from Poland. They are therefore adapted to central European climatic conditions and were characterized by high resistance to net blotch, scald, and steam rust. Therefore, they have the potential to be used in pro-quality breeding. Moreover, naked barley appears as one of the future and climate-smart crops due to its hardiness, ability to grow under low input conditions, and potential for use in climate change adaptation [79]. However, due to accessions heterogeneity, prior selection of pure lines and their evaluation for agronomic traits, yield, and nutrient content will be necessary. By mapping the DArTseq results to the barley reference genome, it was possible to identify the exact chromosomal locations of the loci analyzed. Detailed analysis of chromosome 7H showed that in the region of the *nud* gene, there were multiple SNPs unique to the hulless form and the heterogeneity in this chromosome fragment was significantly reduced. This is consistent with previous findings that indicated both low variability and extensive linkage disequilibrium in the vicinity of the *nud* gene [70,80]. No pleiotropic effect of the *nud* gene was found on agronomic traits such as grain yield and weight, plant height, or heading date [81,82]. In this region, Wang et al. [83] localized a QTL hotspot region underlying traits related to grain size and weight. In addition, numerous previous works indicate the presence of QTLs of grain yield and thousand-grain weight in this region [81,84,85]. Thus, the identified SNPs may also be associated with grain weight and size which may be useful in breeding high-yielding cultivars of hulless barley. Currently, the yield of hulless barley is much lower than that of hulled barley, which is mainly due to the adhering husk on hulled barley, which accounts for about 10–13% of the weight and volume of harvested barley grain [86]. Intensification of breeding work to improve the yield of hulless-grain barley to that of hulled barley, therefore, seems reasonable. The Cas9 RNA-guided endonuclease used to knockout the *nud* gene in hulled cultivars seems to be an interesting tool in this context [82].

Another important aspect is the color of barley grain. The mature barley grain could have different seed coat colors including but not limited to yellow, blue, purple, and black [87]. The color of the seed coat is related to flavonoids and specifically to the synthesis of anthocyanins. It is believed that proanthocyanidins synthesized in the testa are responsible for the yellow color, and purple pigmentation is due to the synthesis of anthocyanins in the pericarp and glumes [88–90]. Four loci of purple seed coat color (*Psc*) were located on chromosome 7H [91]. One of them is in a region where an accumulation of SNPs unique to hulless forms was identified. Its sequence is 95% homologous to Arabidopsis *F3'M* (AK366933.1), which belongs to the cytochrome P450 family and is associated with flavonoid biosynthesis [92]. In crop plants, seed coat color is an important agronomic trait because of its association with unique biological activity and, hence, function in health care [93]. The continuously growing market of functional food may be a trigger for the development of breeding colored barley cultivars containing high levels of natural antioxidants such as phenolic compounds, anthocyanins, and essential amino acids [94]. Using barley flour, which itself has antioxidant properties, in the production of bread, pasta, or confectionery products, is becoming increasingly popular. Thus, the use of colored grains for flour, groats, or whole-grain flakes as a trendy health food will become feasible as soon as more cultivars with colored grains are available in the market. In the collection studied, there were both accessions with very light and almost black grain color. The presence of accessions with dark grain may be the beginning of targeted breeding programs. A detailed further analysis of barley collections in the Polish gene bank will probably provide a larger number of accessions with colored grains, including probably also old varieties and breeding materials. Integrating seed morphometric analysis, genome-wide genotyping, and agronomic traits evaluation supplemented with a selection of pure lines as a part of core activities of the gene bank may significantly contribute to the development of pro-quality barley breeding. Parallel educational and information campaigns will contribute to the production and consumption of healthier food and the reduction of civilization diseases in society.

5. Conclusions

The landraces collection in the Polish gene bank displays both genetic and morphological diversity. Worth noting is the fact of varying levels of genetic diversity depending on both the chromosome and the accession country of origin. The gene pools of Polish and Lithuanian landraces show considerable similarity but are not identical. Accessions originating from Iran are characterized by significant genetic distinctiveness from most of the collection and contain many unique loci. The preserved accessions have a variable level of internal heterogeneity. DArTseq analysis confirmed the genetic distinctiveness of hulless forms indicated by ISSR markers in the previous study. It also enabled the identification of unique SNPs located on chromosome 7H within the region carrying the *nud* gene that determines the hulled/hulless caryopsis phenotype. These SNPs may also be important in the selection for traits related to grain weight and size because their QTLs were found there. To improve landraces use in breeding, it is necessary to create special collections for commercial use containing pure lines representing accessions and to provide them with the most comprehensive characterization possible. Such actions are necessary to inject new variability into modern breeding, which must cope with both global warming and increasing consumer expectations for quality and pro-health traits.

Supplementary Materials: The following are available online at https://www.mdpi.com/article/10.3390/agronomy11112330/s1: Table S1: The list of accessions used in the study with eco-geographic description of collection sites; Figure S1: Summary of the number of loci from the raw data and the number of loci that remained after filtering; Figure S2: The distribution of polymorphic information content (PIC) at the loci studied; Figure S3: 3D plot of results of Principal Coordinate Analysis (PCoA) for 116 spring barley landraces with indication of the country of origin. The accession numbers according to Table S1; Figure S4: 3D plot of results of Principal Coordinate Analysis (PCoA) for 116 spring barley landraces with indication of ear type. The accession numbers according to Table S1; Figure S5: 3D plot of results of Principal Coordinate Analysis (PCoA) for 116 spring barley landraces with indication of grain type. The accession numbers according to Table S1; Figure S6: 3D plot of results of Principal Coordinate Analysis (PCoA) for 116 spring barley landraces with indication disease resistance. The accession numbers according to Table S1.

Author Contributions: Conceptualization, M.B.; methodology, M.B.; formal analysis, M.B.; investigation, J.D., J.G. and G.G.; resources, W.P.; data curation, J.D. and M.B.; writing—original draft preparation, J.D.; writing—review and editing, M.B.; visualization, J.D. and M.B.; supervision, M.B.; project administration, W.P. and M.B.; funding acquisition, W.P. All authors have read and agreed to the published version of the manuscript.

Funding: This work was supported by the Multi-annual program: 2015–2020 "Establishment of a scientific basis for biological progress and preservation of plant genetic resources as a source of innovation to support sustainable agriculture and food security of the country" coordinated by Plant Breeding and Acclimatization Institute (IHAR) National Research Institute and financed by the Ministry of Agriculture and Rural Development of Poland. The calculations were performed at the Interdisciplinary Centre for Mathematical and Computational Modelling, University of Warsaw (ICM UW) within the framework of Computational Grant No. G72–19.

Institutional Review Board Statement: Not applicable.

Informed Consent Statement: Not applicable.

Data Availability Statement: The row data of DArTseq SNP used in this study are openly available on the platform Center for Open Science at https://osf.io/mh4tn/ (accessed on 23 October 2021). (doi 10.17605/OSF.IO/MH4TN).

Acknowledgments: The authors would like to express their gratitude to Boguslaw Lapinski for his constructive comments on the manuscript.

Conflicts of Interest: The authors declare no conflict of interest. The funders had no role in the design of the study; in the collection, analyses, or interpretation of data; in the writing of the manuscript, or in the decision to publish the results.

References

1. Veteläinen, M.; Negri, V.; Maxted, N. *European Landraces: On-Farm Conservation, Management and Use*; Bioversity International: Rome, Italy, 2009; pp. 1–22.
2. Marone, D.; Russo, M.A.; Mores, A.; Ficco, D.; Laidò, G.; Mastrangelo, A.M.; Borrelli, G.M. Importance of Landraces in Cereal Breeding for Stress Tolerance. *Plants* **2021**, *10*, 1267. [CrossRef] [PubMed]
3. FAO. *Voluntary Guidelines for the Conservation and Sustainable Use of Farmers' Varieties/Landraces*; FAO: Rome, Italy, 2017.
4. Maxted, N.; Dulloo, M.E.; Ford-Lloyd, B.V. *Enhancing Crop Genepool Use: Capturing Wild Relative and Landrace Diversity for Crop Improvement*; CABI: Boston, MA, USA, 2016.
5. Doebley, J.F.; Gaut, B.S.; Smith, B.D. The molecular genetics of crop domestication. *Cell* **2006**, *127*, 1309–1321. [CrossRef] [PubMed]
6. Kilian, B.; Özkan, H.; Kohl, J.; von Haeseler, A.; Barale, F.; Deusch, O.; Brandolini, A.; Yucel, C.; Martin, W.; Salamini, F. Haplotype structure at seven barley genes: Relevance to gene pool bottlenecks, phylogeny of ear type and site of barley domestication. *Mol. Genet. Genom.* **2006**, *276*, 230–241. [CrossRef] [PubMed]
7. Boczkowska, M.; Łapiński, B.; Kordulasińska, I.; Dostatny, D.F.; Czembor, J.H. Promoting the use of common oat genetic resources through diversity analysis and core collection construction. *PLoS ONE* **2016**, *11*, e0167855.
8. Ellis, R.P.; Forster, B.P.; Robinson, D.; Handley, L.; Gordon, D.C.; Russell, J.R.; Powell, W. Wild barley: A source of genes for crop improvement in the 21st century? *J. Exp. Bot.* **2000**, *51*, 9–17. [CrossRef]
9. Sharma, S.; Upadhyaya, H.D.; Varshney, R.K.; Gowda, C. Pre-breeding for diversification of primary gene pool and genetic enhancement of grain legumes. *Front. Plant Sci.* **2013**, *4*, 309. [CrossRef]
10. Kumar, A.; Verma, R.P.S.; Singh, A.; Sharma, H.K.; Devi, G. Barley landraces: Ecological heritage for edaphic stress adaptations and sustainable production. *Environ. Sustain. Indic.* **2020**, *6*, 100035. [CrossRef]
11. Pasam, R.K.; Sharma, R.; Walther, A.; Özkan, H.; Graner, A.; Kilian, B. Genetic diversity and population structure in a legacy collection of spring barley landraces adapted to a wide range of climates. *PLoS ONE* **2014**, *9*, e116164. [CrossRef]
12. Ceccarelli, S. Positive interpretation of genotype by environment interactions in relation to sustainability and biodiversity. In *Plant Adaptation and Crop Improvement*; IRRI: Manila, Philippines, 1996; pp. 467–486.
13. Grausgruber, H.; Bointner, H.; Tumpold, R.; Ruckenbauer, P.; Fischbeck, G. Genetic improvement of agronomic and qualitative traits of spring barley. *Plant Breed.* **2002**, *121*, 411–416. [CrossRef]
14. Rasmusson, D.; Phillips, R. Plant breeding progress and genetic diversity from de novo variation and elevated epistasis. *Crop Sci.* **1997**, *37*, 303–310. [CrossRef]
15. Okada, Y.; Kanatani, R.; Arai, S.; Ito, K. Interaction between barley yellow mosaic disease-resistance genes rym1 and rym5, in the response to BaYMV strains. *Breed. Sci.* **2004**, *54*, 319–325. [CrossRef]
16. Steffenson, B.J. Analysis of durable resistance to stem rust in barley. *Euphytica* **1992**, *63*, 153–167. [CrossRef]
17. Bregitzer, P.; Mornhinweg, D.; Hammon, R.; Stack, M.; Baltensperger, D.; Hein, G.; O'neill, M.; Whitmore, J.; Fiedler, D. Registration of 'Burton' barley. *Crop Sci.* **2005**, *45*, 1166–1168. [CrossRef]
18. Burnett, P.; Comeau, A.; Qualset, C. Host plant tolerance or resistance for control of barley yellow dwarf. In *Barley Yellow Dwarf: 40 Years of Progress*; D'Arcy, C.J., Burnett, P.A., Eds.; APS Press: Saint Paul, MN, USA, 1995; pp. 321–343.
19. Piffanelli, P.; Ramsay, L.; Waugh, R.; Benabdelmouna, A.; D'Hont, A.; Hollricher, K.; Jørgensen, J.H.; Schulze-Lefert, P.; Panstruga, R. A barley cultivation-associated polymorphism conveys resistance to powdery mildew. *Nature* **2004**, *430*, 887–891. [CrossRef]
20. Dawson, I.K.; Russell, J.; Powell, W.; Steffenson, B.; Thomas, W.T.; Waugh, R. Barley: A translational model for adaptation to climate change. *New Phytol.* **2015**, *206*, 913–931. [CrossRef] [PubMed]
21. Pietrusińska, A.; Żurek, M.; Piechota, U.; Słowacki, P.; Smolińska, K. Searching for diseases resistance sources in old cultivars, landraces and wild relatives of cereals. A review. *Agron. Sci.* **2018**, *73*, 45–60. [CrossRef]
22. Hernandez, J.; Meints, B.; Hayes, P. Introgression breeding in barley: Perspectives and case studies. *Front. Plant Sci.* **2020**, *11*, 761. [CrossRef]
23. Jaradat, A.; Shahid, M. Population and multilocus isozyme structures in a barley landrace. *Plant Genet. Resour.* **2006**, *4*, 108–116. [CrossRef]
24. Bjørnstad, A.; Demissie, A.; Kilian, A.; Kleinhofs, A. The distinctness and diversity of Ethiopian barleys. *Theor. Appl. Genet.* **1997**, *94*, 514. [CrossRef]
25. Demissie, A.; Bjørnstad, Å.; Kleinhofs, A. Restriction Fragment Length Polymorphisms in Landrace Barleys from Ethiopia in Relation to Geographic, Altitude, and Agro-Ecological Factors. *Crop Sci.* **1998**, *38*, 237–243. [CrossRef]
26. Manjunatha, T.; Bisht, I.; Bhat, K.; Singh, B. Genetic diversity in barley (*Hordeum vulgare* L. ssp. vulgare) landraces from Uttaranchal Himalaya of India. *Genet. Resour. Crop Evol.* **2007**, *54*, 55–65. [CrossRef]
27. Abdellaoui, R.; M'Hamed, H.C.; Naceur, M.B.; Bettaieb-Kaab, L.; Hamida, J.B. Morpho-physiological and molecular characterization of some Tunisian barley ecotypes. *Asian J. Plant Sci.* **2007**, *6*, 26–268. [CrossRef]
28. Yahiaoui, S.; Igartua, E.; Moralejo, M.; Ramsay, L.; Molina-Cano, J.L.; Lasa, J.; Gracia, M.; Casas, A. Patterns of genetic and eco-geographical diversity in Spanish barleys. *Theor. Appl. Genet.* **2008**, *116*, 271–282. [CrossRef] [PubMed]
29. Feng, Z.-Y.; Zhang, L.-L.; Zhang, Y.-Z.; Ling, H.-Q. Genetic diversity and geographical differentiation of cultivated six-rowed naked barley landraces from the Qinghai-Tibet plateau of China detected by SSR analysis. *Genet. Mol. Biol.* **2006**, *29*, 330–338. [CrossRef]

30. Hamza, S.; Hamida, W.B.; Rebaï, A.; Harrabi, M. SSR-based genetic diversity assessment among Tunisian winter barley and relationship with morphological traits. *Euphytica* **2004**, *135*, 107–118. [CrossRef]

31. Van Treuren, R.; Tchoudinova, I.; Van Soest, L.; van Hintum, T.J. Marker-assisted acquisition and core collection formation: A case study in barley using AFLPs and pedigree data. *Genet. Resour. Crop Evol.* **2006**, *53*, 43–52. [CrossRef]

32. Dziurdziak, J.; Bolc, P.; Wlodarczyk, S.; Puchta, M.; Gryziak, G.; Podyma, W.; Boczkowska, M. Multifaceted Analysis of Barley Landraces Collected during Gene Bank Expeditions in Poland at the End of the 20th Century. *Agronomy* **2020**, *10*, 1958. [CrossRef]

33. Varshney, R.; Baum, M.; Guo, P.; Grando, S.; Ceccarelli, S.; Graner, A. Features of SNP and SSR diversity in a set of ICARDA barley germplasm collection. *Mol. Breed.* **2010**, *26*, 229–242. [CrossRef]

34. Varshney, R.; Thiel, T.; Sretenovic-Rajicic, T.; Baum, M.; Valkoun, J.; Guo, P.; Grando, S.; Ceccarelli, S.; Graner, A. Identification and validation of a core set of informative genic SSR and SNP markers for assaying functional diversity in barley. *Mol. Breed.* **2008**, *22*, 1–13. [CrossRef]

35. Russell, J.R.; Ellis, R.P.; Thomas, W.T.; Waugh, R.; Provan, J.; Booth, A.; Fuller, J.; Lawrence, P.; Young, G.; Powell, W. A retrospective analysis of spring barley germplasm development fromfoundation genotypes' to currently successful cultivars. *Mol. Breed.* **2000**, *6*, 553–568. [CrossRef]

36. Nandha, P.S.; Singh, J. Comparative assessment of genetic diversity between wild and cultivated barley using g SSR and EST-SSR markers. *Plant Breed.* **2014**, *133*, 28–35. [CrossRef]

37. Mohammadi, S.A.; Sisi, N.A.; Sadeghzadeh, B. The influence of breeding history, origin and growth type on population structure of barley as revealed by SSR markers. *Sci. Rep.* **2020**, *10*, 19165. [CrossRef]

38. Lister, D.L.; Jones, H.; Jones, M.K.; O'Sullivan, D.M.; Cockram, J. Analysis of DNA polymorphism in ancient barley herbarium material: Validation of the KASP SNP genotyping platform. *Taxon* **2013**, *62*, 779–789. [CrossRef]

39. Close, T.J.; Bhat, P.R.; Lonardi, S.; Wu, Y.; Rostoks, N.; Ramsay, L.; Druka, A.; Stein, N.; Svensson, J.T.; Wanamaker, S. Development and implementation of high-throughput SNP genotyping in barley. *BMC Genom.* **2009**, *10*, 582. [CrossRef] [PubMed]

40. Bayer, M.M.; Rapazote-Flores, P.; Ganal, M.; Hedley, P.E.; Macaulay, M.; Plieske, J.; Ramsay, L.; Russell, J.; Shaw, P.D.; Thomas, W. Development and evaluation of a barley 50k iSelect SNP array. *Front. Plant Sci.* **2017**, *8*, 1792. [CrossRef] [PubMed]

41. Comadran, J.; Kilian, B.; Russell, J.; Ramsay, L.; Stein, N.; Ganal, M.; Shaw, P.; Bayer, M.; Thomas, W.; Marshall, D. Natural variation in a homolog of Antirrhinum CENTRORADIALIS contributed to spring growth habit and environmental adaptation in cultivated barley. *Nat. Genet.* **2012**, *44*, 1388–1392. [CrossRef]

42. Torkamaneh, D.; Laroche, J.; Bastien, M.; Abed, A.; Belzile, F. Fast-GBS: A new pipeline for the efficient and highly accurate calling of SNPs from genotyping-by-sequencing data. *BMC Bioinform.* **2017**, *18*, 5. [CrossRef] [PubMed]

43. Milner, S.G.; Jost, M.; Taketa, S.; Mazón, E.R.; Himmelbach, A.; Oppermann, M.; Weise, S.; Knüpffer, H.; Basterrechea, M.; König, P. Genebank genomics highlights the diversity of a global barley collection. *Nat. Genet.* **2019**, *51*, 319–326. [CrossRef] [PubMed]

44. Wenzl, P.; Carling, J.; Kudrna, D.; Jaccoud, D.; Huttner, E.; Kleinhofs, A.; Kilian, A. Diversity Arrays Technology (DArT) for whole-genome profiling of barley. *Proc. Natl. Acad. Sci. USA* **2004**, *101*, 9915–9920. [CrossRef]

45. Bengtsson, T.; Manninen, O.; Jahoor, A.; Orabi, J. Genetic diversity, population structure and linkage disequilibrium in Nordic spring barley (*Hordeum vulgare* L. subsp. vulgare). *Genet. Resour. Crop Evol.* **2017**, *64*, 2021–2033. [CrossRef]

46. Moragues, M.; Comadran, J.; Waugh, R.; Milne, I.; Flavell, A.; Russell, J.R. Effects of ascertainment bias and marker number on estimations of barley diversity from high-throughput SNP genotype data. *Theor. Appl. Genet.* **2010**, *120*, 1525–1534. [CrossRef] [PubMed]

47. NCPGR. EGISET. Available online: https://wyszukiwarka.ihar.edu.pl/pl (accessed on 15 August 2021).

48. Doyle, J.J.; Doyle, J.L. Isolation ofplant DNA from fresh tissue. *Focus* **1990**, *12*, 39–40.

49. Edwards, K.; Johnstone, C.; Thompson, C. A simple and rapid method for the preparation of plant genomic DNA for PCR analysis. *Nucleic Acids Res.* **1991**, *19*, 1349. [CrossRef] [PubMed]

50. Wójcik-Jagła, M.; Rapacz, M.; Barcik, W.; Janowiak, F. Differential regulation of barley (Hordeum distichon) HVA1 and SRG6 transcript accumulation during the induction of soil and leaf water deficit. *Acta Physiol. Plant.* **2012**, *34*, 2069–2078. [CrossRef]

51. Mascher, M.; Gundlach, H.; Himmelbach, A.; Beier, S.; Twardziok, S.O.; Wicker, T.; Radchuk, V.; Dockter, C.; Hedley, P.E.; Russell, J. A chromosome conformation capture ordered sequence of the barley genome. *Nature* **2017**, *544*, 427–433. [CrossRef]

52. Leberg, P. Estimating allelic richness: Effects of sample size and bottlenecks. *Mol. Ecol.* **2002**, *11*, 2445–2449. [CrossRef]

53. Hubisz, M.J.; Falush, D.; Stephens, M.; Pritchard, J.K. Inferring weak population structure with the assistance of sample group information. *Mol. Ecol. Resour.* **2009**, *9*, 1322–1332. [CrossRef]

54. Evanno, G.; Regnaut, S.; Goudet, J. Detecting the Number of Clusters of Individuals Using the Software Structure: A Simulation Study. *Mol. Ecol.* **2005**, *14*, 2611–2620. [CrossRef]

55. Gower, J.C. Generalized procrustes analysis. *Psychometrika* **1975**, *40*, 33–51. [CrossRef]

56. Peakall, R.; Smouse, P.E. Genalex 6: Genetic analysis in Excel. Population genetic software for teaching and research. *Bioinformatics* **2012**, *28*, 2537–2539. [CrossRef]

57. Kalinowski, S.T. hp-rare 1.0: A computer program for performing rarefaction on measures of allelic richness. *Mol. Ecol. Notes* **2005**, *5*, 187–189. [CrossRef]

58. Kopelman, N.M.; Mayzel, J.; Jakobsson, M.; Rosenberg, N.A.; Mayrose, I. Clumpak: A program for identifying clustering modes and packaging population structure inferences across K. *Mol. Ecol. Resour.* **2015**, *15*, 1179–1191. [CrossRef]

59. Zietkiewicz, E.; Rafalski, A.; Labuda, D. Genome fingerprinting by simple sequence repeat (SSR)-anchored polymerase chain reaction amplification. *Genomics* **1994**, *20*, 176–183. [CrossRef] [PubMed]

60. Villa, T.C.C.; Maxted, N.; Scholten, M.; Ford-Lloyd, B. Defining and identifying crop landraces. *Plant Genet. Resour.* **2005**, *3*, 373–384. [CrossRef]

61. Podyma, W.; Boczkowska, M.; Wolko, B.; Dostatny, D.F. Morphological, isoenzymatic and ISSRs-based description of diversity of eight sand oat (Avena strigosa Schreb.) landraces. *Genet. Resour. Crop Evol.* **2017**, *64*, 1661–1674. [CrossRef]

62. Ávila, C.M.; Requena-Ramírez, M.D.; Rodríguez-Suárez, C.; Flores, F.; Sillero, J.C.; Atienza, S.G. Genome-Wide Association Analysis for Stem Cross Section Properties, Height and Heading Date in a Collection of Spanish Durum Wheat Landraces. *Plants* **2021**, *10*, 1123. [CrossRef]

63. Hahn, V.; Würschum, T. Molecular genetic characterization of Central European soybean breeding germplasm. *Plant Breed.* **2014**, *133*, 748–755. [CrossRef]

64. Maccaferri, M.; Harris, N.S.; Twardziok, S.O.; Pasam, R.K.; Gundlach, H.; Spannagl, M.; Ormanbekova, D.; Lux, T.; Prade, V.M.; Milner, S.G. Durum wheat genome highlights past domestication signatures and future improvement targets. *Nat. Genet.* **2019**, *51*, 885–895. [CrossRef] [PubMed]

65. Bustos-Korts, D.; Dawson, I.K.; Russell, J.; Tondelli, A.; Guerra, D.; Ferrandi, C.; Strozzi, F.; Nicolazzi, E.L.; Molnar-Lang, M.; Ozkan, H. Exome sequences and multi-environment field trials elucidate the genetic basis of adaptation in barley. *Plant J.* **2019**, *99*, 1172–1191. [CrossRef]

66. Amezrou, R.; Gyawali, S.; Belqadi, L.; Chao, S.; Arbaoui, M.; Mamidi, S.; Rehman, S.; Sreedasyam, A.; Verma, R.P.S. Molecular and phenotypic diversity of ICARDA spring barley (Hordeum vulgare L.) collection. *Genet. Resour. Crop Evol.* **2018**, *65*, 255–269. [CrossRef]

67. Chen, F.; Chen, D.; Vallés, M.-P.; Gao, Z.; Chen, X. Analysis of diversity in Chinese cultivated barley with simple sequence repeats: Differences between eco-geographic populations. *Biochem. Genet.* **2010**, *48*, 44–56. [CrossRef]

68. Lasa, J.; Igartua, E.; Ciudad, F.; Codesal, P.; Garcíaa, E.; Gracia, M.; Medina, B.; Romagosa, I.; Molina-Cano, J.; Montoya, J. Morphological and agronomical diversity patterns in the Spanish barley core collection. *Hereditas* **2001**, *135*, 217–225. [CrossRef] [PubMed]

69. Zohary, D.; Hopf, M.; Weiss, E. *Domestication of Plants in the Old World: The Origin and Spread of Domesticated Plants in Southwest Asia, Europe, and the Mediterranean Basin*; Oxford University Press on Demand: Oxford, UK, 2012.

70. Taketa, S.; Kikuchi, S.; Awayama, T.; Yamamoto, S.; Ichii, M.; Kawasaki, S. Monophyletic origin of naked barley inferred from molecular analyses of a marker closely linked to the naked caryopsis gene (nud). *Theor. Appl. Genet.* **2004**, *108*, 1236–1242. [CrossRef] [PubMed]

71. Zeng, X.; Guo, Y.; Xu, Q.; Mascher, M.; Guo, G.; Li, S.; Mao, L.; Liu, Q.; Xia, Z.; Zhou, J. Origin and evolution of qingke barley in Tibet. *Nat. Commun.* **2018**, *9*, 5433. [CrossRef]

72. Kikuchi, S.; Taketa, S.; Ichii, M.; Kawasaki, S. Efficient fine mapping of the naked caryopsis gene (nud) by HEGS (High Efficiency Genome Scanning)/AFLP in barley. *Theor. Appl. Genet.* **2003**, *108*, 73–78. [CrossRef] [PubMed]

73. Taketa, S.; Amano, S.; Tsujino, Y.; Sato, T.; Saisho, D.; Kakeda, K.; Nomura, M.; Suzuki, T.; Matsumoto, T.; Sato, K. Barley grain with adhering hulls is controlled by an ERF family transcription factor gene regulating a lipid biosynthesis pathway. *Proc. Natl. Acad. Sci. USA* **2008**, *105*, 4062–4067. [CrossRef] [PubMed]

74. Yu, S.; Long, H.; Deng, G.; Pan, Z.; Liang, J.; Zeng, X.; Tang, Y.; Tashi, N.; Yu, M. A single nucleotide polymorphism of Nud converts the caryopsis type of barley (Hordeum vulgare L.). *Plant Mol. Biol. Report.* **2016**, *34*, 242–248. [CrossRef]

75. Boros, D.; Rek-Cieply, B.; Cyran, M. A note on the composition and nutritional value of hulless barley. *J. Anim. Feed Sci.* **1996**, *5*, 417–424. [CrossRef]

76. Brennan, C.S. Dietary fibre, glycaemic response, and diabetes. *Mol. Nutr. Food Res.* **2005**, *49*, 560–570. [CrossRef]

77. Jood, S.; Kalra, S. Chemical composition and nutritional characteristics of some hull less and hulled barley cultivars grown in India. *Food/Nahr.* **2001**, *45*, 35–39. [CrossRef]

78. Owen, B.; Farmer, M.; Sosuski, F.; Wu, K. Variation in mineral content of Saskatchewan feed grains. *Can. J. Anim. Sci.* **1977**, *57*, 679–687. [CrossRef]

79. Yadav, R.; Gautam, S.; Palikhey, E.; Joshi, B.; Ghimire, K.; Gurung, R.; Adhikari, A.; Pudasaini, N.; Dhakal, R. Agro-morphological diversity of Nepalese naked barley landraces. *Agric. Food Secur.* **2018**, *7*, 86. [CrossRef]

80. Muñoz-Amatriaín, M.; Cuesta-Marcos, A.; Endelman, J.B.; Comadran, J.; Bonman, J.M.; Bockelman, H.E.; Chao, S.; Russell, J.; Waugh, R.; Hayes, P.M. The USDA barley core collection: Genetic diversity, population structure, and potential for genome-wide association studies. *PLoS ONE* **2014**, *9*, e94688.

81. Barabaschi, D.; Francia, E.; Tondelli, A.; Gianinetti, A.; Stanca, A.M.; Pecchioni, N. Effect of the nud gene on grain yield in barley. *Czech J. Genet. Plant Breed.* **2012**, *48*, 10–22. [CrossRef]

82. Gerasimova, S.V.; Hertig, C.; Korotkova, A.M.; Kolosovskaya, E.V.; Otto, I.; Hiekel, S.; Kochetov, A.V.; Khlestkina, E.K.; Kumlehn, J. Conversion of hulled into naked barley by Cas endonuclease-mediated knockout of the NUD gene. *BMC Plant Biol.* **2020**, *20*, 255. [CrossRef] [PubMed]

83. Wang, Q.; Sun, G.; Ren, X.; Du, B.; Cheng, Y.; Wang, Y.; Li, C.; Sun, D. Dissecting the genetic basis of grain size and weight in barley (Hordeum vulgare L.) by QTL and comparative genetic analyses. *Front. Plant Sci.* **2019**, *10*, 469. [CrossRef]

84. Tondelli, A.; Francia, E.; Visioni, A.; Comadran, J.; Mastrangelo, A.; Akar, T.; Al-Yassin, A.; Ceccarelli, S.; Grando, S.; Benbelkacem, A. QTLs for barley yield adaptation to Mediterranean environments in the 'Nure'×'Tremois' biparental population. *Euphytica* **2014**, *197*, 73–86. [CrossRef]

85. Gong, X.; Wheeler, R.; Bovill, W.D.; McDonald, G.K. QTL mapping of grain yield and phosphorus efficiency in barley in a Mediterranean-like environment. *Theor. Appl. Genet.* **2016**, *129*, 1657–1672. [CrossRef]

86. Rey, J.; Hayes, P.; Petrie, S.; Corey, A.; Flowers, M.; Ohm, J.-B.; Ong, C.; Rhinhart, K.; Ross, A. Production of dryland barley for human food: Quality and agronomic performance. *Crop Sci.* **2009**, *4*, 347–355. [CrossRef]

87. Shoeva, O.Y.; Mock, H.-P.; Kukoeva, T.V.; Börner, A.; Khlestkina, E.K. Regulation of the flavonoid biosynthesis pathway genes in purple and black grains of *Hordeum vulgare*. *PLoS ONE* **2016**, *11*, e0163782.

88. Aastrup, S.; Outtrup, H.; Erdal, K. Location of the proanthocyanidins in the barley grain. *Carlsberg Res. Commun.* **1984**, *49*, 105–109. [CrossRef]

89. Kim, M.-J.; Hyun, J.-N.; Kim, J.-A.; Park, J.-C.; Kim, M.-Y.; Kim, J.-G.; Lee, S.-J.; Chun, S.-C.; Chung, I.-M. Relationship between phenolic compounds, anthocyanins content and antioxidant activity in colored barley germplasm. *J. Agric. Food Chem.* **2007**, *55*, 4802–4809. [CrossRef]

90. Harlan, H.V. *Some Distinctions in Our Cultivated Barleys with Reference to Their Use in Plant Breeding*; US Department of Agriculture: Washington, DC, USA, 1914.

91. Yao, X.; Wu, K.; Yao, Y.; Bai, Y.; Ye, J.; Chi, D. Construction of a high-density genetic map: Genotyping by sequencing (GBS) to map purple seed coat color (Psc) in hulless barley. *Hereditas* **2018**, *155*, 37. [CrossRef]

92. Larson, R.L.; Bussard, J.B. Microsomal flavonoid 3′-monooxygenase from maize seedlings. *Plant Physiol.* **1986**, *80*, 483–486. [CrossRef] [PubMed]

93. Yan, X.; Li, J.; Fu, F.; Jin, M.; Chen, L.; Liu, L. Co-location of seed oil content, seed hull content and seed coat color QTL in three different environments in *Brassica napus* L. *Euphytica* **2009**, *170*, 355–364. [CrossRef]

94. Abdel-Aal, E.-S.M.; Young, J.C.; Rabalski, I. Anthocyanin composition in black, blue, pink, purple, and red cereal grains. *J. Agric. Food Chem.* **2006**, *54*, 4696–4704. [CrossRef] [PubMed]

Article

Exploring the Phenotypic Stability of Soybean Seed Compositions Using Multi-Trait Stability Index Approach

Ahmed M. Abdelghany [1,2,†], Shengrui Zhang [1,†], Muhammad Azam [1], Abdulwahab S. Shaibu [1], Yue Feng [1], Jie Qi [1], Jing Li [1], Yanfei Li [1], Yu Tian [1], Huilong Hong [1], Sobhi F. Lamlom [3], Bin Li [1,*] and Junming Sun [1,*]

[1]   The National Engineering Laboratory for Crop Molecular Breeding, MARA Key Laboratory of Soybean Biology (Beijing), Institute of Crop Sciences, Chinese Academy of Agricultural Sciences, 12 Zhongguancun South Street, Beijing 100081, China; ahmed.abdelghany@agr.dmu.edu.eg (A.M.A.); zhangshengrui@caas.cn (S.Z.); azaamuaf@gmail.com (M.A.); asshuaibu.agr@buk.edu.ng (A.S.S.); 82101179104@caas.cn (Y.F.); qjycyz@outlook.com (J.Q.); lijing02@caas.cn (J.L.); yanfeili2017@gmail.com (Y.L.); 82101181082@caas.cn (Y.T.); hong154778@gmail.com (H.H.)

[2]   Crop Science Department, Faculty of Agriculture, Damanhour University, Damanhour 22516, Egypt

[3]   Plant Production Department, Faculty of Agriculture Saba Basha, Alexandria University, Alexandria 21531, Egypt; dr_faidfaid@yahoo.com

*   Correspondence: libin02@caas.cn (B.L.); sunjunming@caas.cn (J.S.); Tel.: +86-10-82105805 (J.S.); +86-10-82105805(B.L.)

†   These authors contributed equally to this work.

Citation: Abdelghany, A.M.; Zhang, S.; Azam, M.; Shaibu, A.S.; Feng, Y.; Qi, J.; Li, J.; Li, Y.; Tian, Y.; Hong, H.; et al. Exploring the Phenotypic Stability of Soybean Seed Compositions Using Multi-Trait Stability Index Approach. *Agronomy* **2021**, *11*, 2200. https://doi.org/10.3390/agronomy11112200

Academic Editors: Gregorio Barba-Espín and Jose Ramon Acosta-Motos

Received: 30 September 2021
Accepted: 29 October 2021
Published: 30 October 2021

Publisher's Note: MDPI stays neutral with regard to jurisdictional claims in published maps and institutional affiliations.

Abstract: In order to ensure an ongoing and long-term breeding progress of soybean, stable sources of major quality traits across multi-environments need to be identified. Here, a panel of 135 soybean genotypes was tested in three different Chinese environments, including Beijing, Anhui, and Hainan during the 2017 and 2018 growing seasons to identify stable genotypes for cultivation under varying environmental conditions. The weighted average of absolute scores biplot (WAASB) for the best linear unbiased predictions of the genotype-environment interaction and multi-trait stability index (MTSI) were utilized to determine the stability of the soybeans for seven seed composition traits viz; protein content, oil content, and five fatty acids (palmitic, stearic, oleic, linoleic, and linolenic acids). Based on the WAASB index, the following genotypes were identified as stable genotypes for some specific traits; ZDD12828 and ZDD12832 for protein content, WDD01583 and WDD03025 for oil content, ZDD23040 for palmitic acid, WDD00033 for stearic acid, ZDD23822 for oleic acid, ZDD11183 for linoleic acid, and ZDD08489 for linolenic acid. Furthermore, based on MTSI at a selection intensity of 10%, 14 soybean genotypes were selected for their average performance and stability. Overall, the MTSI was shown to be a powerful and simple tool for identifying superior genotypes in terms of both performance and stability, hence, identifying stable soybean genotypes for future breeding programs of quality traits.

Keywords: MTSI; multi-environment; soybean; seed compositions; WAASB

1. Introduction

Soybean is presently acknowledged as one of the leading crops due to its viable source of vegetable protein and oil, making it an additional source of healthy food [1]. Soybean is recognized as the food legume with the greatest protein content (40%) and is second only to peanut regarding oil content (20%). When compared to other vegetable proteins, soy protein has an outstanding amino acid balance [2] and is deemed complete because it contains sufficient levels of amino acids necessary by the body for tissue growth and repair [2]. Moreover, soybean provides 28% of the world's oilseed production for edible oil [3]. Soybean oil is composed of 12% palmitic (16:0), 4% stearic (18:0), 23% oleic (18:1), 53% linoleic (18:2), and 8% linolenic (18:3) acids [4].

Climate change may make it more difficult to cultivate crops in the same way and regions as in the past. The impacts of climate change must be weighed against other changing elements that affect agricultural productivity, such as changes in farming techniques and technology [5,6]. Protein and oil contents in soybeans are two of the most important seed quality factors that soybean breeders, growers, and marketers take into account. It is thus important to investigate the stability of upcoming soybean varieties in terms of these quality parameters under such unprecedented rates of climate change impacts [5,7]. Soybean seed characteristics such as protein and oil are influenced by environmental and genetic variables, as well as their interactions [8]. Importantly, when selecting cultivars for breeding programs, genotype-environment interaction (GEI) is influential and it must be evaluated and taken into account [9]. As a result, to maintain positive trends and rebalance dietary sources of soybean in the future, intensive research is needed to develop such improved and highly stable soybean varieties which would appeal to matching the dietary standards requirement around the globe [10,11].

The impact of GEI on genotypes can be described by particular trait stability. Typically, phenotypic stability of distinct traits, either yield or quality traits, is the ability of a genotype to perform consistently in various environments [12]. It is worth noting that determining the stability pattern of genotypes is a prerequisite for understanding their response to different environments, identification of stable and widely adapted genotypes, and breeding new cultivars that could adapt to the different environments [12,13]. Thus, plant breeders' always aim to select varieties with high performance across different environments. The differences observed in the stability of genotypes are usually a result of GEI [14]. However, the selection of highly stable varieties can be difficult when a breeder has to consider individual GEI of multiple traits under multiple environments. Notably, the GEI of quality characters of soybean has been studied previously by several investigators [12,15–20]. In the selection process, besides choosing the best statistical model to predict genetic values, plant breeders usually handle multiple traits simultaneously [21–23]. However, the simultaneous selection of soybean genotypes with high performance based on multiple traits can be a difficult task. So, the question now is how can we breed superior genotypes quickly enough to meet the expanding world population's food demands?

To address such a question, in the advanced stages of the breeding process, multi-environment trial data are frequently used to make brilliant selections [24–27]. Recently, some improvements in this area include the multi-trait stability index (MTSI) which allows for selection based on both mean performance and stability [27–29]. Multi-environment trials analysis is often conducted with only one single trait [30–32]. However, when several traits are taken into account, the accuracy of genotype recommendations improves. In this regard, a rapid and new technique for analyzing multi-environment trials has been recently developed that integrates the simultaneous selection for the stability of multi traits into a single and easily interpretable index [23,27–29].

The theoretical basis of MTSI [27] depends on mean performance and stability selected simultaneously across multi environments trials. Lower MTSI scores suggest the stability of genotypes based on many variables. This strategy was used in a set of 22 oats (*Avena sativa* L.) genotypes [27] based on the weighted average of absolute scores biplot (WAASB) from the singular value decomposition of the matrix of best linear unbiased predictions index [29,33]. Thus, in the simultaneous selection of genotypes based on their mean performances and stabilities of several traits, the MTSI will be beneficial since it offers a strong and easy-to-use selection method that considers the correlation structure of the features. Moreover, the MTSI has been successfully used in selecting high-performing and stable soybean genotypes under two different conditions (drought and salinity), revealing the efficiency of implementing the method [28]. Furthermore, the MTSI has been used in the identification of stable and superior water-tolerant cassava genotypes [29]. From this literature, it is evident that the MTSI is a valuable tool to plant breeders for the selection of superior genotypes based on multi-traits and multi-environment data. Thus, the main objective of this study is to identify soybean genotype(s) with superior performance based on multiple seed composition traits and high phenotypic stability across multi-environment in China.

2. Materials and Methods

2.1. Plant Materials and Field Experiments

In the present study, 135 soybean accessions from three different countries (87 accessions from China, 43 accessions from the USA, and 5 accessions from Russia) were used. Information about accession number, origin, maturity groups, and other phenotypic characteristics are presented in Table S1. The plant material utilized in this research was provided by the soybean genetic resource research group of the Institute of Crop Sciences, Chines Academy of Agricultural Sciences (CAAS).

Field experiments were conducted at Changping, Beijing (40°13′ N, 116°12′ E), and Sanya, Hainan (18°24′ N, 109°5′ E) in 2017 and 2018, and Hefei, Anhui (33°61′ N 117°0′ E) in 2017. The accessions were planted on 12 June 2017 and 14 June 2018 in Changping; 14 November 2017 and 16 November 2018 in Sanya; and 5 June 2017 in Hefei. The experiment was laid out in a randomized incomplete block design with three replications in each location. The mean monthly temperature, rainfall, and sunshine of the five environments are presented in Table S2. Soybean seeds were sown in a 3-m single row plot at a spacing of 0.1 m within rows and 0.5 m between rows. After emergence, the plants were thinned and only uniform healthy plants were left. Plots were manually harvested when the plants reached physiological maturity. The harvest date of all accessions varied due to the differences in maturity group and location. The growing duration of soybean cultivars was 102–120 days at Changping, 101–119 days in Hefei, and 94–96 days at Sanya.

2.2. Determination of Soybean Seed Protein, Oil, and Fatty Acid Compositions

To evaluate the soybean seed protein, oil, and fatty acid contents, harvested soybean seeds of each cultivar were bulked and around 500 g of seeds were used. Near-infrared spectroscopy was used to determine the protein and oil contents [34]. For each sample, absorption of about 50 g of soybean seeds were determined using the transform near-infrared absorption spectroscopy (Bruker Fourier, Germany). The spectrum value of each sample is the average of triplicate measurements with absorption range between 4000 and 8000 cm^{-1}. The spectra were used to estimate protein and oil contents by the Quant 2 method of Bruker's OPUS 4.2 software. The soybean seeds fatty acid contents were quantified by the derivatization into their methyl esters and determined using gas chromatography (GC) [35]. The procedure for the extraction and determination of fatty acids has been previously reported [36]. The area normalization method was used to quantify the percentages of fatty acids using a GC 2010 workstation [35].

2.3. Data Analysis

2.3.1. Analysis of Variance

Individual analysis of variance (ANOVA) for each environment was conducted, followed by an analysis of joint variance according to the statistical model described in Equation (1):

$$Y_{ijk} = \mu + B/E_{jk} + G_i + E_j + GE_{ij} + \varepsilon_{ijk} \tag{1}$$

where Y_{ijk} represents the ith genotype in the jth environment and the kth block; μ is the overall mean; B/E_{jk} corresponds to the block within the jth environment and in the kth block; G_i is the effect of the ith genotype; E_j is the effect of the jth environment; GE_{ij} is the effect of the interaction of the ith genotype with the jth environment; and ε_{ijk} is the effect of experimental error.

2.3.2. Mean Performance and Stability Indices Based on Multiple Traits

The genotypic stability of each genotype was quantified by the WAASB from the singular value decomposition of the matrix of best linear unbiased predictions for the GEI effects generated by a linear mixed-effect model [27], estimated as indicated in Equation (2):

$$WAASB_i = \sum_{k=1}^{p} |IPCA_{ik}\, EP_k| / \sum_{k=1}^{p} EP_k \tag{2}$$

where $WAASB_i$ is the weighted average of absolute scores of the ith genotype; $IPCA_{ik}$ is the score of the ith genotype in the kth interaction principal component axis (IPCA); and EP_k is the amount of the variance explained by the kth IPCA. The genotype with the lowest WAASB value is considered the most stable, showing the least deviation from the average performance across environments [27].

To estimate the multi-trait stability index (MTSI) [27], Equation (3) below was used as follows:

$$MTSI_i = \left[\sum_{j=1}^{f} (F_{ij} - F_j)^2 \right]^{0.5} \tag{3}$$

where MTSI is the multi-trait stability index for the ith genotype, F_{ij} is the jth score of the i_{th} genotype, and F_j is the jth score of ideotype. The genotype with the lowest MTSI is, therefore, closer to the ideotype and hence has a high mean performance and stability for all variables studied. The stability analyses of the multi-environment trial data using MTSI and WAASB indexes were conducted using the *metan* package [37] of the R 4.0.3 software [38].

3. Results

3.1. Mean Performance of 135 Soybean Accesions for Seed Composition Traits across Five Environments

The performance of the 135 soybean genotypes for seven seed composition traits for individual environment is shown in Table S3. The results showed highly significant differences ($p < 0.001$) among the five environments for the seed composition traits. The highest protein content (46.2%) was observed at Hainan in 2018, whereas the lowest protein content (40.3%) was recorded in Anhui in 2017. In contrast to protein content, the lowest oil content (18.3%) was observed at Hainan in 2018, while the highest oil content (20.3%) was recorded in Hainan in 2017. For fatty acid compositions, the highest contents of palmitic acid (13.5%), stearic acid (5.2%), oleic acid (22.9%), linoleic acid (57%), and linolenic acid (9.72%) were observed at Hainan in 2018, Beijing in 2018, Beijing in 2018, Beijing in 2017, and Hainan in 2018, respectively. The lowest values of palmitic acid (12.4%), stearic acid (4.1%), oleic acid (21.2%), linoleic acid (54.6%) and linolenic acid (7.7%) were recorded at Beijing in 2018, Hainan in 2017, Anhui in 2017, Beijing in 2017, and Anhui in 2017, respectively. The heritability values for the evaluated traits ranged from the lowest ($h^2 = 0.859$) for stearic acid at Hainan in 2017 to the highest for protein content and oleic acid ($h^2 = 0.999$) at Hainan in 2017 (Table S3).

For the mean performance of the genotypes across environments, genotypes ZDD12828 and ZDD11436 had the highest (52%) and lowest (35.7%) protein content, respectively. For oil content, WDD01583 and ZDD12828 recorded the highest and lowest content of 22% and 14.4%, respectively. For fatty acid composition, genotypes with highest value of palmitic, stearic, oleic, linoleic, and linolenic were ZDD09581 (15.2%), ZDD23915 (4.9%), ZDD10100 (28.6%), ZDD11183 (60.2%), and ZDD08489 (10.8%), respectively. In contrast, the lowest levels of palmitic, stearic, oleic, linoleic, and linolenic acids were recorded by WDD00033 (10.6%), WDD01632 (3%), ZDD09581 (13.5%), ZDD11235 (46.9), and ZDD23822 (5.9%), respectively.

3.2. Combined Analysis of Variance

The combined analysis of variance (Table 1) for seed protein and oil components showed that the genotype, environment, and GEI effects were highly significant ($p < 0.001$). The results also indicated that all the fatty acid compositions were significantly affected ($p < 0.05$) by genotype, environment, and GEI, except for palmitic and stearic acids which were not significantly influenced by the environment (Table 1).

Table 1. Combined analysis of variance for seven seed compositions of 135 genotypes across the five environments.

Source	Df	Mean Squares						
		Protein Content	Oil Content	Palmitic Acid	Stearic Acid	Oleic Acid	Linoleic Acid	Linolenic Acid
ENV	4	1970 **	288.00	74.9 ns	66 ns	255 **	404 **	256 **
REP(ENV)	10	49.4 **	49.10	44.7 **	32.6 **	47 **	70.4 **	60.4 **
GEN	134	152 **	54.40	7.78 **	2.16 **	113 **	98.8 **	16.2 **
GEN × ENV	536	18.3 **	4.50	1 **	0.74 **	25.1 **	17 **	2.78 **
Residuals	1340	0.05	0.09	0.09	0.09	0.08	0.08	0.09

ENV: environment; REP: replicate; GEN: genotypes; **: significant at $p < 0.01$ level of probability; ns: not significant.

3.3. AMMI Analysis of Variance for Studied Traits

The AMMI analysis showed highly significant effects ($p < 0.001$) of genotype and GEI for the seed composition traits (Table 2). In addition, the environment showed highly significant effects ($p < 0.001$) on protein, oil, oleic and linoleic acids, and a significant effect ($p < 0.05$) on linoleic acid, while palmitic and stearic acids were not significantly affected. The results further showed that the AMMI model explained the GEI and decomposed it into four interaction principal component axes (IPCAs), accounting for 100% of the total variation for all traits (Table 2). The four IPCAs fitted in the current study were all found to be significant ($p < 0.001$) for all the seed composition traits.

Table 2. Additive main effect and multiplicative interaction (AMMI) analysis of variance for seven seed components for 135 soybean genotypes evaluated in five environments.

Source	Df	Mean Squares						
		Protein	Oil	Palmitic Acid	Stearic Acid	Oleic Acid	Linoleic Acid	Linolenic Acid
ENV	4	1970 ***	282 **	72.2 ns	64.9 ns	253 **	415 **	248 *
REP(ENV)	10	49.4 ***	45.9 ***	47.6 ***	31.6 ***	47 ***	68.8 ***	61.7 ***
GEN	134	152 ***	53.2 ***	7.57 ***	2.16 ***	112 ***	97.8 ***	16.5 ***
GEN × ENV	536	18.3 ***	4.36 ***	0.886 ***	0.63 ***	25.1 ***	16.9 ***	2.62 ***
IPCA1†	137	33.6 ***	7.93 ***	1.53 ***	1.16 ***	43.8 ***	27.9 ***	5.03 ***
IPCA2	135	18.8 ***	4.07 ***	0.825 ***	0.594 ***	22 ***	15.6 ***	2.7 ***
IPCA3	133	11.6 ***	3.09 ***	0.705 ***	0.461 ***	19.8 ***	12.2 ***	1.54 ***
IPCA4	131	8.39 ***	2.2 ***	0.457 ***	0.283 ***	14.1 ***	11.4 ***	1.13 ***
Residuals	1340	0.00321	0.001	0.00576	0.0022	0.00238	0.0000148	0.0092
Total	2560	18.9	5.23	1.07	0.603	16.9	13.1	2.59

ENV: environment; REP: replicate; GEN: genotypes; IPCA, interaction principal component axis. *, **, *** significant at $p < 0.05$, $p < 0.01$, and $p < 0.001$ level of probability, respectively; ns: not significant.

The partitioning of total phenotypic variance due to factors of genotype, environment, and GEI was also estimated. Collectively, the highest contribution to total phenotypic variation (as a relative contribution to the total sum of squares) was captured by genotypes in all measured seed compositions traits except for stearic acid. The genotype effect explained 42.1, 53.1, 36.9, 18.8, 34.6, 39.0, and 33.3% of the total phenotypic variance for protein, oil, palmitic acid, stearic acid, oleic acid, linoleic acid, and linolenic acid, respectively. The lowest environmental effect was captured by oleic acid (2.3%), while stearic acid was highly affected by the environment (16.9%).

3.4. Mean Performance and Stability of Selected Genotypes

The mean vs. WAASB shows the joint interpretation of the mean performance and stability of genotypes for the seed composition traits (Figures 1 and 2). The results showed that genotypes ZDD12828 and ZDD12832 were highly stable (low WAASB index) with high protein contents of 51.9 and 50.6%, respectively, over the grand mean of all genotypes (Figure 1a). As for oil content, WDD01583 (22.9%), WDD03025 (22.7%), and WDD00573 (22.5%) with WAASB index of 0.132, 0.108, and 0.121, respectively were the most stable (Figure 1b).

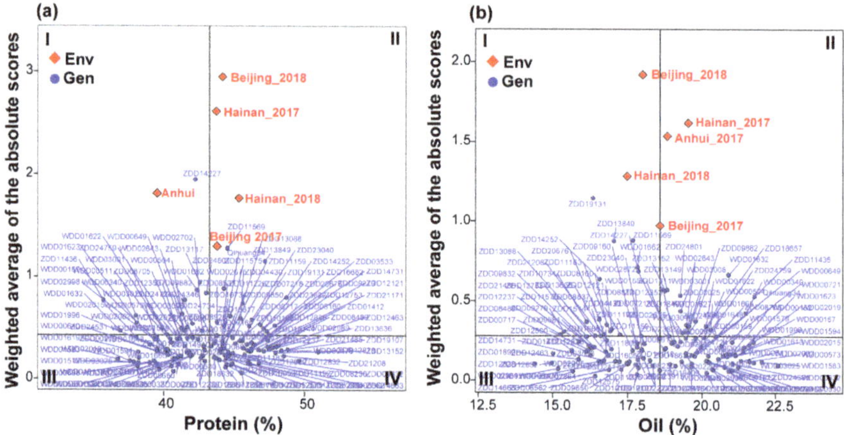

Figure 1. Biplot showing the performance vs. stability of 135 soybean genotypes for protein content (**a**) and oil content (**b**). The x-axis shows the arithmetic mean for each genotype × environment interaction. The y-axis shows the weighted average of absolute scores from the singular value decomposition of the matrix of best linear unbiased predictions for the genotype × environment interaction effects.

For the stability of fatty acid composition (Figure 2a–e), ZDD23040 and ZDD12500 were the most stable genotypes with average palmitic acid contents of 14.2 and 14.1% and WAASB values of 0.119 and 0.108, respectively (Figure 2a). The most stable genotypes with high average stearic acid content were WDD00033 and ZDD24734 with 5.6 and 5.5% and WAASB index values of 0.163 and 0.188, respectively (Figure 2b). Highest levels of oleic acid were recorded by ZDD23822 (28.3%) and ZDD19107 (26.5%) with WAASB values of 0.412 and 0.308, respectively (Figure 2c). The most stable genotypes for linoleic acid were ZDD11183 (61%) and WDD01613 (60.1%) with WAASB values of 0.182 and 0.202, respectively (Figure 2d). For linolenic acid, higher levels than the grand mean coupled with high stability were recorded by ZDD08489 (11.5%), ZDD12463 (11.3%), and ZDD09581 (11.1%) with WAASB values of 0.199, 0.144, and 0.103, respectively (Figure 2e).

3.5. Multi-Trait Stability Index and Genotype Selection

The ranking of the 135 soybean genotypes based on MTSI values is presented in Table S4. The seven seed composition traits, i.e., protein, oil, palmitic acid, stearic acid, oleic acid, linoleic and linolenic acid were used to estimate the MTSI (Figure 3). Overall, the mean of MTSI values across all genotypes was 7.4, where the lowest value was recorded by ZDD12500 (5.42), while the highest value of MTSI (9.9), was recorded by WDD01583 indicating the most and least stable genotypes, respectively. According to the lowest MTSI values at a selection intensity of 10%, 14 soybean genotypes were identified (Figure 3). The selected genotypes with lowest MTSI values were ZDD12500 (5.42), ZDD04430 (5.62), ZDD24734 (5.78), ZDD12463 (5.8), ZDD16617 (5.92), ZDD18657 (5.93), ZDD08812 (6.03), ZDD12832 (6.07), WDD02292 (6.12), ZDD12828 (6.18), ZDD21171 (6.19), ZDD01412 (6.26), WDD00530 (6.31), and ZDD23040 (6.33). These genotypes represent the best soybean materials in terms of high stability and overall performance among the tested panel of 135 soybeans.

The mean of the selected genotypes (Xs) was higher than the original average (Xo) which included all the 135 soybean genotypes for all the measured traits except for oil, oleic, and linoleic components (Table 3). The selection differential (SD) was positive for all traits, except for oil, oleic acid, and linoleic acid compositions. The heritability ranged from 0.66 for stearic acid to 0.92 for oil content (Table 3). Moreover, the selection gain (SG) was positive for all studied traits except for oil, oleic acid, and linoleic acid compositions. The highest positive SG was 3.41% for protein content, whereas palmitic acid had the lowest SG value of 0.98%, while the negative SG ranged from −4.45% for oil content to −0.18% for linoleic acid content.

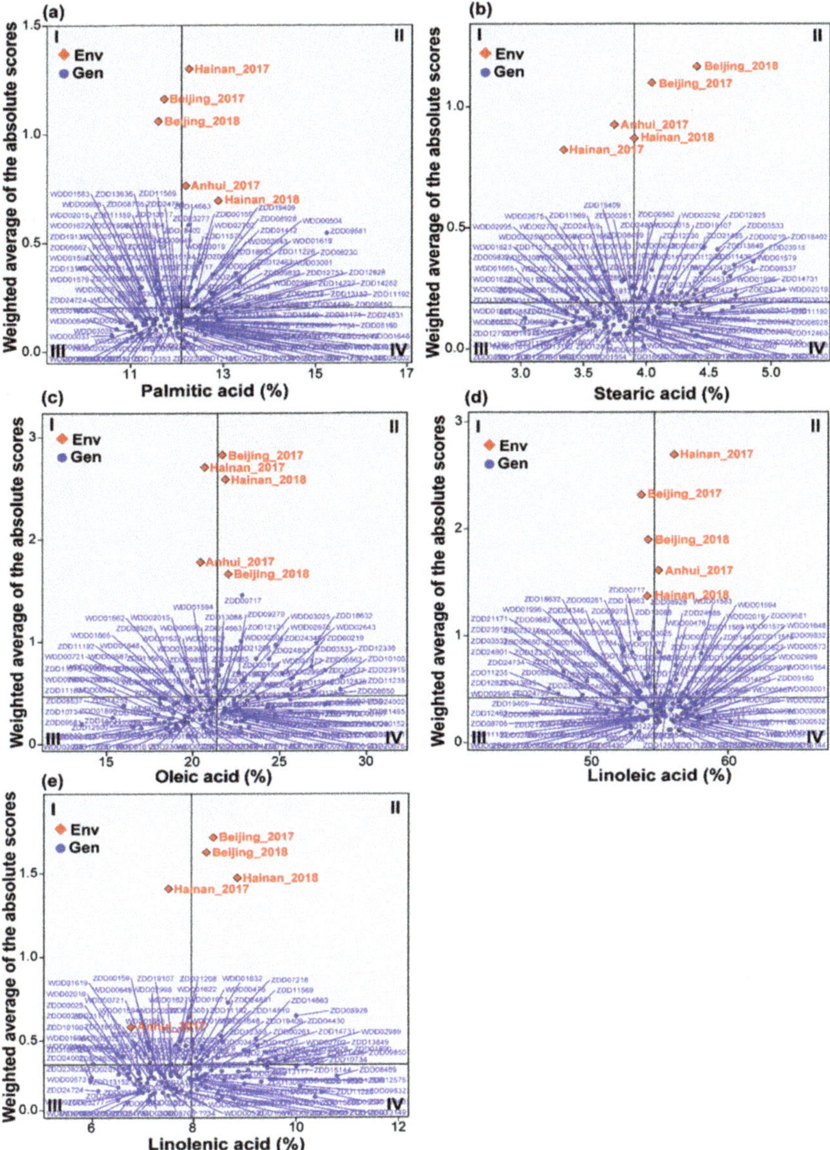

Figure 2. Biplots showing the performance vs. stability of 135 soybean genotypes for palmitic acid (**a**), stearic acid (**b**), oleic acid (**c**), linoleic acid (**d**), and linolenic acid (**e**). The *x*-axis shows the arithmetic mean for each genotype × environment interaction. The *y*-axis shows the weighted average of absolute scores from the singular value decomposition of the matrix of best linear unbiased predictions for the genotype × environment interaction effects.

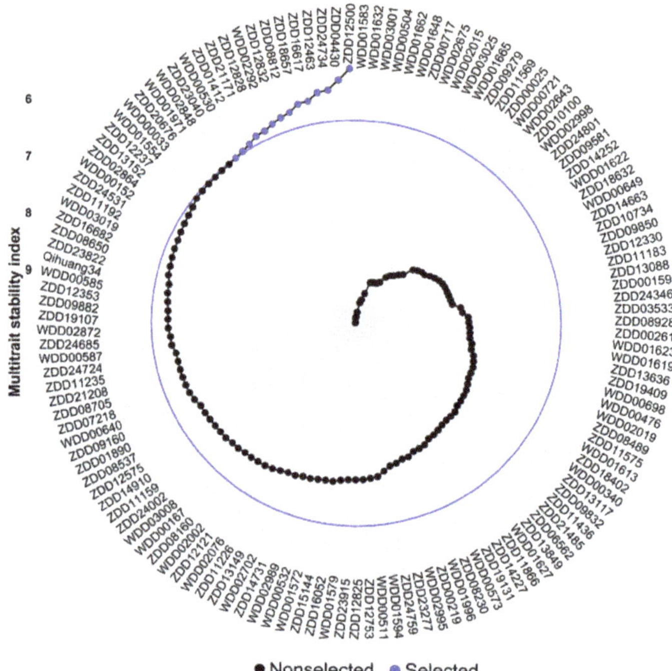

Figure 3. Genotype ranking and selected genotypes for the multi-trait stability index (MTSI) of 135 soybean genotypes based on seven seed composition traits. The selected genotypes are shown in blue color and the blue circle represents the cut-point according to the selection differential of 10%.

Table 3. Estimates of selection differential, selection gain, and heritability based on MTSI for seven seed compositions evaluated for 135 soybean genotypes across five environments.

Trait	Factor	Xo	Xs	SD	SD (%)	SG	SG (%)	h^2
Protein	FA 1	44.1	46	1.93	4.37	1.69	3.84	0.88
Oil	FA 1	19.4	18.4	−0.941	−4.86	−0.86	−4.45	0.92
Palmitic acid	FA 1	12.9	13	0.145	1.13	0.13	0.98	0.87
Linolenic acid	FA 1	8.76	9.01	0.246	2.81	0.20	2.33	0.83
Stearic acid	FA 2	4.69	4.85	0.156	3.33	0.10	2.19	0.66
Oleic acid	FA 2	22.2	21.8	−0.386	−1.74	−0.30	−1.35	0.78
Linoleic acid	FA 3	55.4	55.3	−0.12	−0.216	−0.099	−0.18	0.83

Xo: Overall mean of genotypes; Xs: Mean of the selected genotypes; SD: Selection differential; SG: Selection gain or impact; h^2, heritability.

4. Discussion

Due to the narrow genetic base revealed by most soybean germplasm analyses [39,40], it became a necessity that breeders, researchers, and producers make such genetic diversity information available to secure genetic advance and improvement of such elite soybean germplasms in the future. Soybean's genetic diversity may be successfully and extensively conserved by developing germplasm collections on both local and global scales as vital genetic resources [10]. This endeavor could provide a panel of soybean accessions characterized with not only promising high-seed yield but also high-quality characteristics.

To identify genotypes with wider adaptability and test their stability in terms of yield and quality traits, multi-environmental trials are among major foci for soybean breeding programs and are thus critical for such selection [28,41–47]. The current study showed varied responses of a set of diverse soybean genotypes for seed composition traits under different environments in China, demonstrating the need for determining the magnitude of genotype by environment interactions. The three Chinese agroecological regions where the soybean genotypes were evaluated represent three distinct latitudes with varying climatic variables such as temperature, rainfall, and sunshine duration [36,48]. These ecoregions further demonstrate major growing sites for Chinese soybean germplasm [49]. The current study was conducted using a diverse panel of 135 soybean genotypes to explore their overall stability coupled with high concentrations of seed protein, oil, and five essential fatty acids including palmitic, stearic, oleic, linoleic, and linolenic acids. Several studies have studied the stability of soybean seed compositions across several environments [28,41–47]. Notably, a great change in the protein, oil, and fatty acids compositions across different environments was also reported in other studies [50–53] showing a discrepant response among studied soybean genotypes in protein, oil, saturated and unsaturated fatty acids components for environmental factors including latitudes, temperature, and sunshine duration.

Furthermore, different approaches have been exploited to assess the performance of soybean genotypes across several environments such as AMMI [42,43,45,54], GGE biplot [24,47,55–57], WAASB index [28,58], and MTSI [28]. In this study, highly statistically significant effects were shown by AMMI analysis on G, E, and GEI for all studied traits except for palmitic and stearic acids. The presence of significant GEI revealed a differential performance among the soybean genotypes across the various testing environment. The AMMI analysis has been widely used to select promising genotypes in terms of stability and superiority across multiple environments [59–63]. The high proportion of genotype effect to total phenotypic variance shown in this study for all measured seed compositions traits, except for stearic acid, was also reported in previous studies [64,65]. This indicates that the phenotypic variance of such traits was highly controlled by genotypic variation more than environmental variation and also provides an opportunity for selection gain [65]. Most of the variation for the majority of the seed composition traits was accounted for by the genotype effect followed by genotype × environment interaction, while the environment effect contributed least, which will lead to higher heritability for such traits [42]. This was evident from the high heritability values observed in this study. In contrast, the higher percentage of GEI effect of seed stearic acid over each of genotype and environment effects indicates lower heritability. These findings are in accordance with the previous studies where environmental variance was dominant [12,44,55,58], showing the importance of the GEI effect on some soybean seed composition traits.

Although, genotype effect dominantly characterized the performance of the soybean accessions in this study, however, the higher significant GEI effect could influence the selection efficiency and, consequently, limits the development of adapted genotypes [66]. In this context, searching for ideal and powerful tools such as WAASB and MTSI to identify the most stable and highly performing soybean genotypes became necessary. The WAASB utilized in this study explains the stability based on the WAASB index which considers the entire GEI variance in identifying the stable genotypes [27]. The WAASB which shows the joint interpretation of mean performance and stability in a bi-dimensional plot comprising of four quadrants was recently suggested to provide an easy and robust tool for selecting stable genotypes with high performance [27]. Genotypes within quadrants I and II are assumed to be unstable, while genotypes within quadrants III and IV are highly stable, revealing lesser variation across all environments. This implies that to select for high and stable genotypes, quadrant IV should be considered, whereas quadrant III is preferable to select genotypes with high stability and low content of specific seed composition. Furthermore, genotypes with WAASB values close to zero are identified as the most stable [67]. Based on the WAASB index used in the current study, two genotypes, ZDD12828 and ZDD12832, showed high stability and high seed protein content. Using the same approach, genotypes WDD01583, WDD03025, and WDD00573 were identified as stable accessions with high seed oil content. Similarly, genotypes with desired and stable fatty acids profile were also identified. The superiority of the quantitative measure of the WAASB index was also reported as an important statistical tool for identifying high performing and broadly adapted genotypes in previous studies on oat [27,67], strawberry [23], and cassava [29]. The WAASB shows potential in quantifying the stability in compounded GEI structures because it uses all the estimated IPCA in the computation of WAASB. Thus, WAASB could be successfully applied in the identification of promising genotypes with highly preferable seed compositions and broad adaptability.

More interestingly, the MTSI has been recently used to help in selecting elite genotypes based on the stability and mean performance of multiple traits [27]. MTSI is assumed based on the genotype–ideotype distance (Euclidian) through the scores obtained in factor analysis [27]. With respect to the criteria of MTSI, the genotypes with lower values of MTSI indicate higher stability based on multiple traits measured. In the current study, setting the selection intensity to 10% has resulted in the identification of 14 soybean genotypes, ZDD12500, ZDD04430, ZDD24734, ZDD12463, ZDD16617, ZDD18657, ZDD08812, ZDD12832, WDD02292, ZDD12832, ZDD21171, ZDD01412, WDD00530, and ZDD23040, that were classified as stable or fairly stable for all the traits. Such selection was fairly justified and precisely evaluated under contrasting environment. In term of feasibility of such index, the selection of these genotypes would greatly help in improving the overall mean performance as indicated by the high percentage of selection differentials. In a recent study on soybean, stable soybean genotypes were identified under drought and salinity conditions based on the criterion of MTSI [28]. The MTSI was further utilized to choose the ideal treatment based on twenty-three traits of strawberry [23]. From the current study, based on multiple traits, the MTSI index can provide a unique, robust, and powerful tool to develop better treatment and/or genotypes helping both breeders and agronomists as also reported previously [23,27,28].

5. Conclusions

Several recent issues have constrained the preservation of soybean genotypes with stable seed quality parameters. Therefore, the stability index of genotype performance has the potential to provide reliable estimates of stability in future studies. The WAASB used in this study identified stable genotypes with a high value of seed composition traits, indicating their potential as a source of desirable protein, oil, and fatty acid compositions. Moreover, the MTSI, which is a novel multivariate approach, was used in the current study to discover stable genotypes with numerous characteristics that are appropriate for wider adaptations. The WAASB and MTSI were proven to be efficient tools in evaluating stability and will maximize the use of resources, thereby contributing to the global sustainability of soybean breeding programs.

Supplementary Materials: The following are available online at https://www.mdpi.com/article/10.3390/agronomy11112200/s1, Table S1: Information of the 135 soybean accessions used in this study. Table S2: The monthly average of temperature (°C), precipitation (mm), and sunshine (h) for the experimental sites of Hainan, and Beijing in 2017 and 2018, and Anhui in 2017. Table S3: The performance of 135 soybean genotypes for seven seed components for individual environments. Table S4: The ranking of 135 soybean genotypes based on MTSI values.

Author Contributions: Conceptualization, A.M.A. and J.S.; formal analysis, A.M.A.; funding acquisition, B.L. and J.S.; investigation, J.Q.; methodology, A.M.A., M.A., S.Z., A.S.S., J.L., Y.F., Y.L., Y.T., H.H. and S.F.L.; project administration, B.L. and J.S.; resources, S.Z., M.A. and A.S.S.; supervision, J.S.; visualization, B.L.; writing—original draft, A.M.A.; writing—review and editing, A.M.A., B.L. and J.S. All authors have read and agreed to the published version of the manuscript.

Funding: This study was supported by the National Natural Science Foundation of China (No. 31671716 and 32001574), Ministry of Science and Technology of China (2016YFD0100201) and the Chinese Academy of Agricultural Sciences (CAAS) Innovation Project (2060206-2).

Institutional Review Board Statement: Not applicable.

Informed Consent Statement: Not applicable.

Data Availability Statement: Not applicable.

Acknowledgments: We would like to thank Lijuan Qiu from Institute of Crop Sciences, CAAS, for providing us all of soybean accessions and her helpful suggestions regarding the manuscript. We also would like to thank Xiaobo Wang from Anhui Agricultural University for helping in growing soybean accessions in Anhui province of China.

Conflicts of Interest: The authors declare no competing financial interest in regard to this manuscript.

References

1. Wilcox, J.R.; Shibles, R.M. Interrelationships among seed quality attributes in soybean. *Crop Sci.* **2001**, *41*, 11–14. [CrossRef]
2. Wolf, W.J.; Cowan, J. *Soybeans as a Food Source*; CRC Press: Boca Raton, FL, USA, 1975; ISBN 0878191127.
3. Kim, H.-J.; Ha, B.-K.; Ha, K.-S.; Chae, J.-H.; Park, J.-H.; Kim, M.-S.; Asekova, S.; Shannon, J.G.; Son, C.-K.; Lee, J.-D. Comparison of a high oleic acid soybean line to cultivated cultivars for seed yield, protein and oil concentrations. *Euphytica* **2015**, *201*, 285–292. [CrossRef]
4. Boerma, H.R.; Specht, J.E.; Wilson, R.F. Seed Composition. *Soybeans Improv. Prod. Uses* **2004**, 621–677. [CrossRef]
5. Raza, A.; Razzaq, A.; Mehmood, S.S.; Zou, X.; Zhang, X.; Lv, Y.; Xu, J. Impact of climate change on crops adaptation and strategies to tackle its outcome: A review. *Plants* **2019**, *8*, 34. [CrossRef]
6. Adaawen, S. Understanding climate change and drought perceptions, impact and responses in the rural savannah, West Africa. *Atmosphere* **2021**, *12*, 594. [CrossRef]
7. Chaudhary, K.R.; Wu, J. Stability analysis for yield and seed quality of soybean [*Glycine max* (l.) Merril] across different environments in Eastern South Dakota. *Conf. Appl. Stat. Agric.* **2012**. [CrossRef]
8. Ray, C.L.; Shipe, E.R.; Bridges, W.C. Planting date influence on soybean agronomic traits and seed composition in modified fatty acid breeding lines. *Crop Sci.* **2008**, *48*, 181–188. [CrossRef]
9. Prado, E.E.D.; Hiromoto, D.M.; Godinho, V.D.P.C.; Utumi, M.M.; Ramalho, A.R. Adaptabilidade e estabilidade de cultivares de soja em cinco épocas de plantio no cerrado de Rondônia. *Pesqui. Agropecuária Bras.* **2001**, *36*, 625–635. [CrossRef]
10. Sudarić, A.; Matoša Kočar, M.; Duvnjak, T.; Zdunić, Z.; Markulj Kulundžić, A. Improving Seed Quality of Soybean Suitable for Growing in Europe. In *Soybean for Human Consumption and Animal Feed*; IntechOpen: London, UK, 2020; pp. 416–680.
11. Voss-Fels, K.P.; Stahl, A.; Hickey, L.T. Q&A: Modern crop breeding for future food security. *BMC Biol.* **2019**, *17*, 1–7.
12. Gurmu, F.; Mohammed, H.; Alemaw, G. Genotype x environment interactions and stability of soybean for grain yield and nutrition quality. *Afr. Crop Sci. J.* **2009**, *17*. [CrossRef]
13. Tolorunse, K.D.; Gana, A.S.; Bala, A.; Sangodele, E.A. Yield stability studies of soybean (*Glycine max* (L.) Merrill) under rhizobia inoculation in the savanna region of Nigeria. *Plant Breed.* **2018**, *137*, 262–270. [CrossRef]
14. Becker, H.C.; Leon, J. Stability analysis in plant breeding. *Plant Breed.* **1988**, *101*, 1–23. [CrossRef]
15. Sharma, P.K.; Gupta, P.K.; Govila, O.P. AMMI analysis of a pearl millet yield trial. *Indian J. Genet. Plant Breed.* **1998**, *58*, 183–192.
16. Chandrakar, P.K.; Shrivastava, R.; Agrawal, S.K.; Rao, S.S. Stability analysis of soybean (*Glycine max* L. merrill) varieties in rice zone of Madhya Pradesh. *J. Oilseeds Res.* **1998**, *15*, 247–249.
17. Poysa, V.; Woodrow, L. Stability of soybean seed composition and its effect on soymilk and tofu yield and quality. *Food Res. Int.* **2002**, *35*, 337–345. [CrossRef]
18. Ramana, M.V.; Satyanarayana, A. Stability of yield and its components in soybean (*Glycine max* L. Merrill). *J. Oilseeds Res.* **2005**, *22*, 18–21.
19. Arslanoglu, F.; Aytac, S.; Oner, E.K. Effect of genotype and environment interaction on oil and protein content of soybean (*Glycine max* (L.) Merrill) seed. *Afr. J. Biotechnol.* **2011**, *10*, 18409–18417. [CrossRef]
20. Balešević-Tubić, S.; Đorđević, V.; Miladinović, J.; Đukić, V.; Tatić, M. Stability of soybean seed composition. *Genetika* **2011**, *43*, 217–227. [CrossRef]
21. Akhter, M.; Sneller, C.H. Yield and yield components of early maturing soybean genotypes in the mid-south. *Crop Sci.* **1996**, *36*, 877–882. [CrossRef]
22. Malek, M.A.; Rafii, M.Y.; Afroz, S.S.; Nath, U.K.; Mondal, M. Morphological characterization and assessment of genetic variability, character association, and divergence in soybean mutants. *Sci. World J.* **2014**, *2014*, 968796. [CrossRef]
23. Olivoto, T.; Diel, M.I.; Schmidt, D.; Lúcio, A.D. Multivariate analysis of strawberry experiments: Where are we now and where can we go? *bioRxiv* **2021**, 2012–2020. [CrossRef]
24. Dalló, S.C.; Zdziarski, A.D.; Woyann, L.G.; Milioli, A.S.; Zanella, R.; Conte, J.; Benin, G. Across year and year-by-year GGE biplot analysis to evaluate soybean performance and stability in multi-environment trials. *Euphytica* **2019**, *215*, 1–12. [CrossRef]
25. Jarquin, D.; Howard, R.; Crossa, J.; Beyene, Y.; Gowda, M.; Martini, J.W.R.; Covarrubias Pazaran, G.; Burgueño, J.; Pacheco, A.; Grondona, M. Genomic prediction enhanced sparse testing for multi-environment trials. *G3 Genes Genomes Genet.* **2020**, *10*, 2725–2739.
26. Woyann, L.G.; Meira, D.; Matei, G.; Zdziarski, A.D.; Dallacorte, L.V.; Madella, L.A.; Benin, G. Selection indexes based on linear-bilinear models applied to soybean breeding. *Agron. J.* **2020**, *112*, 175–182. [CrossRef]
27. Olivoto, T.; Lúcio, A.D.C.; da Silva, J.A.G.; Sari, B.G.; Diel, M.I. Mean performance and stability in multi-environment trials II: Selection based on multiple traits. *Agron. J.* **2019**, *111*, 2961–2969. [CrossRef]

28. Zuffo, A.M.; Steiner, F.; Aguilera, J.G.; Teodoro, P.E.; Teodoro, L.P.R.; Busch, A. Multi-trait stability index: A tool for simultaneous selection of soya bean genotypes in drought and saline stress. *J. Agron. Crop Sci.* **2020**, *206*, 815–822. [CrossRef]

29. Koundinya, A.V.V.; Ajeesh, B.R.; Hegde, V.; Sheela, M.N.; Mohan, C.; Asha, K.I. Genetic parameters, stability and selection of cassava genotypes between rainy and water stress conditions using AMMI, WAAS, BLUP and MTSI. *Sci. Hortic.* **2021**, *281*, 109949.

30. Bornhofen, E.; Todeschini, M.H.; Stoco, M.G.; Madureira, A.; Marchioro, V.S.; Storck, L.; Benin, G. Wheat yield improvements in Brazil: Roles of genetics and environment. *Crop Sci.* **2018**, *58*, 1082–1093. [CrossRef]

31. Mohammadi, R.; Armion, M.; Zadhasan, E.; Ahmadi, M.M.; Amri, A. The use of AMMI model for interpreting genotype× environment interaction in durum wheat. *Exp. Agric.* **2018**, *54*, 670–683. [CrossRef]

32. Nowosad, K.; Liersch, A.; Popławska, W.; Bocianowski, J. Genotype by environment interaction for seed yield in rapeseed (*Brassica napus* L.) using additive main effects and multiplicative interaction model. *Euphytica* **2016**, *208*, 187–194. [CrossRef]

33. Olivoto, T. Stability Analysis in R Using the WAASB Index. 2019. Available online: https://doi.org/10.13140/RG.2.2.32425.7536 9/1 (accessed on 21 May 2021).

34. Hymowitz, T.; Dudley, J.W.; Collins, F.I.; Brown, C.M. Estimations of protein and oil concentration in corn, soybean, and oat seed by near-infrared light reflectance. *Crop Sci.* **1974**, *14*, 713–715. [CrossRef]

35. Fan, S.; Li, B.; Yu, F.; Han, F.; Yan, S.; Wang, L.; Sun, J. Analysis of additive and epistatic quantitative trait loci underlying fatty acid concentrations in soybean seeds across multiple environments. *Euphytica* **2015**, *206*, 689–700. [CrossRef]

36. Abdelghany, A.M.; Zhang, S.; Azam, M.; Shaibu, A.S.; Feng, Y.; Li, Y.; Tian, Y.; Hong, H.; Li, B.; Sun, J. Profiling of seed fatty acid composition in 1025 Chinese soybean accessions from diverse ecoregions. *Crop J.* **2020**, *8*, 635–644. [CrossRef]

37. Olivoto, T.; Lúcio, A.D. metan: An R package for multi—Environment trial analysis. *Methods Ecol. Evol.* **2020**, *11*, 783–789. [CrossRef]

38. R Core Team. *R: A Language and Environment for Statistical Computing*; Version 4.0.3; R Foundation for Statistical Computing: Vienna, Austria, 2020.

39. Zhang, G.; Xu, S.; Mao, W.; Hu, Q.; Gong, Y. Determination of the genetic diversity of vegetable soybean [*Glycine max* (L.) Merr.] using EST-SSR markers. *J. Zhejiang Univ. Sci. B* **2013**, *14*, 279–288. [CrossRef]

40. Perić, V.; Nikolić, A.; Babić, V.; Sudarić, A.; Srebrić, M.; Đorđević, V.; Mladenović-Drinić, S. Genetic relatedness of soybean genotypes based on agromorphological traits and RAPD markers. *Genetika-Belgrade* **2014**, *46*, 839–854. [CrossRef]

41. Adie, M.M.; Krisnawati, A. Soybean yield stability in eight locations and its potential for seed oil source in Indonesia. *Energy Procedia* **2015**, *65*, 223–229. [CrossRef]

42. Sudarić, A.; Šimić, D.; Vratarić, M. Characterization of genotype by environment interactions in soybean breeding programmes of southeast Europe. *Plant Breed.* **2006**, *125*, 191–194. [CrossRef]

43. Bhartiya, A.; Aditya, J.P.; Singh, K.; Purwar, J.P.; Agarwal, A. AMMI & GGE biplot analysis of multi environment yield trial of soybean in North Western Himalayan state Uttarakhand of India. *Legum. Res. Int. J.* **2017**, *40*, 306–312.

44. Bhartiya, A.; Aditya, J.P.; Kumari, V.; Kishore, N.; Purwar, J.P.; Agrawal, A.; Kant, L. GGE biplot & ammi analysis of yield stability in multi-environment trial of soybean [*Glycine max* (L.) Merrill] genotypes under rainfed condition of north western Himalayan hills. *Indian J. Genet. Plant Breed.* **2017**, *27*, 227–238.

45. Mwiinga, B.; Sibiya, J.; Kondwakwenda, A.; Musvosvi, C.; Chigeza, G. Genotype x environment interaction analysis of soybean (*Glycine max* (L.) Merrill) grain yield across production environments in Southern Africa. *F. Crop. Res.* **2020**, *256*, 107922. [CrossRef]

46. Amira, J.O.; Ojo, D.K.; Ariyo, O.J.; Oduwaye, O.A.; Ayo-Vaughan, M.A. Relative discriminating powers of GGE and AMMI models in the selection of tropical soybean genotypes. *Afr. Crop Sci. J.* **2013**, *21*, 67–73.

47. Asfaw, A.; Alemayehu, F.; Gurum, F.; Atnaf, M. AMMI and SREG GGE biplot analysis for matching varieties onto soybean production environments in Ethiopia. *Sci. Res. Essays* **2009**, *4*, 1322–1330.

48. Azam, M.; Zhang, S.; Abdelghany, A.M.; Shaibu, A.S.; Feng, Y.; Li, Y.; Tian, Y.; Hong, H.; Li, B.; Sun, J. Seed isoflavone profiling of 1168 soybean accessions from major growing ecoregions in China. *Food Res. Int.* **2020**, *130*. [CrossRef]

49. Sedivy, E.J.; Wu, F.; Hanzawa, Y. Soybean domestication: The origin, genetic architecture and molecular bases. *New Phytol.* **2017**, *214*, 539–553. [CrossRef]

50. Primomo, V.S.; Falk, D.E.; Ablett, G.R.; Tanner, J.W.; Rajcan, I. Genotype x environment interactions, stability, and agronomic performance of soybean with altered fatty acid profiles. *Crop Sci.* **2002**, *42*, 37–44. [CrossRef]

51. Assefa, Y.; Bajjalieh, N.; Archontoulis, S.; Casteel, S.; Davidson, D.; Kovács, P.; Naeve, S.; Ciampitti, I.A. Spatial characterization of soybean yield and quality (amino acids, oil, and protein) for United States. *Sci. Rep.* **2018**, *8*, 14653. [CrossRef]

52. Yaklich, R.W.; Vinyard, B.; Camp, M.; Douglass, S. Analysis of seed protein and oil from soybean northern and southern region uniform tests. *Crop Sci.* **2002**, *42*, 1504–1515. [CrossRef]

53. Wu, T.; Yang, X.; Sun, S.; Wang, C.; Wang, Y.; Jia, H.; Man, W.; Fu, L.; Song, W.; Wu, C. Temporal–spatial characterization of seed proteins and oil in widely grown soybean cultivars across a century of breeding in China. *Crop Sci.* **2017**, *57*, 748–759. [CrossRef]

54. Liu, Z.; Fan, X.; Huang, W.; Yang, J.; Zheng, Y.; Wang, S.; Qiu, L. Stability analysis of seven agronomic traits for soybean [(*Glycine max* (L.) Merr.] Tokachi nagaha and its derived cultivars using the AMMI model. *Plant Prod. Sci.* **2017**, *20*, 499–506. [CrossRef]

55. Nataraj, V.; Pandey, N.; Ramteke, R.; Verghese, P.; Reddy, R.; Onkarappa, T.; Mehtre, S.P.; Gupta, S.; Satpute, G.K.; Mohan, Y. GGE biplot analysis of vegetable type soybean genotypes under multi-environmental conditions in India. *J. Environ. Biol.* **2021**, *42*, 247–253. [CrossRef]

56. Sousa, L.B.; Hamawaki, O.T.; Nogueira, A.P.O.; Batista, R.O.; Oliveira, V.M.; Hamawaki, R.L. Evaluation of soybean lines and environmental stratification using the AMMI, GGE biplot, and factor analysis methods. *Genet. Mol. Res.* **2015**, *14*, 12660–16674. [CrossRef] [PubMed]
57. Cheelo, P.; Lungu, D.; Mwala, M. GGE biplot analysis for identification of ideal soybean [*Glycine max* L. Merrill] test and production locations in Zambia. *J. Exp. Agric. Int.* **2017**, 1–15. [CrossRef]
58. Nataraj, V.; Bhartiya, A.; Singh, C.P.; Devi, H.N.; Deshmukh, M.P.; Verghese, P.; Singh, K.; Mehtre, S.P.; Kumari, V.; Maranna, S.; et al. WAASB based stability analysis and simultaneous selection for grain yield and early maturity in soybean. *Agron. J.* **2021**, *113*, 3089–3099. [CrossRef]
59. Hagos, H.G.; Abay, F. AMMI and GGE biplot analysis of bread wheat genotypes in the northern part of Ethiopia. *J. Plant Breed. Genet.* **2013**, *1*, 12–18.
60. Mahmodi, N.; Yaghotipoor, A.; Farshadfar, E. AMMI stability value and simultaneous estimation of yield and yield stability in bread wheat (*Triticum aestivum* L.). *Aust. J. Crop Sci.* **2011**, *5*, 1837–1844.
61. Yan, W.; Kang, M.S.; Ma, B.; Woods, S.; Cornelius, P.L. GGE biplot vs. AMMI analysis of genotype-by-environment data. *Crop Sci.* **2007**, *47*, 643–653. [CrossRef]
62. Sneller, C.H.; Dombek, D. Comparing soybean cultivar ranking and selection for yield with AMMI and full-data performance estimates. *Crop Sci.* **1995**, *35*, 1536–1541. [CrossRef]
63. Ndhlela, T.; Herselman, L.; Magorokosho, C.; Setimela, P.; Mutimaamba, C.; Labuschagne, M. Genotype × environment interaction of maize grain yield using AMMI biplots. *Crop Sci.* **2014**, *54*, 1992–1999. [CrossRef]
64. Shilpashree, N.; Devi, S.N.; Manjunathagowda, D.C.; Muddappa, A.; Abdelmohsen, S.A.M.; Tamam, N.; Elansary, H.O.; El-Abedin, T.K.Z.; Abdelbacki, A.M.M.; Janhavi, V. Morphological characterization, variability and diversity among vegetable soybean (*Glycine max* L.) genotypes. *Plants* **2021**, *10*, 671. [CrossRef] [PubMed]
65. Whaley, R.; Eskandari, M. Genotypic main effect and genotype-by-environment interaction effect on seed protein concentration and yield in food-grade soybeans (*Glycine max* (L.) Merrill). *Euphytica* **2019**, *215*, 33. [CrossRef]
66. Rao, P.S.; Reddy, P.S.; Rathore, A.; Reddy, B.V.S.; Panwar, S. Application GGE biplot and AMMI model to evaluate sweet sorghum (*Sorghum bicolor*) hybrids for genotype × environment interaction and seasonal adaptation. *Indian J. Agric. Sci.* **2011**, *81*, 438–444.
67. Olivoto, T.; Lúcio, A.D.C.; da Silva, J.A.G.; Marchioro, V.S.; de Souza, V.Q.; Jost, E. Mean performance and stability in multi-environment trials I: Combining features of AMMI and BLUP techniques. *Agron. J.* **2019**, *111*, 2949–2960. [CrossRef]

agronomy

Article

Agromorphological Characterization and Nutritional Value of Traditional Almond Cultivars Grown in the Central-Western Iberian Peninsula

Rodrigo Pérez-Sánchez * and María Remedios Morales-Corts

Faculty of Agricultural and Environmental Sciences, University of Salamanca, Avda. Filiberto Villalobos, 119, 37007 Salamanca, Spain; reme@usal.es
* Correspondence: rodrigopere@usal.es

Abstract: In this study, 24 traditional almond cultivars grown in the central-western Iberian Peninsula, all of them clearly in decline or close to extinction, were characterized from the agromorphological and chemical points of view. A total of 40 agromorphological and chemical descriptors, mainly defined by the IPGRI and the UPOV, were used to describe the flowers, leaves, fruits and the trees themselves over three consecutive years (2015–2017). Some of the cultivars showed distinctive and interesting agronomical characteristics from a commercial point of view, such as high yields and high quality fruit. This was the case of the almond cultivars called "Gorda José" and "Marcelina". Their fruits were quite heavy (nuts: >9.1 g; kernels: >1.9 g), with very low percentages of double kernels (<3%) and high nutritional value (>50% lipids; >21% proteins). The results of the PCA and cluster analysis showed that agromorphological and chemical analysis can provide reliable information on the variability in almond genotypes. This work constitutes an important step in the conservation of genetic almond resources in the central-western Iberian Peninsula.

Keywords: almond descriptors; conservation; endangered cultivars; fruit quality; genetic resources; *Prunus dulcis*

Citation: Pérez-Sánchez, R.; Morales-Corts, M.R. Agromorphological Characterization and Nutritional Value of Traditional Almond Cultivars Grown in the Central-Western Iberian Peninsula. *Agronomy* **2021**, *11*, 1238. https://doi.org/10.3390/agronomy11061238

Academic Editors: Gregorio Barba-Espín, José Ramón Acosta-Motos and Essaid Ait Barka

Received: 5 May 2021
Accepted: 16 June 2021
Published: 18 June 2021

Publisher's Note: MDPI stays neutral with regard to jurisdictional claims in published maps and institutional affiliations.

1. Introduction

The almond (*Prunus dulcis* [Miller] D.A. Webb syn. *P. amygdalus* (L.) Batsch, Rosaceae, 2n = 2x = 16) is one of the oldest and most important nut crops grown commercially worldwide. It originated in the arid mountainous regions of southwestern and central Asia [1] and spread rapidly towards the Mediterranean Basin via seeds carried by caravans along the old Silk Route [2]. The almond is cultivated for its edible seed (the kernel), which is used for direct consumption and for almond-based products and confections [3,4]. In 2019, world production of almonds was 3.49 million metric tons [5]. The United States, Spain, Iran, Australia, Morocco and Syria are the most important almond-producing countries (approximately 80% of world almond production). Concretely, the Iberian Peninsula (the first European producer) has 717,870 ha dedicated to almond production and produces 373,970 metric tons of fruit per year. The main almond-producing regions in the Iberian Peninsula are close to the Mediterranean Sea, such as Andalusia, Murcia and Valencia, but also include inland regions such as Aragón and Castilla–La Mancha. In these regions, most of the almond orchards are not irrigated (92.2%), resulting in very low productivity [6]. Some of the most common cultivars cultured in the Iberian Peninsula are "Marcona", "Desmayo Largueta", "Tuono", "Cristomorto", "Ferragnés", "Ferraduel", "Guara", "Belona", "Soleta", "Mardía", "Masbovera", "Glorieta", "Francolí", "Constantí", "Marinada", "Tarraco", "Vayro", "Parada", "Bonita" and "Casanova". Moreover, many other named cultivars of local origin have evolved from localized ecological niches that are found in different valleys extending inland from the Mediterranean coast [7–9]. The conservation and characterization of these local cultivars is important to avoid the loss of genetic variability and as a potential source of genetic variation for future almond breeding programs. Some of these cultivars show distinctive agronomic characteristics, such as self-compatibility, late blooming, frost tolerance and shell hardness.

Many studies addressing the agromorphological and biochemical characterization of almond cultivars have been undertaken in countries mainly located around the Mediterranean basin [10–24]. In the Iberian Peninsula, studies have been carried out by Montero-Riquelme et al. [25], Felipe [26], Cordeiro et al. [27], Arquero et al. [28], Kodad and Socias i Company [29,30], Vargas et al. [31], Kodad et al. [32–34] and Ramos and Costa [35].

The objective of the present study was to survey, identify and characterize from the agromorphological and nutritional points of view the traditional almond cultivars existing in the central-western region of the Iberian Peninsula in order to avoid their disappearance and so that they can be included in future almond breeding programs.

2. Materials and Methods

2.1. Plant Material

A survey was carried out in the regions known as "Arribes del Duero" (Salamanca, Spain) and "Trás-os-Montes e Alto Douro" (Bragança, Portugal) during 2014. A total of 192 adult almond trees at least ten years old, corresponding to 24 cultivars ("Agapitina", "Bravía", "Cascafina", "Cascona", "Cornicabra", "De Convite", "Desmayo Largueta", "Desmayo Rojo", "Esperanza", "Gorda José", "Marcelina", "Marcona", "Molar de Chavés", "Mollar de Arribes", "Pestaneta de Bragança", "Pestañeta de Arribes", "Peta", "Picuda", "Portuguesa", "Recia", "Redondilla", "Valenciana", "Verdeal" and "Verdinal"), were selected for study at full fruit maturity. Eight trees were evaluated per cultivar.

2.2. Agromorphological Characterization

The agromorphological description of the almond cultivars was based on 19 descriptors established by the UPOV [36] and IPGRI [37] and a further 14 descriptors that were considered relevant for identification. For the determination of some of the descriptors, samples of flowers, fruits and leaves were taken during 2015, 2016 and 2017, using UPOV guidelines. The measurement for each parameter was performed as follows: Flowers were collected at full bloom. Ten flowers were taken from each of the eight trees studied per cultivar and year, and the following quantitative parameters were measured using an electronic digital caliper (COMECTA 5900603, Barcelona, Spain): open flower diameter (cm), petal length (cm), petal width (cm) and pistil length (cm). The numbers of petals, pistils and stamens were also counted. Leaves were collected at the adult stage, at approximately the end of July. Seven leaves were sampled per tree and year, and the following quantitative parameters were measured: petiole length and leaf blade length and width. Anthocyanin pigmentation of the nectaries was indicated by a reddish tonality observed in leaf glands. Two ratios were calculated: the length/width of the leaf blade and the petiole length/leaf blade length. Almond fruits were collected at maturity. The time of maturity was reached when the mesocarp started to dry off. All observations on dry fruits and kernels were made when the ripe fruits had a water content of less than 8%; that is, at least one month after harvesting. A total of 20 fruits per tree and year were taken to determinate each parameter. The three principal dimensions of the nut and kernel, namely length (L), width (W) and thickness (T), were measured using a digital caliper with a sensibility of 0.01 mm. The geometric mean diameter (Dg), sphericity (ø) and surface area (S) were calculated using the following equations: $Dg = (LWT)^{0.333}$, $ø = [(LWT)^{0.333}]/L$ and $S = \pi Dg^2$ [38–41]. Also, following Jain and Bal [42] and Özgüven and Vursavuş [43], the volume (V) was expressed as follows: $V = \pi B^2 L^2 / 6(2L\text{-}B)$, where $B = (WT)^{0.5}$. Shell hardness was evaluated according to the categories established in the guidelines (extremely hard, hard, intermediate, soft and paper). Mass was measured on an electronic balance (Mettler XPR603S, Toledo, Spain) with a sensitivity of ±0.001 g. The percentage of doubles (number of kernels per nut) and kernel taste were also determined. Finally, with regard to whole trees, the vegetative bearing habit of the different cultivars was evaluated by considering growing habits from extremely upright to drooping.

2.3. Chemical Composition

One hundred almonds were randomly selected per cultivar and their shells were removed to obtain the kernel. They were then finely ground by an electric grinder (Moulinex MC3001, Barcelona, Spain) and analyzed, with three replicates, for the following parameters: dry weight, lipids, proteins, dietary fiber, carbohydrates and ash (AOAC procedures [44]). Dry matter (%) was determined in a drying oven (Indelab, mod. IDL.AI 80, Navarra, Spain) at 100 °C using 25 g per sample. The crude protein content (%) of the samples was estimated with the macro-Kjeldahl method, with sulfuric acid digestion (Bloc-Digest 12P SELECTA, Barcelona, Spain) performed prior to the distillation and titration process, which was carried out with IDK132 VELP equipment and a conversion factor (N-Protein) of 5.18. The crude fat/lipids content (%) was determined with a Soxhlet extractor (VELP SER-158 Technilab, Lisses, France) by extraction from a 5 g sample of powdered almond with petroleum ether (boiling point range = 38.2–54.3 °C). The ash content (%) was determined by quantifying the residue after combustion of the dry sample in an analogical muffle furnace (HD230PAD Hobersal, Barcelona, Spain) at 550 ± 15 °C for 6 hours under incineration conditions corresponding to the gravimetric method. Total dietary fiber (%) was measured with an enzymatic–gravimetric method, determining the fiber's hydrolyzed polysaccharides by HPLC and lignin gravimetrically. Total carbohydrates (%) were estimated through the difference between the dry extract and the rest of the components. The energy value (kcal/100 g) was calculated using the general Atwater coefficients: 4 * (% protein) + 9 * (% crude fat) + 4 * (%total carbohydrates).

2.4. Statistical Analyses

Means and standard deviations were calculated for each of the quantitative parameters studied over the 3 years for the 24 traditional almond cultivars. The unit of measurement of each of the parameters studied was based on the individual value of each of the eight trees sampled per cultivar. Finally, based on all the studied parameters, principal component analysis (PCA) was carried out using the SPSS 17.0 program, and a dendrogram of genetic similarities among cultivars was compiled using the furthest neighbor method (Statgraphics Plus 17.0 program).

3. Results

3.1. Flowers

Flower parameters are shown in Table 1. Open flower diameter ranged from 3.48 to 5.51 cm, the cultivars with the largest flowers being "Peta", "Marcelina", "Gorda José" and "Portuguesa". All genotypes had five petals, except "Picuda" and the "Pestañetas" group.

In all cases, the petal length was highly correlated (0.91) with the flower size. Petal width varied from 1.11 to 2.04 cm. The lowest values were observed for "Pestañeta Arribes", "Desmayo Rojo" and "Esperanza". These last two cultivars, together with "Cascafina", "Cascona", "De Convite", "Desmayo Largueta" and "Peta", showed more than one pistil. The length of pistils ranged from 1.26 to 2.05 cm. "Portuguesa" was the cultivar with the largest pistil. The stamen number varied between 28 and 53.

Table 1. Means, standard deviations and ANOVA analyses for some flower parameters in almond cultivars.

Cultivar	Open Flower Diameter (cm)	Number of Petals	Petal Length (cm)	Petal Width (cm)	Number of Pistils	Pistil Length (cm)	Number of Stamens
Agapitina	4.38 (0.18) [defg]	5	1.89 (0.04) [c]	1.27 (0.03) [cd]	1	1.83 (0.03) [hi]	32.28 (1.97) [bcdef]
Bravía	4.35 (0.14) [cdef]	5	1.72 (0.07) [b]	1.61 (0.04) [h]	1	1.26 (0.05) [a]	31.06 (3.01) [abcdef]
Cascafina	4.91 (0.12) [jkl]	5	2.25 (0.05) [j]	1.91 (0.02) [l]	1–2	1.88 (0.04) [jk]	30.84 (1.98) [abcde]
Cascona	3.48 (0.22) [a]	5	1.45 (0.04) [a]	1.44 (0.05) [g]	1–2	1.51 (0.02) [c]	28.12 (2.11) [a]
Cornicabra	4.02 (0.11) [b]	5	1.72 (0.07) [b]	1.23 (0.04) [c]	1	1.73 (0.03) [ef]	30.87 (2.64) [abcde]
De Convite	4.70 (0.15) [hij]	5	2.05 (0.05) [efg]	1.71 (0.02) [ij]	1–2	1.95 (0.02) [lm]	39.38 (1.08) [g]
Desmayo Largueta	4.31 (0.13) [cde]	5	2.02 (0.06) [ef]	1.45 (0.05) [g]	1–2	1.65 (0.04) [d]	29.51 (2.06) [abc]
Desmayo Rojo	4.23 (0.16) [bcd]	5	2.14 (0.04) [hi]	1.17 (0.04) [b]	1–3	1.79 (0.03) [gh]	28.62 (2.35) [a]
Esperanza	4.26 (0.14) [bcde]	5	2.15 (0.05) [hi]	1.18 (0.02) [b]	1–3	1.81 (0.03) [hi]	28.17 (2.54) [a]
Gorda José	5.03 (0.21) [kl]	5	2.25 (0.06) [j]	2.04 (0.03) [m]	1	1.83 (0.03) [hi]	41.28 (2.64) [g]
Marcelina	5.11 (0.17) [l]	5	2.39 (0.05) [k]	1.63 (0.05) [h]	1	1.76 (0.04) [fg]	29.28 (0.82) [ab]
Marcona	4.62 (0.20) [ghi]	5	2.12 (0.04) [gh]	1.35 (0.04) [f]	1	1.48 (0.02) [bc]	28.64 (1.71) [a]
Molar de Chavés	4.60 (0.18) [fghi]	5	2.10 (0.05) [fgh]	1.32 (0.03) [ef]	1	1.45 (0.03) [b]	29.64 (1.99) [abc]
Mollar de Arribes	4.81 (0.17) [ijk]	5	2.22 (0.08) [ij]	1.75 (0.03) [j]	1	1.71 (0.05) [e]	34.02 (1.82) [ef]
Pestaneta de Bragança	4.16 (0.19) [bcd]	5–6	1.79 (0.05) [b]	1.16 (0.04) [b]	1	1.84 (0.03) [ij]	28.09 (1.48) [a]
Pestañeta de Arribes	4.10 (0.20) [bc]	5–6	1.72 (0.07) [b]	1.11 (0.03) [a]	1	1.83 (0.02) [hi]	28.14 (1.05) [a]
Peta	5.51 (0.17) [m]	5	2.49 (0.05) [l]	1.70 (0.02) [i]	1–2	1.92 (0.03) [kl]	31.02 (1.49) [abcdef]
Picuda	4.52 (0.17) [efgh]	5–6	1.99 (0.06) [de]	1.28 (0.04) [de]	1	1.45 (0.04) [b]	53.64 (3.02) [h]
Portuguesa	5.02 (0.22) [kl]	5	2.21 (0.07) [ij]	1.86 (0.05) [k]	1	2.05 (0.02) [n]	33.05 (2.64) [def]
Recia	4.99 (0.14) [kl]	5	2.25 (0.03) [j]	1.62 (0.08) [h]	1	1.98 (0.02) [m]	34.21 (2.19) [f]
Redondilla	4.31 (0.12) [cde]	5	1.93 (0.02) [cd]	1.23 (0.02) [c]	1	1.89 (0.04) [k]	29.15 (1.43) [ab]
Valenciana	4.72 (0.11) [hij]	5	2.02 (0.05) [ef]	1.71 (0.03) [ij]	1	1.95 (0.03) [lm]	32.64 (0.97) [cdef]
Verdeal	4.18 (0.16) [bcd]	5	1.80 (0.04) [b]	1.35 (0.04) [f]	1	1.75 (0.03) [efg]	30.56 (1.99) [abcd]
Verdinal	4.11 (0.10) [bc]	5	1.78 (0.03) [b]	1.31 (0.04) [def]	1	1.72 (0.04) [ef]	30.67 (1.35) [abcd]

ANOVA, analysis of variance; [a–n] Different letters in the same column indicate statistically significant differences between cultivars at the 95% confidence level.

3.2. Leaves

Leaf parameters are summarized in Table 2. Petiole length varied from 1.46 to 3.40 cm, the cultivars with the shortest and the longest leaf peduncles being "Cornicabra" and "Mollar de Arribes", respectively. This last cultivar also showed the highest leaf blade length. At the opposite end for this parameter was the "Bravía" genotype. Its leaves were also the narrowest (2.07 cm). The length/width ratio of the leaf blades ranged from 3.76 cm to 7.91 cm. "Valenciana", "Desmayo Rojo" and "Esperanza" were the cultivars with the highest ratios (5.96, 6.54 and 7.91, respectively). The other ratios calculated (petiole/leaf blade length) had values of around 0.22. The leaf glands showed green or reddish anthocyanin colorations.

3.3. Fruits

Fruit agromorphological parameters are shown in Table 3. Regarding dry fruit, "Gorda José" was the local cultivar that had the largest size and weight parameters, at 4.98 cm length, 3.41 cm width, 2.49 cm thickness and 16.35 g mass. Its nuts had geometric mean diameters close to 3.50 cm and volume and surface area values of around 15.64 cm^3 and 38.02 cm^2, respectively. At the other extreme was the cultivar "Bravía", with mean dry fruit dimensions of 2.82 cm (length), 1.82 cm (width) and 1.31 cm (thickness) and a mean mass value of about 2.70 g. Its nuts also had the lowest values of volume (2.42 cm^3) and surface area (11.17 cm^2). Nut sphericity ranged from 60.31 to 79.00%, the cultivars with the longest and roundest fruits being "Cornicabra" and "Marcona", respectively. It is also important to point out the relevant differences recorded for this parameter between the two "Mollar" cultivars. The "Molar de Arribes" genotype showed more elongated nuts than "Molar de Chavés" (64.22 and 72.29%, respectively). With respect to shell hardness, high resistance to cracking was observed for all the cultivars studied, except for "Desmayo Rojo", "Mollar" and "De Convite". This latter cultivar has dry fruit which can easily be opened by birds. In relation to kernels, "Marcelina" was the traditional cultivar that had the largest size and weight parameters at 2.52 cm length, 1.94 cm width, 0.90 cm thickness and 2.11 g mass. Its kernels recorded geometric mean diameters close to 1.63 cm and volume and surface area values of around 1.56 cm^3 and 8.41 cm^2, respectively. With respect to sphericity, the kernels of this last cultivar were, together with those of the cultivars "Marcona" and "Agapitina", the roundest (values close to 65%). The "Marcelina" genotype also recorded low values of double kernels (3%). This tendency to produce only a single, well-formed kernel is a highly desirable cultivar trait. On the other hand, "Desmayo Largueta", "De Convite", "Cascona", "Cascafina", "Peta", "Esperanza" and "Desmayo Rojo" registered double kernel values higher than 14%. With respect to the kernel/dry fruit mass ratio, "Cascafina" and "De Convite" were the cultivars with the highest yield values (0.34). It could be interesting to use these cultivars in future almond breeding programs. With respect to taste, all the cultivars were sweet except for the "Bravía" and "Recia" genotypes.

Table 2. Means, standard deviations and ANOVA analyses for some leaf parameters in almond cultivars.

	1	2	3	4	5	6
Cultivar	Petiole Length (cm)	Leaf Blade Length (cm)	Leaf Blade Width (cm)	2/3 Ratio	1/2 Ratio	Anthocyanin Coloration of Leaf Glands
Agapitina	2.18 (0.31) [cdefgh]	9.27 (1.27) [abcd]	1.91 (0.34) [bcdefg]	4.85 (0.62) [abcd]	0.24 (0.03) [def]	Green
Bravía	2.06 (0.55) [bcdefgh]	7.79 (1.19) [a]	2.07 (0.38) [cdefgh]	3.76 (0.60) [a]	0.26 (0.05) [f]	Reddish
Cascafina	2.56 (0.39) [ghij]	10.15 (1.18) [bcdefg]	1.89 (0.21) [bcdefg]	5.37 (0.58) [bcd]	0.25 (0.03) [ef]	Green
Cascona	2.29 (0.23) [defghi]	10.90 (1.19) [cdefgh]	2.33 (0.34) [efgh]	4.68 (0.65) [abc]	0.21 (0.02) [bcde]	Green
Cornicabra	1.46 (0.38) [a]	9.60 (1.93) [abcde]	2.18 (0.50) [efgh]	4.40 (0.69) [ab]	0.15 (0.03) [a]	Green
De Convite	3.06 (0.33) [jk]	8.42 (0.45) [ab]	1.88 (0.11) [bcdef]	4.48 (0.10) [ab]	0.36 (0.02) [g]	Reddish
Desmayo Largueta	1.95 (0.32) [abcd]	10.21 (1.15) [bcdefg]	1.93 (0.43) [bcdefg]	5.29 (0.90) [bcd]	0.19 (0.02) [abc]	Green
Desmayo Rojo	2.43 (0.40) [efghi]	12.09 (1.21) [ghi]	1.85 (0.36) [bcde]	6.54 (1.15) [e]	0.20 (0.02) [bcd]	Reddish
Esperanza	2.03 (0.39) [abcdefg]	10.52 (0.92) [cdefgh]	1.33 (0.16) [a]	7.91 (0.90) [f]	0.19 (0.03) [abc]	Reddish
Gorda José	2.47 (0.22) [fghi]	11.89 (0.98) [fgh]	3.11 (0.35) [i]	3.82 (0.82) [a]	0.21 (0.03) [bcde]	Green
Marcelina	2.46 (0.39) [efghi]	10.11 (1.02) [bcdef]	2.17 (0.20) [defgh]	4.66 (0.42) [abc]	0.24 (0.02) [def]	Green
Marcona	2.21 (0.27) [cdefghi]	11.07 (0.94) [defgh]	2.47 (0.33) [h]	4.48 (0.58) [ab]	0.20 (0.04) [bcd]	Reddish
Molar de Chavés	2.30 (0.38) [defghi]	11.52 (2.01) [efgh]	2.51 (0.37) [h]	4.58 (0.72) [abc]	0.19 (0.03) [abc]	Reddish
Mollar de Arribes	3.40 (0.55) [k]	14.03 (2.34) [i]	2.36 (0.39) [fgh]	5.94 (0.79) [de]	0.24 (0.02) [def]	Reddish
Pestaneta de Bragança	2.03 (0.21) [abcdefg]	8.51 (0.85) [ab]	1.68 (0.23) [abcd]	5.06 (0.55) [bcd]	0.23 (0.03) [cdef]	Green
Pestañeta de Arribes	1.94 (0.10) [abcdef]	8.37 (0.50) [ab]	1.60 (0.16) [abc]	5.23 (0.62) [bcd]	0.23 (0.01) [cdef]	Green
Peta	2.76 (0.41) [ij]	11.10 (1.10) [defgh]	2.38 (0.30) [gh]	4.66 (0.56) [abc]	0.25 (0.02) [ef]	Reddish
Picuda	2.63 (0.54) [hij]	11.46 (1.10) [efgh]	3.04 (0.40) [i]	3.77 (0.59) [a]	0.23 (0.04) [cdef]	Reddish
Portuguesa	1.89 (0.28) [abcde]	11.39 (1.04) [efgh]	2.26 (0.28) [efgh]	5.04 (0.51) [bcd]	0.17 (0.02) [ab]	Green
Recia	2.45 (0.43) [efghi]	12.49 (1.66) [hi]	2.37 (0.58) [fgh]	5.27 (0.90) [bcd]	0.20 (0.03) [bcd]	Green
Redondilla	2.01 (0.31) [abcdefg]	9.09 (1.31) [abc]	1.60 (0.26) [abc]	5.68 (1.18) [cde]	0.22 (0.02) [cdef]	Reddish
Valenciana	1.74 (0.19) [abcd]	8.28 (0.96) [ab]	1.96 (0.43) [bcdefg]	5.96 (0.83) [de]	0.21 (0.03) [bcde]	Green
Verdeal	1.59 (0.27) [ab]	8.03 (1.08) [a]	1.49 (0.38) [ab]	5.38 (0.53) [bcd]	0.19 (0.02) [abc]	Green
Verdinal	1.64 (0.36) [abc]	8.12 (1.21) [a]	1.50 (0.29) [ab]	5.41 (0.86) [bcde]	0.20 (0.03) [bcd]	Green

ANOVA, analysis of variance; [a–k] Different letters in the same column indicate statistically significant differences between cultivars at the 95% confidence level.

Table 3. Means, standard deviations and ANOVA analyses for some fruit parameters in almond cultivars.

Cultivar	Nut					
	Weight (g)	Length (cm)	Width (cm)	Thickness (cm)	Volume (cm³)	Geometric Mean Diameter (mm)
Agapitina	6.49 (1.31) [fg]	3.10 (0.25) [abc]	2.60 (0.15) [efg]	1.64 (0.17) [cd]	5.18 (1.02) [bcdef]	23.62 (2.16) [bcd]
Bravía	2.72 (0.37) [a]	2.82 (0.15) [a]	1.82 (0.28) [a]	1.31 (0.06) [a]	2.42 (0.67) [a]	18.86 (1.94) [a]
Cascafina	3.17 (2.02) [ab]	3.38 (0.19) [bcdefgh]	2.42 (0.15) [cde]	1.56 (0.08) [bc]	4.68 (1.15) [bcde]	23.34 (2.00) [bcd]
Cascona	5.50 (1.15) [cdefg]	3.45 (0.27) [cdefghi]	2.40 (0.22) [cde]	1.61 (0.14) [bcd]	4.88 (1.09) [bcdef]	23.69 (2.08) [bcd]
Cornicabra	5.11 (1.28) [cdef]	3.83 (0.45) [ij]	2.07 (0.17) [ab]	1.56 (0.08) [bc]	4.23 (0.98) [bc]	23.10 (1.96) [bc]
De Convite	3.86 (0.30) [abc]	3.70 (0.09) [ghij]	2.37 (0.21) [bcde]	1.45 (0.10) [ab]	4.44 (1.12) [bcd]	23.32 (2.28) [bcd]
Desmayo Largueta	4.31 (1.33) [abad]	3.56 (0.51) [efghi]	2.21 (0.20) [bcd]	1.34 (0.13) [a]	3.63 (0.94) [ab]	21.90 (1.83) [ab]
Desmayo Rojo	5.00 (0.77) [cdef]	3.24 (0.18) [bcdef]	2.46 (0.13) [cde]	1.64 (0.14) [cd]	4.95 (1.22) [bcdef]	23.53 (2.07) [bcd]
Esperanza	4.80 (0.59) [bcde]	3.38 (0.08) [bcdefgh]	2.52 (0.13) [de]	1.64 (0.11) [cd]	5.22 (1.41) [bcdef]	24.08 (1.96) [bcd]
Gorda José	16.35 (0.87) [j]	4.98 (0.23) [k]	3.41 (0.15) [j]	2.49 (0.11) [g]	15.64 (3.83) [j]	34.79 (2.41) [f]
Marcelina	9.14 (0.52) [i]	3.48 (0.15) [cdefghi]	2.88 (0.10) [ghi]	1.85 (0.10) [ef]	7.26 (1.42) [hi]	26.44 (2.17) [de]
Marcona	5.77 (1.15) [defg]	3.13 (0.20) [abcd]	2.65 (0.20) [efg]	1.83 (0.16) [ef]	6.13 (1.00) [efgh]	24.73 (2.09) [bcde]
Molar de Chavés	5.65 (0.99) [defg]	3.30 (0.39) [bcdefg]	2.54 (0.20) [ef]	1.62 (0.14) [cd]	5.13 (1.08) [bcdef]	23.85 (1.99) [bcd]
Mollar de Arribes	5.43 (0.90) [cdefg]	3.79 (0.60) [hij]	2.40 (0.22) [cde]	1.59 (0.12) [bc]	5.10 (1.14) [bcdef]	24.34 (1.86) [bcd]
Pestaneta de Bragança	6.87 (1.01) [gh]	3.20 (0.21) [abcdef]	2.64 (0.19) [efg]	1.64 (0.10) [cd]	5.37 (1.20) [cdefg]	23.99 (1.99) [bcd]
Pestañeta de Arribes	6.80 (0.56) [gh]	3.13 (0.15) [abcd]	2.63 (0.06) [efg]	1.63 (0.06) [cd]	5.24 (1.05) [bcdef]	23.74 (2.11) [bcd]
Peta	8.22 (1.04) [hi]	4.09 (0.17) [j]	2.67 (0.12) [efg]	1.99 (0.12) [f]	7.92 (1.52) [i]	27.87 (2.30) [e]
Picuda	4.90 (0.85) [cdef]	3.54 (0.21) [defghi]	2.85 (0.20) [fgh]	1.60 (0.06) [bcd]	6.05 (1.29) [defgh]	25.24 (2.04) [cde]
Portuguesa	6.51 (0.92) [fg]	3.61 (0.22) [fghi]	2.68 (0.16) [efg]	1.76 (0.09) [de]	6.37 (1.04) [fghi]	25.70 (2.15) [cde]
Recia	5.43 (0.69) [cdefg]	2.98 (0.24) [ab]	2.19 (0.22) [bc]	1.62 (0.26) [cd]	4.04 (0.97) [abc]	21.93 (1.78) [ab]
Redondilla	6.18 (1.81) [efg]	3.17 (0.47) [abcde]	2.50 (0.33) [cde]	1.72 (0.15) [cde]	5.30 (1.11) [cdefg]	23.86 (1.96) [bcd]
Valenciana	6.51 (0.98) [fg]	3.30 (0.24) [bcdefg]	2.57 (0.18) [efg]	1.85 (0.12) [ef]	6.13 (1.26) [efgh]	25.01 (2.07) [bcde]
Verdeal	6.23 (1.02) [efg]	3.40 (0.20) [cdefgh]	3.17 (0.25) [hij]	1.63 (0.11) [cd]	6.90 (1.47) [ghi]	25.97 (2.13) [cde]
Verdinal	6.38 (0.78) [efg]	3.42 (0.16) [cdefghi]	3.20 (0.28) [ij]	1.65 (0.09) [cd]	7.11 (1.33) [hi]	26.21 (2.24) [cde]

Table 3. *Cont.*

Cultivar	Nut				Kernel	
	Sphericity (%)	Surface Area (cm^2)	Shell Hardness	Weight (g)	Length (cm)	Width (cm)
Agapitina	76.19 (6.22) [ef]	17.52 (1.82) [bcd]	Hard	1.47 (0.23) [ef]	2.26 (0.22) [ab]	1.69 (0.22) [ijk]
Bravía	66.87 (5.16) [abcde]	11.17 (2.03) [a]	Hard	0.89 (0.14) [a]	2.26 (0.11) [ab]	1.10 (0.18) [a]
Cascafina	69.05 (5.62) [abcde]	17.11 (1.96) [bc]	Hard	1.09 (0.14) [abcd]	2.43 (0.14) [bcdef]	1.53 (0.07) [fghi]
Cascona	68.66 (5.47) [abcde]	17.63 (2.15) [bcd]	Hard	1.40 (0.28) [def]	2.43 (0.17) [bcdef]	1.48 (0.13) [defg]
Cornicabra	60.31 (4.98) [a]	16.76 (1.85) [bc]	Hard	1.27 (0.14) [cdef]	2.65 (0.23) [def]	1.35 (0.20) [bcde]
De Convite	63.02 (5.29) [abc]	17.08 (2.31) [bc]	Soft	1.35 (0.12) [cdef]	2.67 (0.12) [ef]	1.50 (0.06) [efg]
Desmayo Largueta	61.51 (5.73) [ab]	15.06 (1.67) [b]	Hard	1.05 (0.37) [abc]	2.60 (0.24) [cdef]	1.21 (0.13) [ab]
Desmayo Rojo	72.62 (6.00) [cdef]	17.39 (1.95) [bc]	Intermediate	1.18 (0.14) [abcdef]	2.30 (0.15) [abc]	1.33 (0.09) [bcd]
Esperanza	71.25 (6.43) [bcdef]	18.21 (2.04) [bcde]	Hard	0.92 (0.07) [ab]	2.23 (0.06) [ab]	1.30 (0.08) [bc]
Gorda José	69.85 (5.48) [abcdef]	38.02 (3.52) [h]	Hard	1.94 (0.18) [g]	2.61 (0.18) [cdef]	1.80 (0.11) [kl]
Marcelina	75.97 (6.11) [ef]	21.96 (2.01) [fg]	Hard	2.11 (0.23) [g]	2.52 (0.22) [bcdef]	1.94 (0.05) [l]
Marcona	79.00 (5.87) [f]	19.21 (1.93) [cdefg]	Hard	1.30 (0.14) [cdef]	2.08 (0.10) [a]	1.58 (0.14) [ghi]
Molar de Chavés	72.29 (5.00) [cdef]	17.87 (1.87) [bcd]	Intermediate	1.33 (0.16) [cdef]	2.32 (0.21) [abc]	1.55 (0.16) [fghi]
Mollar de Arribes	64.22 (5.42) [abcd]	18.61 (2.26) [cdef]	Intermediate	1.39 (0.13) [def]	2.51 (0.24) [bcdef]	1.52 (0.14) [fgh]
Pestaneta de Bragança	74.98 (6.31) [ef]	18.08 (2.38) [bcde]	Hard	1.37 (0.11) [cdef]	2.32 (0.16) [abc]	1.76 (0.09) [jk]
Pestañeta de Arribes	75.84 (6.19) [ef]	17.70 (1.68) [bcd]	Hard	1.36 (0.04) [cdef]	2.30 (0.09) [abc]	1.75 (0.07) [jk]
Peta	68.14 (5.87) [abcde]	24.40 (2.49) [g]	Hard	1.39 (0.22) [def]	2.72 (0.15) [f]	1.67 (0.09) [hijk]
Picuda	71.29 (6.01) [bcdef]	20.01 (2.09) [cdefg]	Hard	1.21 (0.11) [abcdef]	2.25 (0.15) [ab]	1.50 (0.08) [efg]
Portuguesa	71.19 (5.99) [bcdef]	20.74 (2.14) [defg]	Hard	1.48 (0.22) [f]	2.55 (0.13) [bcdef]	1.62 (0.08) [ghij]
Recia	73.59 (6.14) [def]	15.10 (1.52) [b]	Hard	1.23 (0.17) [bcdef]	2.29 (0.13) [abc]	1.40 (0.08) [cdef]
Redondilla	75.26 (5.62) [ef]	17.07 (1.76) [bc]	Hard	1.19 (0.21) [abcdef]	2.34 (0.14) [abad]	1.55 (0.16) [fghi]
Valenciana	75.78 (5.76) [ef]	19.65 (1.64) [cdefg]	Hard	1.15 (0.19) [abcde]	2.24 (0.14) [ab]	1.55 (0.09) [fghi]
Verdeal	76.38 (6.20) [ef]	21.18 (2.37) [defg]	Hard	1.30 (0.17) [cdef]	2.39 (0.17) [abcde]	1.46 (0.11) [cdefg]
Verdinal	76.63 (6.07) [ef]	21.58 (2.43) [efg]	Hard	1.31 (0.21) [cdef]	2.41 (0.20) [bcdef]	1.47 (0.13) [defg]

Table 3. *Cont.*

Cultivar	Kernel						Taste
	Thickness (cm)	Volume (cm^3)	Geometric Mean Diameter (mm)	Sphericity (%)	Surface Area (cm^2)	Doubles (%)	
Agapitina	0.85 (0.08) [bc]	1.16 (0.21) [c]	14.80 (1.02) [cd]	65.49 (4.94) [hi]	6.88 (0.62) [de]	2	Sweet
Bravía	0.84 (0.06) [abc]	0.69 (0.23) [a]	12.77 (1.11) [a]	56.54 (3.21) [bcd]	5.12 (0.59) [ab]	6	Bitter
Cascafina	0.73 (0.06) [ab]	0.91 (0.20) [abc]	13.94 (0.96) [abc]	57.38 (4.89) [bcdef]	6.10 (0.70) [abcd]	17	Sweet
Cascona	0.85 (0.08) [bc]	1.04 (0.19) [bc]	14.50 (1.20) [bc]	59.70 (3.93) [cdefghi]	6.60 (0.71) [abcd]	16	Sweet
Cornicabra	0.80 (0.06) [abc]	0.93 (0.20) [abc]	14.19 (0.99) [abc]	53.55 (3.65) [abc]	6.32 (0.68) [abcd]	3	Sweet
De Convite	0.70 (0.06) [ab]	0.91 (0.22) [abc]	14.09 (1.14) [abc]	52.79 (4.07) [ab]	6.23 (0.67) [abcd]	15	Sweet
Desmayo Largueta	0.68 (0.09) [a]	0.68 (0.18) [a]	12.88 (0.90) [ab]	49.54 (4.62) [a]	5.21 (0.62) [abc]	14	Sweet
Desmayo Rojo	0.80 (0.06) [abc]	0.83 (0.21) [ab]	13.47 (0.94) [abc]	58.57 (3.94) [bcdefg]	5.70 (0.66) [abcd]	21	Sweet
Esperanza	0.70 (0.07) [ab]	0.68 (0.19) [a]	12.66 (1.03) [a]	56.77 (3.82) [bcde]	5.03 (0.60) [a]	20	Sweet
Gorda José	0.70 (0.04) [ab]	1.10 (0.23) [bc]	14.86 (1.00) [cd]	56.95 (3.81) [bcde]	6.93 (0.72) [de]	0	Sweet
Marcelina	0.90 (0.06) [c]	1.56 (0.21) [d]	16.37 (1.28) [d]	64.99 (4.83) [ghi]	8.41 (0.78) [e]	3	Sweet
Marcona	0.78 (0.06) [abc]	0.92 (0.22) [abc]	13.68 (0.97) [abc]	65.77 (5.02) [i]	5.87 (0.70) [abcd]	4	Sweet
Molar de Chavés	0.80 (0.07) [abc]	0.99 (0.18) [abc]	14.22 (1.09) [abc]	61.30 (4.15) [defghi]	6.35 (0.75) [abcd]	7	Sweet
Mollar de Arribes	0.82 (0.10) [abc]	1.05 (0.20) [bc]	14.61 (1.04) [c]	58.24 (4.41) [bcdef]	6.70 (0.69) [bcd]	5	Sweet
Pestaneta de Bragança	0.80 (0.06) [abc]	1.15 (0.23) [bc]	14.83 (1.11) [cd]	63.92 (4.87) [efghi]	6.90 (0.71) [cde]	3	Sweet
Pestañeta de Arribes	0.80 (0.07) [abc]	1.13 (0.19) [bc]	14.76 (1.13) [cd]	64.17 (5.06) [fghi]	6.84 (0.70) [de]	4	Sweet
Peta	0.69 (0.09) [ab]	1.02 (0.21) [bc]	14.62 (0.98) [c]	53.78 (3.94) [abc]	6.71 (0.73) [bcd]	17	Sweet
Picuda	0.85 (0.06) [bc]	1.00 (0.22) [abc]	14.20 (1.17) [abc]	63.12 (4.08) [defghi]	6.33 (0.69) [abad]	4	Sweet
Portuguesa	0.79 (0.05) [abc]	1.10 (0.20) [bc]	14.82 (1.20) [cd]	58.14 (4.61) [bcdef]	6.89 (0.72) [de]	2	Sweet
Recia	0.80 (0.08) [abc]	0.87 (0.23) [abc]	13.68 (0.99) [abc]	59.75 (4.19) [cdefghi]	5.87 (0.67) [abcd]	8	Intermediate
Redondilla	0.79 (0.06) [abc]	0.98 (0.21) [abc]	14.19 (1.02) [abc]	60.67 (3.76) [defghi]	6.32 (0.70) [abcd]	4	Sweet
Valenciana	0.70 (0.08) [ab]	0.83 (0.22) [ab]	13.44 (0.97) [abc]	60.00 (4.99) [cdefghi]	5.67 (0.68) [abcd]	3	Sweet
Verdeal	0.79 (0.06) [abc]	0.93 (0.19) [abc]	14.01 (1.06) [abc]	58.64 (4.52) [bcdefgh]	6.16 (0.73) [abcd]	5	Sweet
Verdinal	0.80 (0.05) [abc]	0.96 (0.21) [abc]	14.14 (1.10) [abc]	58.70 (3.84) [bcdefgh]	6.28 (0.69) [abcd]	5	Sweet

ANOVA, analysis of variance; [a–l] Different letters in the same column indicate statistically significant differences between cultivars at the 95% confidence level.

Fruit chemical parameters are summarized in Table 4. The kernel dry weight value was quite similar in all the cases (95.54–96.56%). "Mollar de Arribes" was the cultivar with the lowest water level in kernels. Greater differences were recorded for the rest of the chemical parameters. Lipid content varied from 48.82% to 56.80%, "Mollar de Arribes", "Cascafina" and "Peta" being the cultivars with the most oleaginous kernels. Lipids content is highly correlated with energy levels (r = 0.95). On the other hand, it is also important to note that the "Molar de Chavés" cultivar, the other genotype called "Mollar" by the local growers, recorded a very low lipids level, close to 49%. Protein content ranged between 15.54% and 23.39%, "Peta" and "Esperanza" being the cultivars with the lowest and the highest values, respectively. It can be observed that protein content is inversely correlated with the lipid fraction (r = −0.87). Dietary fiber content varied from 14.03 to 17.99%. "Picuda", "Gorda José" and "Peta" were the cultivars that had the highest values. Carbohydrates content varied from 2.86% to 4.82%, "Recia", "Desmayo Rojo" and "Esperanza" being the cultivars with the highest levels. Finally, ash content ranged from 3.02% to 4.17%. It can be observed that the results were very similar for all the almond cultivars.

3.4. Vegetative Tree Habits

Very different vegetative habits were observed, ranging from very upright or upright to completely drooping. The habit of the "Agapitina" cultivar was between upright and very upright. On the opposite side, "Desmayo Largueta', "Verdeal" and "Verdinal" were the only almond genotypes that had drooping growth habits. The rest of the cultivars showed vegetative habits between medium and spreading. This was the case for "Mollar de Arribes", "Pestaneta de Bragança", "Pestañeta de Arribes" and "Valenciana" (medium-upright); "Cascafina", "Desmayo Rojo", "Esperanza", "Mollar de Arribes" and "Peta" (medium); "Bravía" (medium-spreading); "Cascona", "De Convite", "Gorda José" and "Portuguesa" (spreading); and "Cornicabra", "Marcelina", "Marcona", "Picuda", "Recia" and "Redondilla" (spreading-drooping). Finally, it is also important to note that the two genotypes called "Mollar" by the local growers, "Molar de Chavés" and "Mollar de Arribes", showed medium-upright and medium growth habits, respectively.

3.5. Statistical Analyses

Principal component analysis (PCA) was used to identify the traits with the highest variation between cultivars and the greatest impact on separation of them in the dataset [45]. PCA results based on flower, leaf, fruit and tree traits showed that more than 54% of the variability observed was explained by the first three components (PC1–PC3). The first component (PC1), accounting for 27.65% of the total variance, was influenced by nut weight and size parameters such as thickness, volume, geometric mean diameter and surface area. The second component (PC2) accounted for 15.27% of total variation and was mainly explained by leaf petiole length and nut and kernel sphericity. Finally, the third principal component (PC3), explaining 11.72% of total variation, was integrated by the kernel thickness and the lipid content. Figure 1 shows a scatterplot of the first two principal components (PCs) for the 24 traditional almond cultivars based on agromorphological and chemical characteristics. It can be observed that there was high variation between genotypes, indicating that the studied germplasm is a good gene pool candidate for breeding programs.

Table 4. Means, standard deviations and ANOVA analyses for some parameters of nutritive composition in almond cultivars.

Cultivar	Dry Weight (%)	Lipids (%)	Proteins (%)	Dietary Fiber (%)	Carbohydrates (%)	Ash (%)	Energy (kcal/100 g)
Agapitina	95.82 (0.71) [ab]	52.76 (2.06) [cd]	20.81 (0.92) [cd]	15.03 (0.96) [ab]	3.82 (0.42) [cde]	3.55 (0.31) [bcdef]	573.28 (15.07) [bcdef]
Bravía	95.69 (0.86) [ab]	50.87 (2.92) [abc]	23.06 (1.04) [efg]	14.84 (1.13) [ab]	4.04 (0.58) [cdef]	3.03 (0.29) [a]	566.15 (20.11) [abcd]
Cascafina	96.14 (0.79) [ab]	56.13 (2.00) [fg]	18.57 (0.87) [b]	14.81 (1.35) [ab]	3.69 (0.50) [c]	3.16 (0.34) [abcd]	593.83 (12.79) [ef]
Cascona	95.67 (0.69) [ab]	52.67 (2.76) [cd]	21.09 (0.96) [cd]	15.19 (1.27) [ab]	3.97 (0.52) [cde]	3.09 (0.39) [abc]	573.59 (18.34) [bcdef]
Cornicabra	95.76 (0.74) [ab]	52.94 (3.01) [cdef]	20.75 (0.83) [cd]	14.78 (1.76) [ab]	3.76 (0.57) [cd]	3.77 (0.30) [efghi]	574.26 (13.92) [bcdef]
De Convite	96.33 (0.77) [ab]	52.83 (2.61) [cde]	21.63 (0.91) [def]	14.97 (0.92) [ab]	3.91 (0.45) [cde]	3.11 (0.33) [ab]	577.59 (14.67) [cdef]
Desmayo Largueta	95.69 (0.80) [ab]	50.77 (2.42) [abc]	21.64 (0.94) [def]	15.56 (1.51) [ab]	4.05 (0.51) [cdef]	3.82 (0.36) [efghi]	559.49 (16.82) [abc]
Desmayo Rojo	96.21 (0.74) [ab]	52.83 (2.38) [cde]	21.02 (1.02) [cd]	14.33 (1.10) [a]	4.63 (0.47) [ef]	3.55 (0.29) [bcdef]	577.95 (11.30) [bcdef]
Esperanza	95.80 (0.79) [b]	49.08 (1.97) [ab]	23.39 (1.11) [g]	15.07 (1.60) [ab]	4.56 (0.49) [def]	4.09 (0.38) [ghi]	553.28 (17.83) [ab]
Gorda José	96.55 (0.68) [ab]	50.26 (2.49) [abc]	21.64 (0.99) [def]	16.26 (1.09) [b]	4.47 (0.60) [cdef]	4.17 (0.28) [i]	556.50 (14.12) [abc]
Marcelina	95.94 (0.83) [ab]	52.52 (2.61) [cd]	20.95 (0.86) [cd]	14.74 (1.54) [ab]	3.71 (0.58) [c]	4.12 (0.33) [hi]	571.28 (15.99) [abcde]
Marcona	96.03 (0.78) [ab]	52.47 (2.46) [cd]	21.16 (0.94) [cd]	15.03 (1.49) [ab]	4.10 (0.52) [cdef]	3.46 (0.37) [abcdef]	573.27 (21.67) [abcde]
Molar de Chavés	95.54 (0.84) [a]	49.91 (1.95) [abc]	23.21 (0.97) [fg]	14.96 (0.90) [ab]	3.99 (0.43) [cde]	3.64 (0.30) [defgh]	557.63 (17.22) [abc]
Mollar de Arribes	96.56 (0.80) [ab]	56.80 (2.81) [g]	19.67 (0.87) [bc]	15.08 (1.82) [ab]	2.13 (0.46) [a]	3.02 (0.28) [a]	598.28 (10.65) [f]
Pestaneta de Bragança	96.08 (0.75) [ab]	52.73 (3.02) [cd]	20.73 (0.85) [cd]	15.36 (1.74) [ab]	4.04 (0.52) [cdef]	3.36 (0.37) [abcde]	573.49 (18.34) [bcdef]
Pestañeta de Arribes	95.91 (0.87) [ab]	52.97 (2.43) [cdef]	20.57 (0.92) [cd]	15.14 (1.66) [ab]	3.80 (0.44) [cd]	3.64 (0.33) [defgh]	574.21 (13.96) [bcdef]
Peta	95.82 (0.78) [ab]	56.06 (3.07) [efg]	15.54 (0.97) [a]	17.99 (1.04) [c]	2.86 (0.47) [ab]	3.62 (0.30) [defg]	577.90 (19.57) [bcdef]
Picuda	95.59 (0.69) [a]	48.82 (2.09) [a]	23.35 (0.90) [g]	15.64 (0.98) [ab]	4.07 (0.50) [cdef]	3.85 (0.27) [fghi]	548.78 (12.83) [a]
Portuguesa	95.54 (0.68) [a]	50.06 (2.31) [abc]	23.33 (1.00) [g]	14.93 (1.16) [ab]	3.83 (0.53) [cde]	3.53 (0.35) [bcdef]	559.06 (15.71) [abc]
Recia	95.96 (0.81) [ab]	54.77 (2.73) [defg]	18.78 (0.83) [b]	14.04 (1.07) [a]	4.82 (0.56) [f]	3.79 (0.37) [efghi]	587.25 (19.80) [def]
Redondilla	95.94 (0.75) [ab]	51.86 (2.64) [abcd]	21.56 (1.06) [de]	14.82 (0.93) [ab]	3.86 (0.49) [cde]	4.04 (0.33) [ghi]	568.18 (11.82) [abcd]
Valenciana	96.23 (0.70) [ab]	52.24 (2.82) [bcd]	21.82 (0.98) [defg]	14.99 (1.02) [ab]	3.71 (0.47) [c]	3.61 (0.40) [cdefg]	572.24 (21.23) [abcde]
Verdeal	96.17 (0.83) [ab]	52.73 (2.71) [cd]	20.73 (1.11) [cd]	14.88 (1.14) [ab]	3.93 (0.60) [cde]	4.05 (0.36) [ghi]	573.09 (16.79) [bcdef]
Verdinal	96.02 (0.75) [ab]	53.05 (2.00) [cdef]	22.14 (0.94) [defg]	14.03 (0.91) [a]	3.65 (0.54) [bc]	3.38 (0.31) [abcdef]	580.41 (14.27) [cdef]

ANOVA, analysis of variance; [a–i] Different letters in the same column indicate statistically significant differences between cultivars at the 95% confidence level.

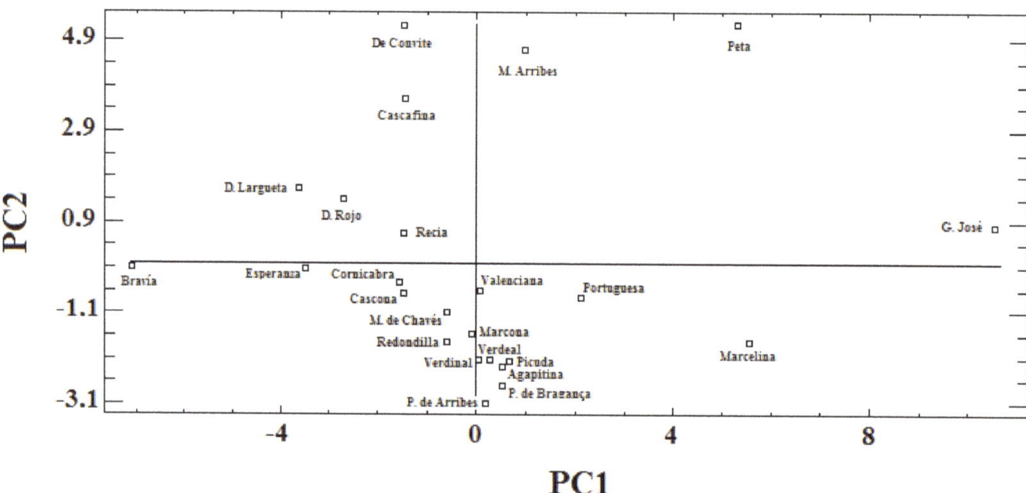

Figure 1. Scatterplot of the first two principal components (PCs) for the 24 traditional almond cultivars based on agromorphological and chemical characteristics.

Figure 2 shows a dendrogram of the relationships among the almond cultivars from the analysis of all the agromorphological and chemical parameters studied. It can be observed that the "Gorda José" cultivar showed the greatest differences compared to the rest of cultivars included in the study. Its nuts showed a large size in comparison with the others.

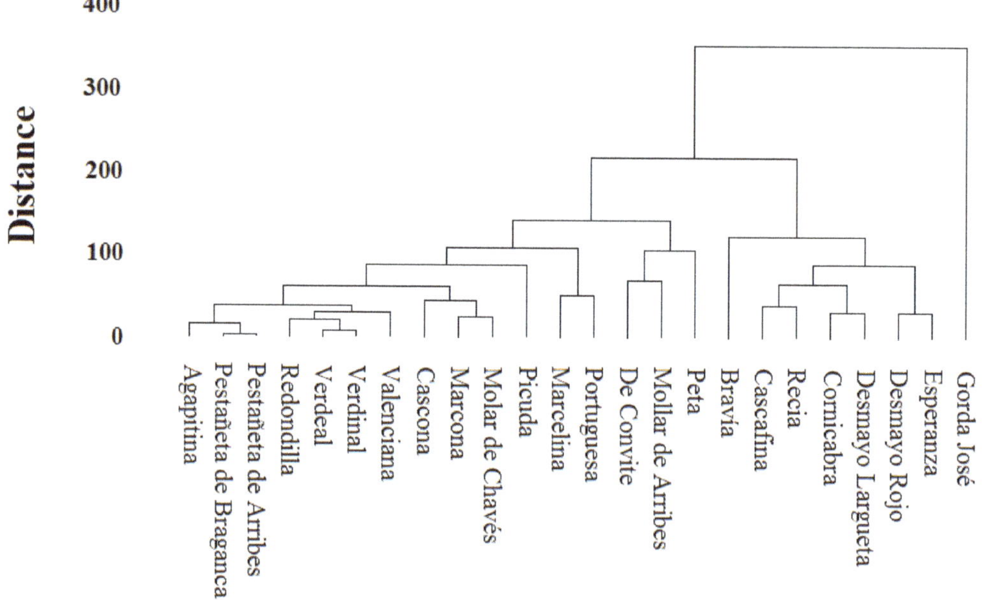

Figure 2. Dendrogram produced using the furthest neighbor method (Euclidean) from the agromorphological and chemical characteristics of traditional almond cultivars included in the study.

4. Discussion

4.1. Agromorphological and Chemical Characterization

The traditional almond cultivars from the central-western Iberian Peninsula showed great variability from an agromorphological point of view. Many of them, despite being practically unknown in the scientific literature, presented very interesting productive characters. Some previous researchers have provided information about some of the cultivars included in the study ("Marcona" and "Desmayo Largueta").

With respect to the flowers, Socias i Company et al. [46] also observed that there are important variations among almond cultivars. They indicated that the flower size is frequently related to the nut size and that the numbers of petals and pistils can range in relation to the type flower. With regard to the stamens, they established that their number can oscillate between 20 and 30, but may reach 40. In our study, a reduced number of cultivars showed six petals or two/three pistils and the number of stamens varied between 28.09 and 53.64. Arquero et al. [28] defined the "Marcona" flowers as small and elongated. In our case, the flowers of this last cultivar had a medium size in relation to the rest of cultivars included in the study.

According to Felipe [26], despite there being important differences among cultivars, almond leaves are generally narrow, long and pointed. Leaf characteristics are also affected by environmental conditions and the general health and vigor of the plant [47]. Moreover, leaf size also varies with position, with shoot leaves tending to be large and spur leaves small [48]. Moreover, Grasselly [49] have pointed out that wild cultivars generally have very small leaves, a probable adaptation to the xerophytic conditions under which these cultivars evolved. This was probably the case for the "Bravía" cultivar. Its leaves had the lowest leaf blade length/width ratio (3.76).

Regarding fruit parameters, Socias i Company et al. [46] also observed that the almond parameters are highly variable depending on the cultivar. Moreover, they indicated that traditional cultivars may produce dry fruits of only 2–3 g. This was the case with the cultivars "Bravía" (2.72 g), "Cascafina" (3.17 g) and "De Convite" (3.86 g). It is important to note that the first cultivar showed small nuts (geometric mean diameter = 18.86 mm, surface area = 11.17 cm^2 and volume = 2.42 cm^3) and the other two had thin and soft shells, respectively. Sorkheh et al. [50] considered small fruit size as one of the most common obstacles to the use of the bitter cultivars for breeding. The weight registered by the almonds of the "Gorda José" cultivar was much higher (16.35 g). It might be said that the nuts of this traditional cultivar are among the heaviest (usual maximum weight of 20 g [46]). Other researchers who have also characterized "Marcona" almonds include Melhaoui et al. [24]. Their productive results were slightly lower than those recorded in this work (almond weight = 3.58 g, geometric diameter = 21.54 mm, volume = 5.28 cm^3). These results could have been due to the arid conditions of the cultivation area. With respect to the sphericity, all the nuts were more or less elongated (mean value = 71.24%). Aydin [51] also reported mean sphericity values of almond nuts close to 70%. The "Marcona" cultivar showed the roundest nuts (79%). This result agrees with that obtained by Melhaoui et al. [24] (79.24%). Muncharaz [52] also identified "Desmayo Largueta" nuts as elongated. Finally, an important number of cultivars showed hard or very hard stony shells. This is due to the fact that hard shells are preferred in the Mediterranean region since the cultivars then seem to be more adapted to non-irrigated conditions, and more resistant to depredation by birds and penetration by insect larvae damaging the kernel. Furthermore, the nuts can be stored for a long time, with reduced problems of becoming rancid or excessively dry, thus allowing their sale throughout the year [46]. Moreover, Aydin [51] also stated that the rupture strength of almond nuts decreased with increasing moisture content. Felipe [26], Mañas et al. [53], Muncharaz [52], Arquero et al. [28] and Batlle et al. [47] also reported that the dry fruits of "Marcona", "Desmayo Largueta" and "Pestañeta" were very hard.

With regard to the kernel, its size and weight are cultivar traits. In general, it can be said that the average weight and size of the almond kernels were quite high (1.36 g and 0.97 cm^3). "Marcelina" and "Gorda José" were the cultivars that registered the largest

(>1.10 cm^3) and heaviest (>1.94 g) kernels. Socias i Compañy et al. [46] indicated that the range of kernel weight varies between 0.5 to 1.5 g, those that exceed 1.2 g being preferred for most uses. They also commented that the general trend in the industry is the preference for large kernels in order to facilitate and cheapen the processes of cracking and blanching. Nonetheless, for some special confectioneries very small sizes are chosen, as well as those with definite shapes. For sugared almonds ("peladillas" or dragées) and for chocolate almonds, large kernels are selected, preferably round to reduce the layer of sugar or chocolate covering the kernel. For chocolate bars, small and particularly flat kernels are preferred to ensure that they are covered by the chocolate and that the thickness of the bar is maintained. In addition, there is a strong environmental and seasonal effect on size, including crop load, tree vigor, soil moisture and weather patterns [47]. With respect to kernel yield, the average value was around to 23%. Concretely, "Marcona" almonds had a shelling percentage of 22%. Identical results for nuts of this cultivar were recorded by Melhaoui et al. [24]. Muncharaz [52] also indicated that the shelling percentages for "Marcona" and "Desmayo Largueta" almonds ranged between 22–28% and 24–28%, respectively. With regard to the sphericity, the kernel values (average data = 59.10%) were lower than those recorded for the nuts. In this sense, Gradziel and Lampinen [54] indicated that very large-sized kernels often bring increased market prices but appear to be associated with lower final tree yields. Felipe [26], Muncharaz [52], Arquero et al. [28] and Melhaoui et al. [24] also defined the "Marcona" kernel as globular and rounded. In addition, a significant number of the almond cultivars showed low percentages of double kernels, among them, the "Marcona" nuts, with a double kernels ratio of 4%. Similar results (0–5%) were obtained for the fruits of this cultivar by Melhaoui et al. [24] and Arquero et al. [28]. However, there is a slight difference between the value recorded in this study and that reported by Muncharaz [52] for "Desmayo Largueta" almonds (14% and 2%, respectively). In this regard, Batlle et al. [47] have noted that, although double kernels are considered to be a negative trait, lowering crop value, organoleptic quality does not appear to be affected. In this sense, Grasselly and Crossa-Reynaud [55] have pointed out a possible dominance effect with strong seasonal differences. Low temperatures before blooming [56] or at flowering time [57,58] have been mentioned as promoting higher percentages of double kernels. In relation to taste, an important number of cultivars had sweet kernels. According to Batlle et al. [47], the main trend under almond domestication was selection of types with sweet, non-poisonous seeds. Most cultivated almonds produce sweet kernels, but some have a slightly bitter flavor. This was probably the case with the "Recia" cultivar. A mild bitter flavor can be detected in some cultivars and can be considered pleasing in some special confectioneries. It is also important to point out that the sweet or bitter taste depends on the cultivar, so all the fruits of each cultivar will be sweet or bitter, regardless of the genotype of the pollen parent. Finally, regarding all these agromorphological fruit parameters, Socias i Company et al. [59] have pointed out that, although the physical traits of the almond do not affect the sensory characteristics of the almond kernel, they are very important for the processing industry and must be taken into account in the ensemble of requirements for any cultivar, together with the production and consumer sectors.

With respect to chemical characterization of the almonds, there were important differences among cultivars. In general, almond kernels are a rich lipid source, essentially composed of mono- and polyunsaturated fatty acids. Concretely, the average value of lipid fractions for the almonds analyzed was 52.42%. This result is in agreement with that reported by Kodad [60] for Spanish almond cultivars (40–67% oil content of the kernel dry weight). As previously mentioned, "Peta", "Cascafina" and "Mollar de Arribes" were the cultivars that presented the highest levels of lipids in the kernels (>56%). This lipid content is a very important factor in the confectionery industry because higher oil contents result in less water absorption by the almond paste [61]. Kernels with a high percentage of oil can be used to produce nougat or to extract oil, which is used in the cosmetic and pharmaceutical industries. However, kernels with a low percentage of oil are required to produce almond milk [62], almond flour and several kinds of food because of their correlated high protein content [63].

Protein was the second major chemical component of the almond kernels after the lipid fraction. Its average content was 21.13%, with values above 23% in the cultivars "Bravía", "Molar de Chavés", "Portuguesa", "Picuda" and "Esperanza". This average result is in agreement with that reported by Kodad [60] for Spanish almond cultivars (15.7–21.1% protein content of the kernel dry weight). According to Alessandroni [61], protein content is inversely correlated with the lipid fraction and the ratio between these two components (R1: % lipids/% protein) is important in the preparation of some processed products because the absorption of water by almond paste is dependent on the balance between these two components. For marzipan production, a low index would be desired, whereas for nougat production, a high index would be preferred.

Dietary fiber was the third major chemical component of the almond kernels. Its average content was 15.10%. Ruggeri et al. [64] reported slightly lower dietary fiber contents than those registered in this study for Italian almond cultivars (11–14% dietary fiber content of the kernel dry weight). These contents have positively effects on colonic health and cholesterol levels [65].

The average content of carbohydrates was 3.89%. It can be observed that these results for kernel carbohydrates composition again agree with those reported by Kodad [60] for Spanish almond cultivars (1.8–7.6% carbohydrates content of the kernel fresh weight). In this sense, it is also important to point out that sugars, starch and some sugar alcohols are the only carbohydrate forms present in the almond kernels that can be digested, absorbed and metabolized by humans to provide a source of energy [60]. Moreover, soluble sugars, while present in relatively low amounts, are sufficient to make kernels sweet-tasting [66].

In addition, the average content of ash was 3.60%. The almond kernel is considered a good source of mineral elements, playing an important role in human health [60]. Regarding this, Romojaro et al. [67] and Saura-Calixto et al. [65] also reported low variability for this parameter in Spanish almond cultivars (3.05–3.45%).

In relation to energy, the high nutritive value of almond kernels arises mainly from their high lipid content, which constitutes an important source of calories [68]. The average caloric content of the kernels analyzed was 571.70 kcal per 100 g fresh weight of edible portion. A similar result (578 kcal/100 g FW) was reported by Gradziel [69].

Finally, the growth habits were highly variable among cultivars, ranging from very upright or upright (the "Agapitina" cultivar) to completely drooping (the "Desmayo Largueta" cultivar, among others). Socias i Company et al. [46], Mañas et al. [53], Felipe [26] and Muncharaz [52] have also observed that "Desmayo Largueta" has a fairly common drooping growth habit. Arquero et al. [28] classified the vegetative habits of "Marcona" and "Desmayo Largueta" cultivars as spreading. In this regard, Espada and Connell [70] have indicated that upright trees have better shaking efficiency than trees with roundish or hanging canopies. However, when the habit is very erect, the tree may have a canopy with insufficient spread. This makes orchard management difficult but may be useful in new developments of high density plantings. Generally an upright to spreading habit is preferable, as it facilitates formation of the tree and mechanization of the different operations [46]. Thus, the tree habit of modern European cultivars is generally spreading [47].

4.2. Statistical Analyses

Principal component analysis (PCA) results showed that more than 54% of the variability observed was explained by the first three components. These results agree with those obtained by Gouta et al. [71], Čolić et al. [72] and Khadivi-Khub and Etemadi-Khah [73] for almond cultivars of the Mediterranean area. PCA revealed that the first three components explained comparable values (from 34% to 57%) of the total variation, based on morphological and biochemical traits. Furthermore, Lansari et al. [74], Talhouk et al. [11], and Sorkheh et al. [50], who used a similar analysis to compare kernel, nut and leaf characters in different almond collections, found that the variables contributing to nut and kernel size were more important than leaf traits.

By analyzing the dendrogram, and taking into account all the results shown above, a series of synonymies among the almond cultivars can also be detected. Such is the case with "Pestaneta de Bragança" and "Pestañeta de Arribes", and with "Verdeal" and "Verdinal". Significant similarities were observed between the cultivars for these two cases of synonymies for all agromorphological and chemical traits. By contrast, a homonym was also detected: "Molar de Chavés" and "Mollar de Arribes". Despite their major agromorphological and chemical differences, both names are often used interchangeably by some growers.

The results of the PCA and cluster analysis showed that agromorphological and chemical analysis can provide reliable information on the variability in almond genotypes. In correspondence with our findings, Ledbetter and Shonnard [75], Talhouk et al. [11], Sorkheh et al. [76], Zeinalabedini et al. [77] and Khadivi-Khub and Etemadi-Khah [73] have also shown that morphological evaluation is an efficient tool for characterization of almond germplasm and for species distinction. The overall analysis of all traits illustrates a wide diversity that may have important implications for management of genetic resources.

5. Conclusions

Twenty-four traditional almond cultivars grown in the central-western Iberian Peninsula, all of them clearly in decline or close to extinction, were characterized from the agromorphological and chemical points of view. Some of the cultivars showed distinctive and interesting agronomical characteristics from a commercial point of view, such as high yields and high quality fruit. This was the case for the almond cultivars called "Gorda José" and "Marcelina". Their fruits were quite heavy (nuts: >9.1 g; kernels: >1.9 g), with very low percentages of double kernels (<3%) and high nutritional value (>50% lipids; >21% proteins). The results of PCA and cluster analysis showed that agromorphological and chemical analysis can provide reliable information on the variability in almond genotypes. Two synonymies ("Pestaneta de Bragança" and "Pestañeta de Arribes"; "Verdeal" and "Verdinal") and one homonym ("Molar de Chavés" and "Mollar de Arribes") were also detected. This work constitutes an important step in the conservation of genetic almond resources in the central-western Iberian Peninsula.

Author Contributions: Conceptualization, R.P.-S. and M.R.M.-C.; Methodology, R.P.-S. and M.R.M.-C.; Software, R.P.-S.; Validation, R.P.-S. and M.R.M.-C.; Investigation, R.P.-S. and M.R.M.-C.; Data curation, R.P.-S.; Writing—original draft preparation, R.P.-S.; Writing—review and editing, R.P.-S. and M.R.M.-C.; Project administration, R.P.-S.; Funding acquisition, R.P.-S. All authors have read and agreed to the published version of the manuscript.

Funding: This research was supported by the Biodiversity Foundation of the Spanish Ministry of Environment and Rural and Marine Affairs (MARM) and the Spanish Federation of Municipalities and Provinces (FEMP). (Project reference: MARM-FB-2008-012).

Institutional Review Board Statement: Not applicable.

Acknowledgments: The authors thank the almond growers of the "Arribes del Duero" (Salamanca, Spain) and "Trás-os-Montes e Alto Douro" (Bragança, Portugal) for their major contributions to this work.

Conflicts of Interest: The authors declare no conflict of interest.

References

1. Watkins, R. Cherry, plum, peach, apricot and almond. *Prunus* spp. In *Evolution of Crop Plants*; Simmonds, N.W., Ed.; Cambridge University Press: London, UK, 1976; pp. 242–247.
2. Mohamed, B.S. The contribution of *Prunus webbii* to almond evolution. *PGR Newsl.* **2004**, *140*, 9–13.
3. Kester, D.E.; Gradziel, T.M. Almonds. In *Fruit Breeding—Volume III: Nuts*; Janick, J., Moore, J.N., Eds.; Wiley-Blackwell: New York, NY, USA, 1996; pp. 1–96.
4. Duke, J.A. *Handbook of Nuts*; CRC Press: Boca Ratón, FL, USA, 2000.

5. Food and Agriculture Organization (FAO). Faostat: Agriculture Data. 2019. Available online: http://www.fao.org/faostat/es/#data/QC (accessed on 28 April 2021).
6. Gradziel, T.M.; Curtis, R.; Socias i Company, R. Production and growing regions. In *Almonds. Botany, Production and Uses*; Socias i Company, R., Gradziel, T.M., Eds.; CABI: Wallingford, UK, 2017; pp. 70–86.
7. Vargas, F.J. *El Almendro en la Provincia de Tarragona*; Excma. Diputación Provincial de Tarragona, Servicio Agropecuario Provincial: Tarragona, Spain, 1975.
8. García, J.E.; Egea, L.; Egea, J.; Berenguer, T. Programme d'amélioration de l'amandier au C.E.B.A.S. de Murcie. *Options Méditerranéennes* **1985**, *85*, 7–8.
9. García, J.E.; Egea, L.; Egea, J.; Berenguer, T. Étude et sélection de variétés d'amandier dans le sudest de l'Espagne. *Rapp. EUR* **1988**, *11557*, 33–46.
10. Nieddu, G.; Chessa, I.; Pala, M.; Lovicu, G. Evaluation of almond germplasm in Sardinia: Further observations. *Acta Hortic.* **1994**, *373*, 135–140. [CrossRef]
11. Talhouk, S.N.; Lubani, R.T.; Baalbaki, R.; Zurayk, R.; Al Khatib, A.; Parmaksizian, L.; Jaradat, A.A. Phenotypic diversity and morphological characterization of Amygdalus L. species in Lebanon. *Genet. Resour. Crop Evol.* **2000**, *47*, 93–104. [CrossRef]
12. De Giorgio, D.; Polignano, G.B. Evaluation the biodiversity of almond cultivars from a germplasm collection field in Southern Italy. In *Sustaining the Global Farm*; Stott, D.E., Mohtar, R.H., Steinhardt, G.C., Eds.; ISCO: West Lafayette, IN, USA, 2001; pp. 305–311.
13. Kaska, N.; Kafkas, S.; Padulosi, S.; Wassimi, N.; Ak, B.E. Characterization of nuts species of Afghanistan: I—Almond. *Acta Hortic.* **2006**, *726*, 147–156. [CrossRef]
14. Chalak, L.; Chehade, A.; Kadri, A. Morphological characterization of cultivated almonds in Lebanon. *Fruits* **2007**, *62*, 177–186. [CrossRef]
15. Arslan, S.; Vursavuş, K.K. Physico-mechanical properties of almond nut and its kernel as a function of variety and moisture content. *Philipp. Agric. Sci.* **2008**, *91*, 171–179.
16. Jahanban, A.; Jahanban, R.; Jamei, R.; Jahanban, A. Morphology and physicochemical properties of 40 genotypes of almond (*Amygdalus communis* L.) fruits. *Eur. J. Exp. Biol.* **2012**, *2*, 2456–2464.
17. Hanine, H.; Zinelabidine, L.H.; Hssaini, H.; Nablousi, A.; Ennahli, S.; Latrache, H.; Zahir, H. Pomological and biochemical characterization of almond cultivars in Morocco. *Turk. J. Agric. Nat. Sci.* **2014**, *1*, 743–753.
18. Hanine, H.; Zinelabidine, L.H.; H'ssaini, H.; Ennahli, S.; Latrache, H.; Hmid, I. Phenotypic, morphological diversity and biochemical characterization of 14 almond cultivars from Morocco. *Moroc. J. Chem.* **2015**, *3*, 394–406.
19. Ardjmand, A.; Piri, S.; Imani, A.; Piri, S. Evaluation of morphological and pomological diversity of 62 almond cultivars and superior genotypes in Iran. *J. Nuts* **2014**, *5*, 39–50.
20. Kodad, O.; Estopañán, G.; Fagroud, M.; Juan, T.; Socias i Company, R. Physical and chemical traits of almond kernels of the local almond populations in Morocco: Commercial and industrial end-uses. *Acta Hortic.* **2014**, *1028*, 233–238. [CrossRef]
21. Kodad, O.; Lebrigui, L.; El-Amrani, L.; Socias i Company, R. Physical fruit traits in Moroccan almond seedlings: Quality aspects and post-harvest uses. *Int. J. Fruit Sci.* **2015**, *15*, 36–53. [CrossRef]
22. Sepahvand, E.; Khadivi-Khub, A.; Momenpour, A.; Fallahi, E. Evaluation of an almond collection using morphological variables to choose superior trees. *Fruits* **2015**, *70*, 53–59. [CrossRef]
23. Giordani, E.; Berti, M.; Rauf Yaqubi, M. Phenotypic characterisation of almond accessions collected in Afganistan. *Adv. Hortic. Sci.* **2016**, *30*, 207–216.
24. Melhaoui, R.; Addi, M.; Houmy, N.; Abid, M.; Mihamou, A.; Serghini-Caid, H.; Sindic, M.; Elamrani, A. Pomological characterization of main almond cultivars from the North Eastern Morocco. *Int. J. Fruit Sci.* **2019**, *19*, 413–422. [CrossRef]
25. Montero-Riquelme, F.J. *Caracterización Morfológica del Almendro*; Servicio de Publicaciones de la Universidad de Castilla-La Mancha: Murcia, Spain, 1993.
26. Felipe, A.J. *El Almendro. Volume I. El Material Vegetal*; Mira Editores: Zaragoza, Spain, 2000.
27. Cordeiro, V.; Monteiro, A.; Oliveira, M.; Ventura, J. Study of some physical characters and nutritive composition of the Portuguese's (local) almond varieties. *Options Méditerranéennes* **2001**, *56*, 333–337.
28. Arquero, O.; Lovera, M.; Navarro, A.; Viñas, M.; Salguero, A.; Serrano, N. *Hábitos de Vegetación y Respuesta a la Intensidad de la Poda de Formación de las Principales Variedades de Almendro*; Consejería de Agricultura y Pesca, Junta de Andalucía: Sevilla, Spain, 2008.
29. Kodad, O.; Socias i Company, R. Variability of oil content and of major fatty acid composition in almond (*Prunus amygdalus* Batsch) and its relationship with kernel quality. *J. Agric. Food Chem.* **2008**, *56*, 4096–4101. [CrossRef] [PubMed]
30. Kodad, O.; Socias i Company, R. Fruit quality in almond as related to the type of pollination in self-compatible genotypes. *J. Am. Soc. Hortic. Sci.* **2008**, *133*, 320–326. [CrossRef]
31. Vargas, F.J.; Romero, M.; Clavé, J.; Alegre, S.; Miarnau, X. *Variedades de Almendro IRTA*; Ibercaja: Madrid, Spain, 2009.
32. Kodad, O.; Mamouni, A.; Lahlo, M.; Socias i Company, R. Implicación commercial e industrial de la variabilidad del contenido en aceite y proteína y de los caracteres físicos del fruto y de la pepita del almendro en las condiciones climáticas mediterráneas. *Inf. Técnica Económica Agrar.* **2011**, *107*, 300–314.
33. Kodad, O.; Estopañán, G.; Juan, T.; Socias i Company, R. Genetic diversity in an almond germplasm collection: Application of a chemometric approach. *Acta Hortic.* **2013**, *976*, 237–242. [CrossRef]

34. Kodad, O.; Estopañán, G.; Juan, T.; Alonso, J.M.; Espiau, M.T.; Socias i Company, R. Oil content, fatty acid composition and tocopherol concentration in the Spanish almond genebank collection. *Sci. Hortic.* **2014**, *177*, 99–107. [CrossRef]

35. Ramos, M.A.; Costa, M.G. *Characteristics of the Fruits of Some Varieties of Almond Grown in the Algarve Region*; Faculty of Sciences and Technology, University of Algarve: Faro, Portugal, 2014.

36. UPOV. *Guidelines for the Conduct of Test for Distinctness, Homogeneity and Stability. Almond (Prunus amygdalus Batsch)*; International Union for the Protection of New Varieties of Plants: Geneva, Switzerland, 1978.

37. IPGRI. *Descriptors List for Almond (Prunus amygdalus) (Revised)*; International Board for Plant Genetic Resources: Rome, Italy, 1985.

38. Mohsenin, N.N. *Physical Properties of Plant and Animal Materials*; Gordon and Breach Science Publishers: New York, NY, USA, 1978.

39. Baryeh, E.A. Physical properties of Bambara groundnuts. *J. Food Eng.* **2001**, *47*, 321–326. [CrossRef]

40. Güner, M.; Dursun, E.; Dursun, I.G. Mechanical behavior of the hazelnut under compression loading. *Biosyst. Eng.* **2003**, *4*, 485–491. [CrossRef]

41. Olajide, J.O.; Igbeka, J.C. Some physical properties of groundnut kernels. *J. Food Eng.* **2003**, *58*, 201–204. [CrossRef]

42. Jain, R.K.; Bal, S. Physical properties of pearl millet. *J. Agric. Eng. Res.* **1997**, *66*, 85–91. [CrossRef]

43. Özgüven, F.; Vursavuş, K. Some physical, mechanical and aerodynamic properties of pine (*Pinus pinea*) nuts. *J. Food Eng.* **2005**, *68*, 191–196. [CrossRef]

44. AOAC. *Official Methods of Analysis*, 16th ed.; Association of Official Analytical Chemists: Arlington, TX, USA, 1995.

45. Iezzoni, A.F.; Pritts, M.P. Applications of principal components analysis to horticultural research. *HortScience* **1991**, *26*, 334–338. [CrossRef]

46. Socias i Company, R.; Ansón, J.M.; Espiau, M.T. Taxonomy, botany and physiology. In *Almonds. Botany, Production and Uses*; Socias i Company, R., Gradziel, T.M., Eds.; CABI: Wallingford, UK, 2017; pp. 1–42.

47. Batlle, I.; Dicenta, F.; Socias i Company, R.; Gradziel, T.M.; Wirthensohn, M.; Duval, H.; Vargas, F.J. Classical genetics and breeding. In *Almonds. Botany, Production and Uses*; Socias i Company, R., Gradziel, T.M., Eds.; CABI: Wallingford, UK, 2017; pp. 111–148.

48. Gradziel, T.M. Almond (*Prunus dulcis*). In *Breeding of Plantation Crops*; Priyadarshan, M., Jain, S.M., Eds.; Springer: Berlin, Germany, 2008; pp. 1–33.

49. Grasselly, C. Les espèces sauvages d'amandiers. *Options Méditerranéennes* **1976**, *32*, 28–43.

50. Sorkheh, K.; Shiran, B.; Rouhi, V.; Asadi, E.; Jahanbazi, H.; Moradi, H.; Gradziel, T.M.; Martínez, P. Phenotypic diversity within native Iranian almond (*Prunus* spp.) species and their breeding potential. *Genet. Resour. Crop Evol.* **2009**, *56*, 947–961. [CrossRef]

51. Aydin, C. Physical properties of almond nut and kernel. *J. Food Eng.* **2003**, *60*, 315–320. [CrossRef]

52. Muncharaz, M. *El Almendro—Manual Técnico*; Mundi-Prensa: Madrid, España, 2004.

53. Mañas, F.; López, P.; Rodríguez, A. Mejora genética de material autóctono de almendro en Castilla-La Mancha para floración tardía y calidad. *Mem. ITAP* **2000**, *2000*, 293–325.

54. Gradziel, T.; Lampinen, B. Defining the limits of almond productivity to facilitate marker assisted selection and orchard management. *Acta Hortic.* **2011**, *912*, 47–52. [CrossRef]

55. Grasselly, C.; Crossa-Raynaud, P. *L'Amandier*; G.P. Maisonneuve et Larose: Paris, France, 1980.

56. Egea, J.; Burgos, L. Double kernel fruits in almond (*Prunus dulcis* Mill.) as related to pre-blossom temperatures. *Ann. Appl. Biol.* **1994**, *126*, 163–168. [CrossRef]

57. Rikhter, A.A. Ways and methods of almond breeding (in Russian). *Tr. Gos. Nikitsk. Bot. Sada* **1969**, *43*, 81–94.

58. Spiegel-Roy, P.; Kochba, J. The inheritance of bitter and double kernel characters in the almond (*Prunus amygdalus* Batsch). *Z. Pflanz.* **1974**, *71*, 319–329.

59. Socias i Company, R.; Alonso, J.M.; Kodad, O. Fruit quality in almond: Physical aspects for breeding strategies. *Acta Hortic.* **2009**, *814*, 475–480. [CrossRef]

60. Kodad, O. Chemical composition of almond nuts. In *Almonds. Botany, Production and Uses*; Socias i Company, R., Gradziel, T.M., Eds.; CABI: Wallingford, UK, 2017; pp. 428–448.

61. Alessandroni, A. Le mandorle. *Panifazione Pastic.* **1980**, *8*, 67–71.

62. Ramusino, F.C.; Intonti, R.; Stachini, A. Analisi del latte di mandorle e dello sciroppo di orzata. *Bolletino Dei Lab. Chem. Prov.* **1961**, *12*, 491–504.

63. Longhi, S. High-Protein Animal Feed. *Ital. Pat.* **1952**, *470*, 433.

64. Ruggeri, S.; Cappelloni, M.; Gambelli, L.; Nicoli, S.; Carnovale, E. Chemical composition and nutritive value of nuts grown in Italy. *J. Food Sci.* **1998**, *3*, 243–251.

65. Saura-Calixto, F.; Bauzá, M.; Martínez, F.; Argamentería, A. Amino acids, sugars, and inorganic elements in the sweet almond. *J. Agric. Food Chem.* **1981**, *29*, 509–511. [CrossRef] [PubMed]

66. Schirra, M. Postharvest technology and utilization of almonds. *Hortic. Rev.* **1997**, *20*, 267–292.

67. Romojaro, F.; García, J.E.; López-Andreu, F.J. Estudio sobre la composición química de variedades de almendra del sureste español. *An. Edafol. Agrobiol.* **1977**, *36*, 121–131.

68. Sabate, J.; Hook, D.G. Almonds, walnuts and serum lipids. In *Lipids in Human Nutrition*; Spiller, G.A., Ed.; CRC Press: Boca Raton, FL, USA, 1996; pp. 137–144.

69. Gradziel, T.M. History of cultivation. In *Almonds. Botany, Production and Uses*; Socias i Company, R., Gradziel, T.M., Eds.; CABI: Wallingford, UK, 2017; pp. 43–69.

70. Espada, J.L.; Connell, J.H. Almond harvesting. In *Almonds. Botany, Production and Uses*; Socias i Company, R., Gradziel, T.M., Eds.; CABI: Wallingford, UK, 2017; pp. 406–427.
71. Gouta, H.; Mars, M.; Gouiaa, M.; Ghrab, M.; Zarrouk, M.; Mliki, A. Genetic diversity of almond (*Prunus amygdalus* Batsch) in Tunisia: A morphological traits analysis. *Acta Hortic.* **2011**, *912*, 351–358. [CrossRef]
72. Čolić, S.; Rakonjac, V.; Zec, G.; Nicolić, D.; Fotirić, M. Morphological and biochemical evaluation of selected almond [*Prunus dulcis* (Mill.) D.A.Webb] genotypes in northern Serbia. *Turk. J. Agric. For.* **2012**, *36*, 429–438.
73. Khadivi-Khub, A.; Etemadi-Kah, A. Phenotypic diversity and relationships between morphological traits in selected almond (*Prunus amygdalus*) germplasm. *Agrofor. Syst.* **2015**, *89*, 205–216. [CrossRef]
74. Lansari, A.; Iezzoni, A.F.; Kester, D.E. Morphological variation within collection of Moroccan almond clones and Mediterranean and North American cultivars. *Euphytica* **1994**, *78*, 27–41.
75. Ledbetter, C.A.; Shonnard, C.B. Evaluation of selected almond (*Prunus dulcis* (Miller) D.A. Webb) germplasm for several shell and kernel characteristics. *Fruit Var. J.* **1992**, *46*, 79–82.
76. Sorkheh, K.; Shiran, B.; Khodambashi, M.; Moradi, H.; Gradziel, T.M.; Martinez-Gómez, P. Correlations between quantitative tree and fruit almond traits and their implications for breeding. *Sci. Hortic.* **2010**, *125*, 323–331. [CrossRef]
77. Zeinalabedini, M.; Sohrabi, S.; Nikoumanesh, K.; Imani, A.; Mardi, M. Phenotypic and molecular variability and genetic structure of Iranian almond cultivars. *Plant Syst. Evol.* **2012**, *298*, 1917–1929. [CrossRef]

Article

Assessing the Genetic Diversity and Population Structure of a Tunisian Melon (*Cucumis melo* L.) Collection Using Phenotypic Traits and SSR Molecular Markers

Hela Chikh-Rouhou [1,*,†], Najla Mezghani [2,3,†], Sameh Mnasri [2], Neila Mezghani [4] and Ana Garcés-Claver [5]

1   Regional Research Centre on Horticulture and Organic Agriculture (CRRHAB/IRESA), LR21AGR03, University of Sousse, 4042 Sousse, Tunisia
2   National Gene Bank of Tunisia, Boulevard Leader Yasser Arafat, 1080 Tunis, Tunisia; najla_mezghani@yahoo.fr (N.M.); mnasrisameh@yahoo.fr (S.M.)
3   Research Unit of Organic and Conventional Vegetable Crops UR13AGR03, High Institute of Agronomy of Chott Mariem, University of Sousse, 4042 Sousse, Tunisia
4   LICEF Research Center, TELUQ University, 5800 Rue Saint-Denis, Quebec, QC H2S3L5, Canada; neila.mezghani@teluq.ca
5   Centro de Investigación y Tecnología Agroalimentaria de Aragón, Instituto Agroalimentario de Aragón-IA2 (CITA-Universidad de Zaragoza), 50059 Zaragoza, Spain; agarces@cita-aragon.es
*   Correspondence: hela.chikh.rouhou@gmail.com
†   These authors contributed equally to this work.

Citation: Chikh-Rouhou, H.; Mezghani, N.; Mnasri, S.; Mezghani, N.; Garcés-Claver, A. Assessing the Genetic Diversity and Population Structure of a Tunisian Melon (*Cucumis melo* L.) Collection Using Phenotypic Traits and SSR Molecular Markers. *Agronomy* 2021, 11, 1121. https://doi.org/10.3390/agronomy11061121

Academic Editors: Gregorio Barba-Espín and Jose Ramon Acosta-Motos

Received: 23 April 2021
Accepted: 28 May 2021
Published: 31 May 2021

Publisher's Note: MDPI stays neutral with regard to jurisdictional claims in published maps and institutional affiliations.

Abstract: The assessment of genetic diversity and structure of a gene pool is a prerequisite for efficient organization, conservation, and utilization for crop improvement. This study evaluated the genetic diversity and population structure of 24 Tunisian melon accessions, by using 24 phenotypic traits and eight microsatellite (SSR) markers. A considerable phenotypic diversity among accessions was observed for many characters including those related to agronomical performance. All the microsatellites were polymorphic and detected 30 distinct alleles with a moderate (0.43) polymorphic information content. Shannon's diversity index (0.82) showed a high degree of polymorphism between melon genotypes. The observed heterozygosity (0.10) was less than the expected heterozygosity (0.12), displaying a deficit in heterozygosity because of selection pressure. Molecular clustering and structure analyses based on SSRs separated melon accessions into five groups and showed an intermixed genetic structure between landraces and breeding lines belonging to the different botanical groups. Phenotypic clustering separated the accessions into two main clusters belonging to sweet and non-sweet melon; however, a more precise clustering among *inodorus*, *cantalupensis*, and *reticulatus* subgroups was obtained using combined phenotypic–molecular data. The discordance between phenotypic and molecular data was confirmed by a negative correlation ($r = -0.16$, $p = 0.06$) as revealed by the Mantel test. Despite these differences, both markers provided important information about the diversity of the melon germplasm, allowing the correct use of these accessions in future breeding programs. Together they provide a powerful tool for future agricultural and conservation tasks.

Keywords: genetic diversity; breeding lines; landraces; phenotypic traits; molecular markers

1. Introduction

Melon (*Cucumis melo* L., 2n = 24) is a morphologically diverse and outcrossing horticultural crop of economic importance that belongs to the Cucurbitaceae family. The species is subdivided into two subspecies, namely, subsp. *agrestis* (Naud.) and subsp. *melo*, on the basis of vegetative morphological characteristics and fruit variation, length, and distribution of hairs on the ovary and young fruit [1]. Melons have been grouped into several horticultural groups [2], with the *cantalupensis*, *inodorus*, and *reticulatus* market types being the most economically important ones in American, Asian, and Mediterranean countries [3].

Melon diversity is spread across primary and secondary centers. Whereas the primary centers are located in central and southern Asia, the Far East Asian and Mediterranean regions comprise the secondary centers [2]. Because of its strategic position and the diversity of ecosystems and climatic conditions, Tunisia has been a crossroad of several civilizations, and itis considered one of the richest Mediterranean countries in terms of plant genetic resources including landraces and wild relatives. Being genetically diverse and well adapted to local environmental conditions, landraces are considered as an important reservoir of useful genes that could be exploited in crop breeding programs.

Tunisian melon landraces have been identified as highly tolerant to many biotic stresses such as Powdery mildew [4], Fusarium wilt [5], aphids [6], and viruses [7]. Although some Tunisian landraces are still cultivated in rural areas through traditional farming systems, the adoption of modern varieties at the expense of autochthonous germplasm has reduced the genetic diversity. This has contributed to genetic erosion resulting in significant loss of valuable genetic diversity for technological quality, adaptation to low inputs, and tolerance to stress conditions [8,9]. Moreover, little attention has been paid to the conservation of this germplasm. The extensive collection, preservation, and genetic diversity assessment of Tunisian melon landraces are vital in order to prevent genetic erosion, increase genetic variability for melon breeding, and introduce new traits into modern melon cultivars.

Several studies have been carried out to estimate the genetic variation within the melon germplasm through different approaches such as morphological descriptors [10,11] and molecular markers [12]. The evaluation of morphological traits has been frequently combined with agronomical [8,13], physiological, and biochemical data such as pH, total soluble solids, polysaccharide content, organic acids, and vitamins [14,15]. However, the expressions of most of these morphological traits are generally influenced by environmental factors and cultivation practices. Molecular markers that reveal polymorphism at the DNA level have been considered a powerful tool for the estimation of plant genetic diversity characterization and for the discrimination of different morphological individuals from different sources [16]. Various molecular markers have been successfully used to characterize the melon germplasm including restriction fragment length polymorphisms (RFLPs) [17], random amplified polymorphic DNA (RAPDs) [18], amplified fragment length polymorphisms (AFLPs) [19], and simple sequence repeats, also called microsatellites (SSRs) [12]. Specifically, SSRs have proven to be useful marker systems in recent melon genetic diversity and population structure studies [14,20,21] due to their abundance in the genome, high polymorphism, reliability, and codominant nature [20,21].

To the best of our knowledge, few studies have addressed the genetic diversity of Tunisian melon landraces using either morphological [8,22] or molecular markers [23,24]. This investigation is the first report on the assessment of the genetic diversity and structure of the Tunisian melon germplasm by combining both morphological descriptors and SSR markers. Landraces and local breeding lines belonging to different botanical groups were included to evaluate the intra and inter variation among accessions and among botanical groups and to examine the level of untapped genetic variation in the local germplasm that could be exploited in future melon improvement programs.

2. Materials and Methods

2.1. Plant Material and Experimental Design

A melon collection including 14 landraces and 10 breeding lines was considered in this study (Table 1). Native landraces adapted to their growing conditions represent an opportunity to explore their variability and identify promising traits. The landraces, named as indicated by the farmers, were open-pollinated landraces collected during 2016 and 2017 from different geographic regions of Tunisia. The breeding lines, derived from landrace individuals, were selected at the CRRHAB Tunisia research center for their resistance to fungal diseases (Fusarium wilt or Powdery mildew) and maintained by self-pollination for 4 years. Characterization of both landraces and breeding lines is necessary for an efficient breeding program and to release new elite cultivars. The accessions were previously morphologically classified into five botanical groups of *C. melo* subsp. *melo* (*inodorus*, *reticulatus*, *cantalupensis*, *flexuosus*, and *chate*; 16, 3, 2, 2, and 1 accessions, respectively) according to Pitrat [2].

Table 1. Description of 24 melon accessions used in the study.

Code	Local Name	Locality Origin	Horticultural Group	Type	Description
Maaz1	Maazoun	Chott Mariem-Sousse	*inodorus*	Landrace	Sweet fruit without aroma; yellow skin
Maaz2	Maazoun	Menzel Chaker-Sfax	*inodorus*	Landrace	Sweet fruit without aroma, yellow-green skin
Maaz3	Maazoun	Wardanine-Monastir	*inodorus*	Landrace	Sweet fruit without aroma; yellow-green skin
Maaz4	Maazoun	Sidi Bouzid	*inodorus*	Landrace	Sweet fruit without aroma; yellow-green skin
Arbi1	Arbi	Jammel-Monastir	*inodorus*	Landrace	Sweet fruit without aroma, yellow skin
Arbi2	Arbi	Jendouba	*inodorus*	Landrace	Fruit without aroma; creamy-white skin
Arbi3	Arbi	Jammel-Monastir	*inodorus*	Landrace	Sweet fruit without aroma; orange skin
Arbi4	Arbi	Jammel-Monastir	*inodorus*	Landrace	Sweet fruit without aroma; yellow skin
Arbi5	Arbi	Kairouan	*reticulatus*	Landrace	Fruit with aroma, netted skin
Arbi6	Arbi	El Jem-Mahdia	*inodorus*	Landrace	Sweet fruit without aroma, yellow skin
Dz1	Dziri	Manzel Kamel-Monastir	*inodorus*	Landrace	Sweet fruit without aroma; light-green skin
Dz2	Dziri-arbi	Sidi Banour-Mahdia	*inodorus*	Landrace	Sweet fruit without aroma of light-green color
FL	Fakous Salem	Wardanine- Monastir	*flexuosus*	Landrace	Elongated fruit, nonsweet, eaten raw before ripeness
Horch	Harchay	Wardanine- Monastir	*chate*	Landrace	Fruit small-sized, nonsweet, eaten raw before ripeness
L1-Maaz	Maazoun	-	*inodorus*	Breeding line	Sweet fruit without aroma, skin of yellow-green color
L2-FL	Fakous	-	*flexuosus*	Breeding line	Elongated fruit, nonsweet, eaten raw before ripeness
L3-Trab	Trabelsi	-	*inodorus*	Breeding line	Sweet fruit without aroma of orange color
L4-Gal	Galaoui	-	*reticulatus*	Breeding line	Sweet fruit with aroma, netted/corked skin
L5-Dz	Dziri	-	*inodorus*	Breeding line	Sweet fruit without aroma; light-green skin
L7-Sarac	Sarachika	-	*inodorus*	Breeding line	Sweet Fruit without aroma; yellow skin
L8-Ru	RD	-	*cantalupensis*	Breeding line	Sweet fruit with aroma and grooves
L9-Ra	Rupa	-	*cantalupensis*	Breeding line	Sweet fruit with aroma and grooves
L10-Anan	Ananas	-	*reticulatus*	Breeding line	Sweet fruit with aroma, netted skin
L13-Raf	V4	-	*inodorus*	Breeding line	Sweet fruit without aroma; orange skin

The experiment was carried out from February to July during two seasons, 2018 and 2019, at the experimental station of Sahline-CRRHAB located in the Central East Region of Tunisia (35°45'02″ N, 10°42'44″ E). Accessions were initially sown in compost, and seedlings at the three-leaf stage were transplanted into a greenhouse. Three replications containing 10 plants of each accession were arranged in a randomized complete block design with a row spacing of 80 cm and a within-row spacing of 40 cm. During culture, agronomic practices including irrigation, weeding, and fertilization were conducted uniformly as required in all plots.

2.2. Phenotypic Characterization

Melon accessions were scored for 12 quantitative and 12 qualitative traits related to leaf, stem, flower, fruit, and seed. Five central plants of each accession in each replication were selected for sampling. The traits were selected following the descriptor lists of the International Union for the Protection of New Varieties of Plants (UPOV). Quantitative data were recorded on (1) days to maturity, from sowing to harvest, (2) stem diameter and length, (3) number of fruits per vine, (4) fruit weight, length, diameter, and thickness, (5) cavity diameter, (6) total soluble solids, and (7) 100-seed weight. Qualitative data concerned (1) sex expression (andromonoecious or monoecious), (2) leaf color and blade size, (3) separation of peduncle from fruit, (4) fruit grooves and netting, (4) fruit shape, (5) fruit skin and flesh color, (6) fruit firmness andshelf life, and (7) seed color.

The quantitative traits (length, width, and diameter) were measured with a ruler or caliper, fruit weight was measured with an electronic balance, and fruit firmness was measured with a penetrometer, while qualitative traits were evaluated by attributing a code to each character states mentioned in UPOV guidelines. Total soluble sugars (TSS), expressed as degree Brix (°Brix) in fruit juice, were determined using a digital refractometer (Atago, Tokyo, Japan). Skin and flesh colors of marketable fruits were assessed using the Royal Horticultural Society Color Chart.Observations on leaf blade were made on fully developed but not old leaves, and those related to fruits were made on fully ripe ones. Fruit skin and flesh colors were assessed using the Royal Horticultural Society Color Chart.

Analysis of variance (ANOVA) followed by the Duncan test was performed for quantitative traits to test the significance of variation between the accessions ($p < 0.05$) using the statistical procedures in SAS software version 9.1 [25].

Pearson correlation analysis was also carried out to estimate the relationship between all quantitative traits and two qualitative traits (fruit firmness and fruit shelf life) using R Studio software version 1.1.456 [26].

To evaluate the levels of phenotypic variation among accessions, Euclidean similarity coefficients were calculated using the SimInt procedure implemented in NTSYSpc software version 2.1 [27] and served for dendrogram construction using the unweighted pair-group method of averages (UPGMA).

2.3. DNA Extraction and Microsatellite Analysis

Six to seven plants per accession were randomly selected for molecular characterization. Genomic DNA was extracted from young leaves of individual plants using a modified CTAB method [28]. The quality and quantity of DNA were determined using a NanoDrop ND-10000 spectrophotometer (NanoDrop Technologies, Wilmington, DE, USA), and the diluted DNA (10 ng/µL) was stored at −20 °C until PCR analysis.

A set of eight SSRs, previously proven to be highly performant for genetic melon characterization, were used to assess the genetic diversity between and among accessions in our collection [29,30]. These SSRs were selected for their high polymorphism, for their equitable distribution throughout the genome, for their similar annealing temperature (55 °C) to facilitate the multiplexing of several loci into one capillary electrophoresis run, and because they were mapped on the consensus genetic map in melon published by Diaz et al. [31].

PCR amplification reactions were carried out according to Mallor et al. [32]. Amplifications were performed in a total volume of 20µL containing 12.5 ng of genomic DNA, 1 × PCR buffer (20 mM Tris–HCl pH 8.4 + 50 mM KCl), 2 mM $MgCl_2$, 65 µM of each dNTP (Invitrogen, Carlsbad, CA, USA), 0.0625 µM of forward primer extended by 18 nt M13 tail at its 5′ end, 0.25 µM of each reverse and M13-forward primer (5′-CACGACGTTGTAAAACGAC-3′) labeled with one of the four fluorescent dyes (6-FAM, VIC, PET, or NED; Applied Biosystems, Foster City, CA, USA), and 0.2 U of Taq DNA polymerase (Invitrogen). Amplification reactions were performed in a Perkin-Elmer 9700 thermocycler (Norwalk, CT, USA) with the following program: 5 min denaturation at 94 °C followed by 35 cycles at 94 °C for 15 s, 55 °C for 15 s, and 72 °C for 15 s, with a final extension at 72 °C for 10 min. Amplified fragments were separated by capillary electrophoresis on an ABI 3130XL Genetic Analyzer (Applied Biosystems, Foster City, CA, USA). The raw data produced and the size of the SSR alleles were analyzed using Peak Scanner software (Applied Biosystems).

The polymorphism of each locus was scored as presence (1) versus absence (0) of a specific allele, and data were transformed into a biallelic matrix using allele size. Genetic parameters (number of alleles per locus (N), number of genotypes per locus (G), observed heterozygosity (Ho), expected heterozygosity (He), Shannon's information index (I), fixation index (Fis), or inbreeding coefficient) were determined using GenAlEx software version 6.5 [33]. The polymorphism information content (PIC) value was calculated using the formula described by Botstein et al. [34].

The genetic distance between accessions based on the simple matching (SM) coefficient was estimated using the software NTSYSpc 2.1 [27]. The resulting matrix served as input data for the cluster analysis using the UPGMA to generate a dendrogram.

Moreover, analysis of the molecular variance (AMOVA) among and within populations was performed using GenAlEx 6.5 program.

A joint analysis based on a combination of phenotypic and genotypic similarity matrix was also conducted. Quantitative traits were converted into three discrete classes as reported in Yildiz et al. [35].

To measure the goodness of fit for the phenotypic and molecular cluster analysis, cophenetic correlation values between the original similarity matrices and the cophenetic matrices given by the UPGMA clustering process were calculated by a Mantel test procedure [36]. Correlation between morphological characters and molecular markers was also tested using the same procedure.

2.4. Population Structure

The software package STRUCTURE version 2.3.1 [37] was used to provide the most reliable grouping of the 24 melon accessions, which was analyzed using a Bayesian method (100,000 burn-ins, 100,000 Markov Chain Monte Carlo) under the admixture model. To determine the proper number of clusters (K), criteria set by Evanno et al. [38] were followed.

Several population numbers (from K = 1 to 10) were tested, and the logarithm posterior probability for each K was recorded. The total number of populations was set when the probability reached a plateau for higher K. Genotypes were assigned to defined populations if the value of the corresponding membership probability was higher than 0.8; otherwise, they were considered to be admixed.

3. Results

3.1. Phenotypic Characterization

The melon accessions under study showed a wide range of variability for almost all the phenotypic traits studied. The frequency distribution for 12 qualitative characters (discontinuous variables) is shown in Figure 1, in which fruit characters showed the highest level of polymorphism. Six fruit shapes were observed in the studied accessions, whereby ovate (33.3%) and elliptical (29.2%) shapes were the most abundant followed by round shape (20.8%); elongated fruits were observed at 8.4% for the accessions of *flexuosus* group, while flattened and obovate fruits were observed at 4.2%.

Fruit skin color was also distributed into six classes including yellow (37.5%), yellow-green (20.8%), light green, green, and orange with the same frequency (12.5%), and creamy (4.2%). Flesh color was distributed into four classes; the majority had orange flesh (37.5%). The majority of accessions were found to be non-sutured (75%), without netting (75%), and presented a strong attachment of the peduncle to the fruit (58.3%).

For improving the shelf life of melon fruits, firmness is an important trait for maintaining the quality of fruits; 54.2% of the accessions were firm, 20.8% were moderately firm, and 25% were soft. A short (15 days) to medium (25–30 days) shelf life was observed in 66.6% of accessions; 12.5% of accessions had a very short (approximately 1 week) or long shelf life (55 days), and 8.3% of the accessions showed a very long shelf life (approximately 3 months) for some accessions of *inodorus* group. Melon accessions were andromonoecious (87.5%) for those belonging to *inodorus*, *cantalupensis*, and *reticulatus* groups, and monoecious (12.5%) for those belonging to *flexuosus* and *chate* groups (Figure 1).

Analysis of variance applied to quantitative characteristics (Table 2) showed significant ($p < 0.05$) to highly significant ($p < 0.0001$) differences between accessions for all recorded traits. The coefficient of variation ranged from 5.5% (lowest) to 46.5% (highest) for the date to maturity (DM) and fruit length (FL), respectively. Accessions of the non-sweet group (FL, L2-FL (*flexuosus*), and Horch (*chate*)) presented the lowest values for date to maturity (103–105 days), fruit weight (84–204.9 g), fruit diameter (3–4.4 cm), fruit thickness (0.5–2 cm), and total soluble solids (4.7–7.4 °Brix), whereas accessions L10-Anan and Arbi5 belonging to the *reticulatus* group had the highest fruit weight (1765.2 g and 1489.6 g, respectively) followed by L13-Raf (1373 g, *inodorus*). This accession was also distinguished by its latest maturity (DM = 159 days), as well as its largest and thickest fruits (FD and FT = 13.9 cm and 3.8 cm, respectively), whereas the accession Arbi2 of the *inodorous* group showed the lowest TSS (8.4 °Brix) among the sweet group (*reticulatus, cantalupensis,* and *inodorus*).

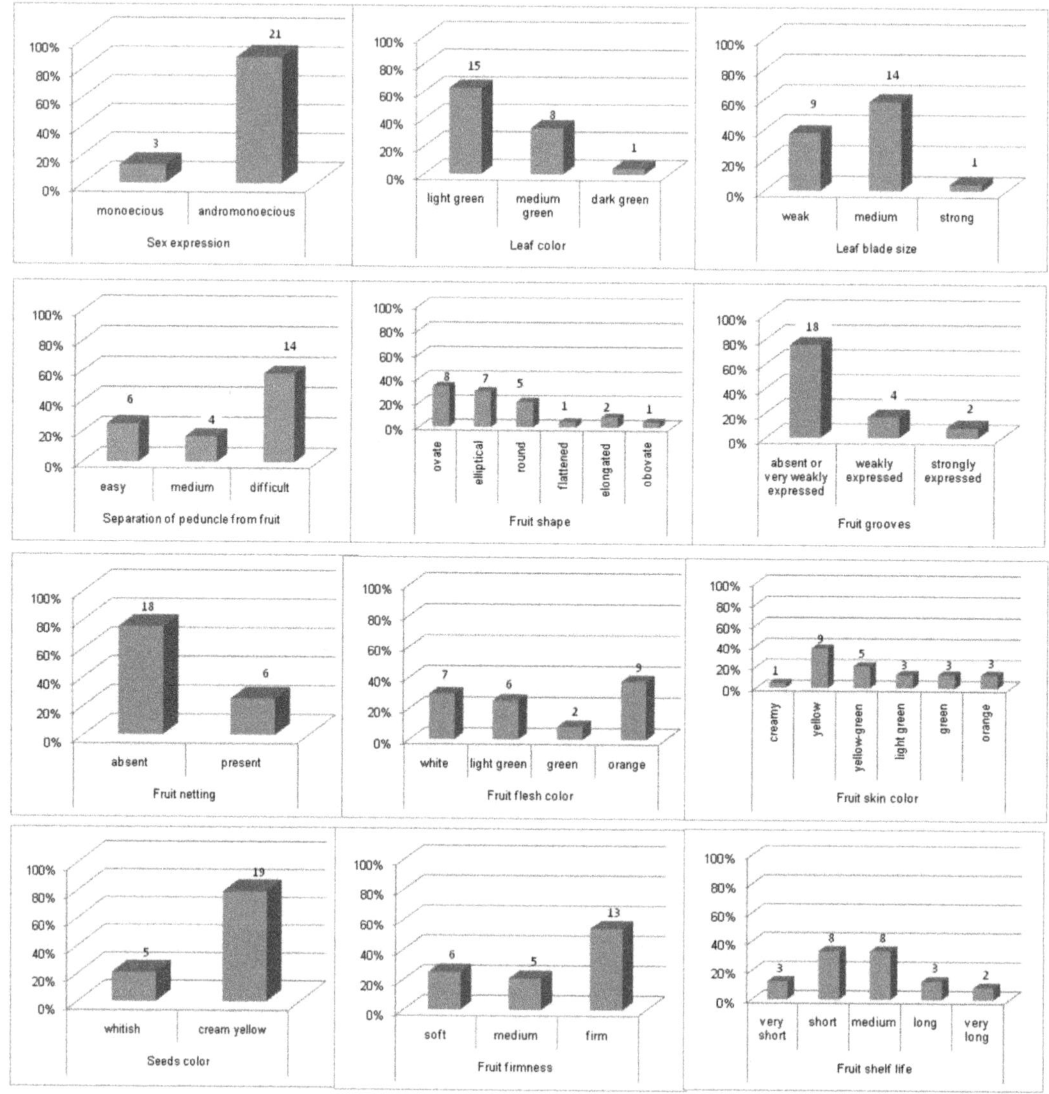

Figure 1. Frequency distribution of phenotypic qualitative traits in melon accessions. Values on each bar represent the number of accessions.

Table 2. Quantitative traits of 24 Tunisian melon accessions belonging to different botanical groups.

	DM (days)	SD (mm)	FN	SL (cm)	FW (g)	FL (cm)	FD (cm)	CD (cm)	FT (cm)	TSS (°Brix)	SN	100-SW (g)
inodorus group												
Maaz1	139.3 ± 4.6	9.4 ± 1.1	3.2 ± 0.2	178.3 ± 17.5	881.0 ± 190.7	13.0 ± 2.1	11.1 ± 1.4	4.9 ± 0.7	2.7 ± 0.5	9.8 ± 0.6	410.7 ± 85.9	5.5 ± 2.5
Maaz2	150.0 ± 6.92	7.3 ± 0.5	2.4 ± 0.5	168.7 ± 28.2	1063.4 ± 101.8	14.8 ± 1.1	13.5 ± 0.8	5.4 ± 1.5	4.5 ± 1.1	10.2 ± 1.1	767.7 ± 193.4	6.5 ± 0.3
Maaz3	139.3 ± 4.61	7.0 ± 1.1	3.2 ± 0.5	112.3 ± 6.8	735.8 ± 56.6	11.2 ± 0.8	10.6 ± 1.1	4.5 ± 0.5	3.2 ± 0.7	10.9 ± 0.5	393.3 ± 84.1	6.0 ± 0.5
Maaz4	133.0 ± 1.73	7.5 ± 0.9	3.1 ± 0.4	158.0 ± 15.8	845.3 ± 75.1	12.0 ± 1.2	9.8 ± 0.2	4.2 ± 0.8	2.6 ± 0.2	9.5 ± 0.6	404.7 ± 53.86	5.1 ± 0.6
Arbi1	137.3 ± 8.08	9.0 ± 2.5	2.0 ± 1.0	180.3 ± 6.5	915.5 ± 220.9	14.1 ± 2.2	10.2 ± 1.9	5.7 ± 0.9	2.4 ± 0.6	9.8 ± 0.1	693.3 ± 110.1	7.5 ± 0.3
Arbi2	136.7 ± 4.61	12.6 ± 0.5	2.1 ± 0.8	197.7 ± 17.7	544.7 ± 62.4	11.9 ± 1.6	9.4 ± 0.5	5.3 ± 0.8	2.2 ± 0.4	8.4 ± 0.2	428.3 ± 143.5	7.3 ± 0.5
Arbi3	135.3 ± 12.7	11.3 ± 1.2	2.7 ± 0.3	172.3 ± 14.7	831.8 ± 40.5	16.8 ± 6.2	10.7 ± 0.5	5.6 ± 0.4	2.5 ± 0.2	11.2 ± 1.1	571.7 ± 110.1	6.9 ± 0.8
Arbi4	146.0 ± 10.5	8.9 ± 1.5	2.7 ± 1.2	146.0 ± 20.1	1035.1 ± 111.9	15.9 ± 2.5	11.3 ± 1.7	5.6 ± 0.2	2.8 ± 0.3	11.1 ± 1.0	529.3 ± 52.5	6.4 ± 1.2
Arbi6	130.0 ± 3.46	9.5 ± 1.1	3.0 ± 0.3	136.0 ± 22.5	955.5 ± 102.3	17.1 ± 2.6	11.4 ± 0.9	5.8 ± 0.5	2.6 ± 1.0	11.5 ± 1.5	500.0 ± 97.7	7.2 ± 0.2
Dz1	154.0 ± 3.26	10.3 ± 0.8	2.4 ± 0.5	171.0 ± 28.8	823.2 ± 67.5	18.5 ± 4.5	11.1 ± 2.1	5.1 ± 0.5	3.1 ± 0.7	11.3 ± 1.4	415.0 ± 99.1	5.1 ± 0.9
Dz2	150.0 ± 6.92	8.5 ± 1.2	2.3 ± 0.3	177.3 ± 11.6	702.0 ± 70.1	11.5 ± 0.3	9.8 ± 1.6	5.5 ± 0.4	2.3 ± 0.7	9.0 ± 1.1	414.7 ± 53.5	6.6 ± 0.8
L1-Maaz	150.3 ± 14.8	9.9 ± 0.8	2.1 ± 0.7	179.7 ± 6.5	971.7 ± 204.5	12.3 ± 1.6	12.0 ± 1.4	5.1 ± 1.4	3.7 ± 0.6	11.0 ± 1.9	596.7 ± 66.3	6.6 ± 0.4
L3-trab	134.7 ± 7.02	9.3 ± 1.1	2.4 ± 0.7	213.3 ± 18.4	801.4 ± 200.9	20.0 ± 9.2	9.0 ± 1.6	4.4 ± 0.8	2.5 ± 0.5	9.7 ± 0.3	622.0 ± 80.0	6.7 ± 1.5
L5-Dz	147.3 ± 11.5	10.4 ± 0.7	2.7 ± 0.6	158.3 ± 17.5	1050.9 ± 38.9	16.9 ± 3.2	10.6 ± 2.1	5.8 ± 1.2	2.6 ± 1.0	9.6 ± 0.9	955.3 ± 171.7	4.1 ± 0.9
L7-Sarac	149.0 ± 12.1	7.7 ± 1.2	2.7 ± 0.1	127.3 ± 8.5	790.6 ± 124.8	12.1 ± 2.8	11.1 ± 2.0	4.5 ± 0.5	3.1 ± 0.3	9.8 ± 0.5	466.0 ± 66.9	7.5 ± 1.0
L13-Raf	159.0 ± 3.4	8.7 ± 0.6	2.3 ± 0.3	136.0 ± 12.3	1373.0 ± 110.7	13.9 ± 1.8	13.90.8 ±	6.7 ± 0.6	3.8 ± 0.3	11.5 ± 1.3	485.3 ± 77.1	9.5 ± 1.2
cantalupensis group												
L8-Ru	132.0 ± 3.5	6.8 ± 1.0	1.9 ± 1.0	92.7 ± 8.7	436.1 ± 67.4	9.8 ± 0.6	9.7 ± 0.7	4.2 ± 0.2	3.3 ± 0.4	9.7 ± 0.7	384.7 ± 86.4	6.0 ± 0.5
L9-Ra	139.3 ± 9.2	10.5 ± 1.3	2.2 ± 0.9	175.7 ± 25.01	1101.1 ± 48.5	13.1 ± 0.9	11.2 ± 2.0	5.7 ± 0.6	2.9 ± 0.3	11.8 ± 1.1	382.3 ± 67.1	6.9 ± 0.8
reticulatus group												
Arbi5	130.0 ± 3.4	6.5 ± 1.0	2.4 ± 0.9	195.7 ± 15.2	1489.6 ± 238.0	29.4 ± 9.2	12.9 ± 3.0	6.5 ± 1.8	2.9 ± 0.6	8.8 ± 0.4	421.3 ± 82.6	10.7 ± 1.5
L4-Gal	142.0 ± 8.0	12.4 ± 2.2	2.7 ± 0.8	133.7 ± 14.16	1025.4 ± 123.1	14.1 ± 0.7	11.41.3 ±	5.9 ± 0.6	2.7 ± 0.4	10.1 ± 0.4	674.0 ± 68.8	6.7 ± 0.4
L10-Anan	144.7 ± 9.2	9.4 ± 1.5	1.8 ± 0.5	196.7 ± 21.3	1765.2 ± 259.7	16.7 ± 3.9	12.4 ± 3.5	5.5 ± 1.5	2.9 ± 0.8	9.5 ± 0.2	410.7 ± 70.6	5.9 ± 1.2
flexuosus group												
L2-FL	103.3 ± 4.1	11.0 ± 0.6	35.0 ± 5.0	190.0 ± 25.8	204.9 ± 32.7	29.3 ± 3.1	4.2 ± 1.1	4.0 ± 1.0	1.9 ± 0.5	4.9 ± 0.6	301.0 ± 35.5	5.6 ± 0.6
FL	105.3 ± 5.5	11.0 ± 0.5	36.7 ± 6.0	226.7 ± 25.2	203.3 ± 15.3	38.3 ± 3.5	4.4 ± 0.5	3.7 ± 1.1	2.0 ± 0.5	4.7 ± 0.5	213.0 ± 24.1	5.3 ± 0.5
chate group												
Horch	103.3 ± 3.0	7.3 ± 0.6	8.0 ± 1.0	112.3 ± 6.8	84.0 ± 16.4	4.3 ± 0.5	3.0 ± 0.3	1.9 ± 0.5	0.5 ± 0.1	7.4 ± 0.4	197.3 ± 18.3	2.7 ± 0.3
CV	5.5	11.9	37.6	18.6	41.9	46.5	17.7	22.5	23.1	10.6	36.3	25.4
F value	11.5 **	7.4 **	92.4 **	3.6 **	3.9 **	9.61 **	6.8 **	2.4 *	4.4 **	9.3 **	2.46 *	2.66 *

DM: days to maturity; SD: stem diameter, FN: fruit number, SL: stem length, FW: fruit weight, FL: fruit length, FD: fruit diameter, CD: cavity diameter, FT: fruit thickness, TSS: total soluble solids, SN: seed number, 100-SW: 100-seed weight. Values are Means ± standard deviation. CV: coefficient of variation, * significant at $p < 0.05$, ** highly significant at $p < 0.0001$.

Pearson correlation coefficients (*r*) were calculated to determine the relationships between all quantitative parameters and two qualitative parameters (fruit firmness and fruit shelf life). A total of 20 features were correlated at a $p < 0.05$ significance level (Figure 2). Date to maturity (DM) was significantly and positively correlated with fruit diameter (FD; $r = 0.87$), fruit thickness (FT; $r = 0.75$), cavity diameter (CD; $r = 0.68$), and total soluble solids (TSS; $r = 0.68$) but negatively correlated with fruit number (FN; $r = -0.75$). Fruit diameter was positively and significantly correlated with fruit weight (FW; $r = 0.86$), fruit thickness (FT; $r = 0.83$), and cavity diameter (CD; $r = 0.83$) but negatively correlated with fruit number (FN, $r = -0.75$). Significant and positive correlations were also observed between fruit weight (FW) and cavity diameter (CD; $r = 0.81$) and between CD and 100-seed weight (100-SW; $r = 0.66$), whereas TSS had a significant and negative correlation with fruit number FN ($r = -0.83$) and stem length (SL; $r = -0.54$).

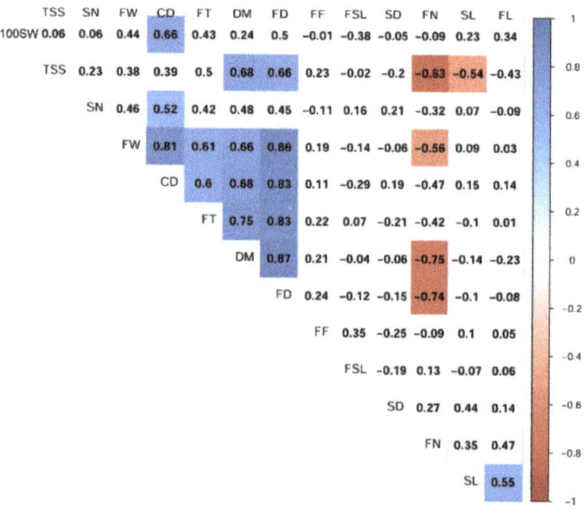

Figure 2. Pearson correlation coefficients between quantitative phenotypic traits and two qualitative traits (fruit firmness and fruit shelf life). DM: days to maturity, SD: stem diameter, SL: stem length; FN: fruit number, FW: fruit weight, FL: fruit length, FD: fruit diameter, CD: cavity diameter, FT: fruit thickness, FF: fruit firmness, FSL: fruit shelf life, TSS: total soluble solids, SN: seed number, 100-SW: 100-seed weight. Only the statistically significant correlation coefficients ($p < 0.05$) are colored. Among the significant coefficients, the color type and intensity indicate the direction and the strength of the association.

A dendrogram combining quantitative and qualitative traits was generated to evaluate the general pattern of variance and to establish the relationship among melon accessions (Figure 3). Accessions were discriminated into two main clusters. The first cluster CL1 included three accessions of *flexuosus* and *chate* botanical groups characterized by a monoecious sex expression, the lowest fresh weights, and a non-sweet taste. The second cluster CL2 was formed by the remaining 21 accessions belonging to three groups (*cantalupensis*, *reticulatus*, and *inodorus*) characterized by an andromonoecious sex expression, an intermediate to high fresh weight, and a sweet taste. CL2 was subdivided into two subclusters, the first one (CL2-1) containing four accessions characterized by the highest fresh weights and the second one (CL2-2) containing 17 accessions characterized by intermediate fresh weights.

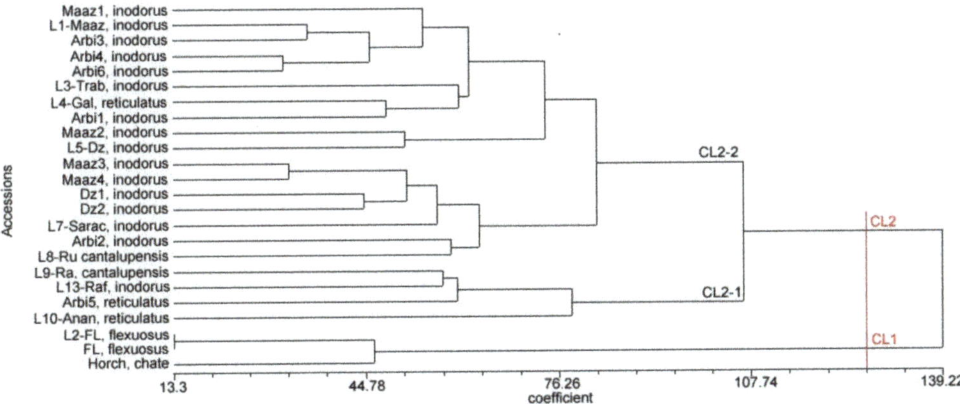

Figure 3. Dendrogram of the Tunisian melon accessions based on UPGMA analysis using Euclidean similarity coefficient for quantitative and qualitative phenotypic traits.

3.2. Molecular Characterization

All SSR markers were polymorphic, and there were no duplicates in the collection. An example of a capillary SSR profile using CM38- and ECM58- labeled primers is shown in Figure 4. A total of 30 alleles were identified in 166 plants representing 24 accessions. The number of identified alleles ranged from two for ECM204, ECM132, and CMCT160 to seven for MU7194, with an average of 3.75 alleles per locus. The length of the amplified fragments ranged from 117 to 346 bp. A total of 53 genotypes were generated (Table 3). Allele frequencies ranged from 0.003 for locus CSWCT10 allele 206 to 0.96 for locus ECM204 allele 330 (Figure 5). Allele 206 (CSWCT10; 0.003) and allele 153 (MU7194; 0.024) were specific to the Arbi4 and Horchay accessions, respectively.

Table 3. Genetic features of eight microsatellite markers used for the assessment of genetic diversity in 24 melon accessions. Ch: chromosome, N: number of alleles, G: number of genotypes, Ho and He: observed and expected heterozygosity, respectively, I: Shannon's diversity index, Fis: inbreeding coefficient, PIC: polymorphic information content.

SSR *Locus*	Ch	Primer Sequences	N	Size Range (bp)	G	Ho	He	I	Fis	PIC
CM38	10	F:TAGCATCTGATCGGAAAACC R:CAACTTCATCCGCCAAGAAT	3	134–144	6	0.17	0.19	0.96	0.06	0.57
ECM58	1	F:TTGAAGCTTCTTCACCTTCTCTTT R:CACCCCACAAGGGTTCAATA	5	132–157	11	0.19	0.23	1.50	0.18	0.76
CSWCT10	3	F:AGATCGGAATTGAAAAAG R:AAAGGGGCTTCCTCTCTA	3	197–206	4	0.05	0.03	0.26	−0.68	0.13
ECM204	7	F:CTCTCTTCATTTCCCCTCGTT R:TGGCCTGGAAAGTAAGGGTAT	2	330–346	3	0.01	0.02	0.14	0.21	0.06
MU7194	4	F:TACTCCTCGCTGATCTTCCC R:AAAGGAAGAAGCGCACAAAA	7	117–153	12	0.13	0.22	1.48	0.42	0.71
ECM132	6	F:CCATTCGTCAAGCAAAGCTAC R:CAAAGGGTCCCTCAATTTCTC	2	292–295	3	0.10	0.09	0.48	−0.06	0.30
CMN04_19	9	F:TTCTTCCCACCAAACCTACG R:AAATGGCAGAGAGCGAGAAA	6	238–262	11	0.08	0.14	1.31	0.48	0.66
CMCT160a	11	F:GTCTCTCTCCCTTATCTTCCA R:ACGGTGTTTGGTGTGAGAAG	2	231–233	3	0.04	0.04	0.43	−0.27	0.26
Total			30	——	53	—	—	—	—	——
Mean			3.75	——	—	0.10	0.12	0.82	0.04	0.43

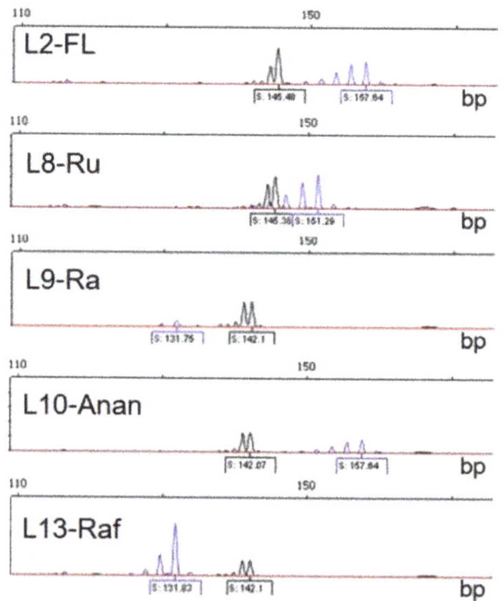

■ CM38 labeled with the NED fluorescence dye
■ ECM58 labeled with the FAM fluorescence dye

Figure 4. Phenogram image of molecular polymorphism for five melon accessions (L2-FL, L8-Ru, L9-Ra, L10-Anan, and L13-Raf) detected by CM38 and ECM58 SSR markers labeled with the NED and FAM fluorescence dyes, respectively.

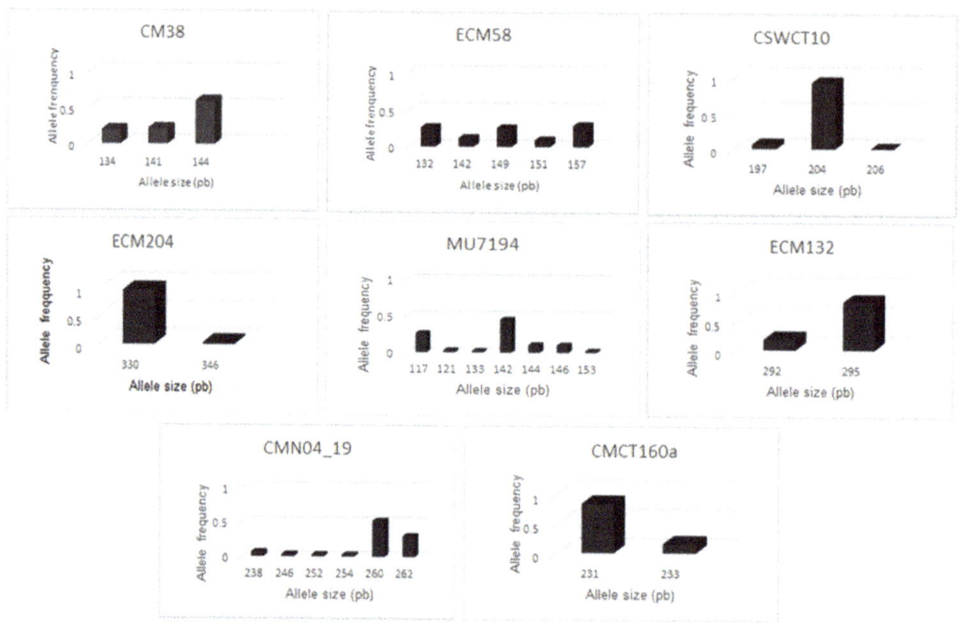

Figure 5. Distribution of allele frequencies for the eight SSR loci studied in melon accessions.

The PIC value, estimating the discriminatory power of loci, ranged from 0.06 (ECM204) to 0.76 (ECM58) with an average of 0.43 (Table 3). Shannon's information index (I) was between 0.14 (ECM204) and 1.5 (ECM58) with an average of 0.82. The observed heterozygosity (H_o) ranged between 0.01 for ECM204 and 0.19 for ECM58 with an average of 0.1, being lower than expected (H_e = 0.12). H_e was lower for the loci with a low number of alleles and higher for the loci with a high number of alleles. The low level of heterozygosity within accessions wasconfirmed by a positive inbreeding coefficient (Fis = 0.04).

The pattern of relationships among the 24 accessions is depicted in the UPGMA dendrogram based on simple matching (SM) coefficient (Figure 6). There was no clear clustering between the accessions in relation to the collection site and botanical group. However, two major clusters could be defined by cutting the dendrogram at the lowest range of similarity value (0.57). The first cluster (CL1) included two landraces FL and Arbi2. The second one (CL2) contained the remaining 12 landraces intermixed with the 10 breeding lines. At about a 0.70 similarity level, CL2 was subdivided into four groups; the first group (G1) consisted of 13 accessions (Maaz1, Maaz2, L1-Maaz, L3-Trab, Dz1, L9-Ra, L10-Anan, L4-Gal, Arbi5, L2-FL, L5-Dz, L7-Sara, and L8-Ru), with most of them (nine) being breeding lines. The remaining nine accessions (Maaz3, Maaz4, Horch, Arbi6, L13-Raf, Dz2, Arbi1, Arbi3, and Arbi4) were spread by pairs or triplets into the three other groups (G2–G4). Intermixing of landrace and breeding line accessions belonging to different taxa indicated a genetic resemblance with each other.

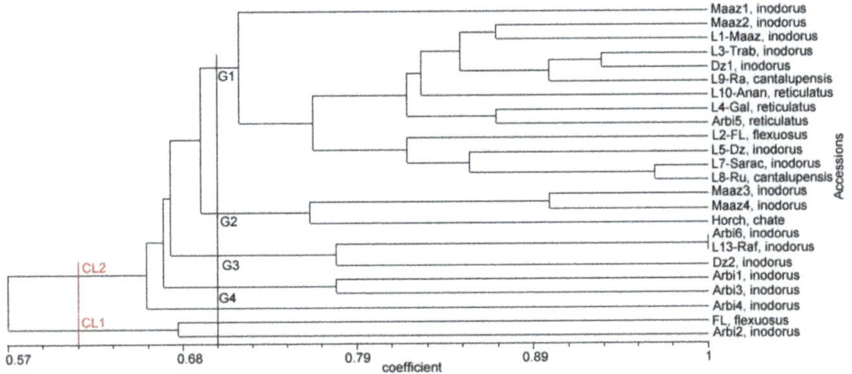

Figure 6. Dendrogram of the Tunisian melon accessions based on UPGMA analysis using the simple matching coefficient after amplification with SSR primers.

According to the genetic distance matrix (data not shown), the most similar accessions were Arbi6 and L13-Ra (SM = 0.99) followed by L7-Sarac and L8-Ru (0.96) and L9-Ra and L3-Trab or Dz1 (0.90); the most dissimilar ones were FL and Maaz3 (0.38) followed by FL and Dz1 or Dz2, Horch, and Dz2 with asimilar coefficient of 0.48.

Analysis of molecular variance (AMOVA) was used to estimate the partitioning of genetic variance among and within populations (Table 4). AMOVA results based on SSR data revealed that the largest proportion (75%) of the total genetic variance occurred among accessions, and only 25% occurred within accessions.

Table 4. Analysis of molecular variance (AMOVA) of 24 melon accessions based on eight microsatellite markers. SV: source of variation, df: degrees of freedom, SS: sum of squares, MS: mean squares, Est.var: estimated variance component, PV: proportion of variance.

SV	df	SS	MS	Est.var	PV (%)	*p*-Value
Among populations	23	794.763	34.555	4.763	75	0.01
Within populations	142	229.333	1.615	1.615	25	0.01
Total	165	1024.096		6.378	100	

3.3. Combined Analysis of Phenotypic and Genotypic Data

At the lowest range of similarty 0.62, the combined phenotypic–molecular dendrogram (Figure 7) separated melon accessions into two main clusters, CL1 (non sweet melon) and CL2 (sweet melon), as shown by the phenotypic dendrogram, but with a more precise grouping of accessions in CL2, according to the botanical group. Thus, this cluster was formed by four groups (G1–G4) with G1 containing most (12) of the *inodorus* accessions except for Arbi1 (G2), L7-Sarac (G2), Dz2 (G3), and Arbi2 (G4), whereas *cantalupensis* accessions (L8-Ru and L9-Ra) and *reticulatus* accessions (L4-Gal, L10-Anan, and Arbi5) were clustered together in G2 with L10-Anan and Arbi5 and L8-Ru and L9-Ra being located close to each other.

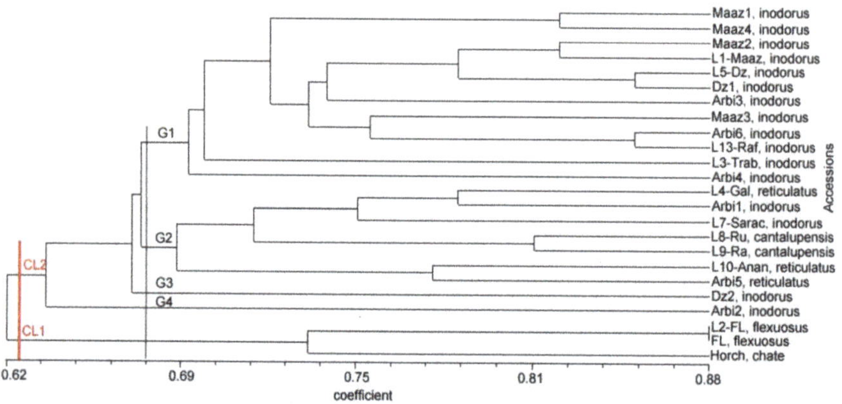

Figure 7. Dendrogram of the Tunisian melon accessions based on UPGMA analysis using combined phenotypic and molecular SM similarity coefficient.

3.4. Relationships and Concordance between Phenotypic and Molecular Markers

The cophenetic correlation coefficient value between the dendrogram and the original distance matrix estimated from the phenotypic and SSR markers was rmorph = 0.71 and rSSR = 0.75, respectively (data not shown), indicating a good fit for both data. A negative and non significant correlation between morphological characters and molecular markers ($r = -0.16$, $p = 0.06$) was found, confirming the disparity between morphological variation and genetic polymorphism of the studied accessions.

3.5. Population Structure

Genetic structure analysis of the individual samples using STRUCTURE was used to provide the most reliable discrimination of the melon accessions. The analysis indicated an intermixed genetic structure between landraces and breeding lines, as well as among botanical groups. Evanno's test indicated that the most informative number of subpopulations (K) was 5 (Figure 8) suggesting the existence of five major groups in the collection, as previously revealed by the molecular cluster analysis. The different groups were defined by five colors (Figure 9). The yellow color predominated the genetic profile of the tested accessions, followed by the red, pink, green, and blue colors. The first group, with a genetic profile dominated by a yellow color, included accessions Maaz2, L1-Maaz, L2-FL, Dz1, L3-Trab, L9-Ra, L10-Ana, and Arbi5 grouped in the UPGMA tree (G1). The second group, with a red color, contained breeding lines L5-Dz, L7-Sarac, and L8-Ru (G1). The third group, dominated by blue color, included landraces Arbi6 and L13-Raf (G3 with SM coefficient = 0.99). The fourth group, with a pink color, included landraces FL, Arbi2 (cluster 1), and Arbi3. Lastly, the fifth group, dominated by a green color, was formed by Arbi4, Dz2, and Maaz4. Accessions belonging to these groups had all individuals with a membership higher than 0.8, indicating that they were strongly assigned to subpopulations, except for Arbi 4, Arbi 3, L9-Ra, and Dz2, each displaying one to three individuals with an admixture allelic form (membership < 0.8). The remaining accessions, with membership probabilities lower than 0.8 for almost all individuals, were classified into an admixture group. This group included five accessions: Maaz1, Maaz3, L4-Gal, Horch, and Arbi1 (Figure 9).

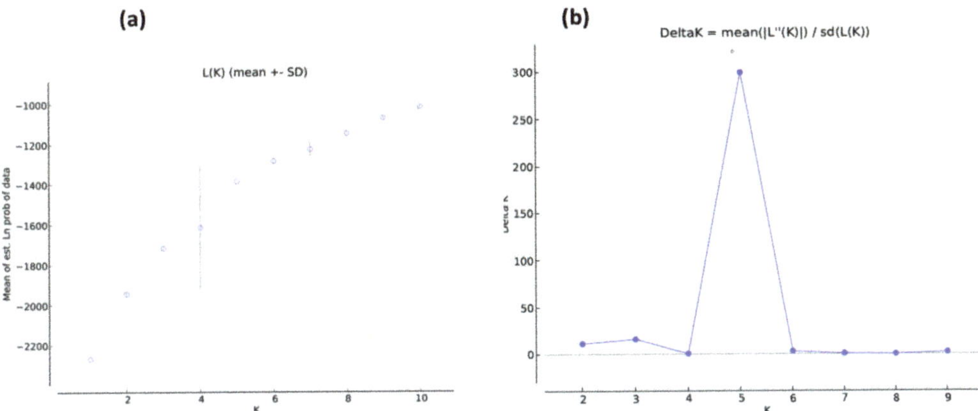

Figure 8. Estimation of the optimum number of clusters for melon accession according to Evanno's method. (**a**) The graph displays the log likelihood (LnP(D)) for each K value. (**b**) The graph displays the ΔK for each K value. The sharp peak of ΔK at K = 5 suggests five subpopulations.

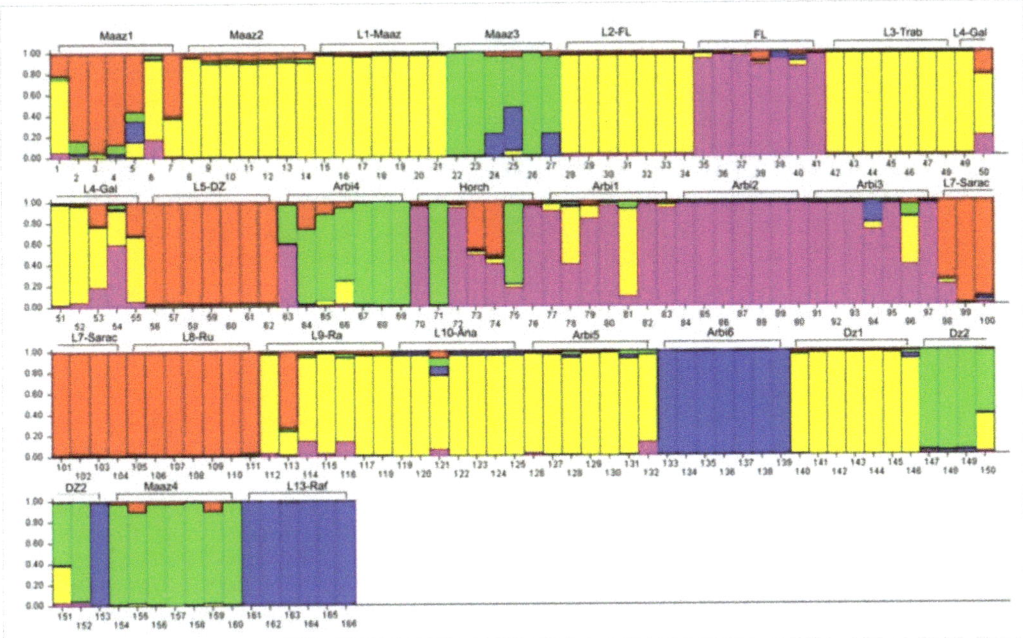

Figure 9. Model-based cluster membership of 24 melon accessions into five subpopulations identified with STRUCTURE analysis using SSR primers. The corresponding membership probability is presented in the vertical axis. Vertical bars represent each individual analyzed in this study, and bars are divided into several colors when there is evidence of admixture.

4. Discussion

Landraces are a valuable repository of gene pool for breeding in a changing environment. They often harbor rich genetic diversity, important for traditional agriculture sustainability under climatic change conditions [39]. The Tunisian melon germplasm is a reservoir of genes for unique traits related to biotic stress resistance [4–6,40] and fruit quality [15] which can be transferred to modern cultivars through hybridization. The assessment of genetic diversity and structure of such a unique gene pool is a prerequisite for its efficient organization, conservation, and utilization for melon improvement, new cultivar deployment, and hybrid seed production.

In the present study, we investigated the genetic diversity and population structure for 24 Tunisian melon accessions including landraces and local breeding lines. Phenotypic characterization showed a wide range of variability among accessions of different botanical groups for almost all the phenotypic traits studied, and each group presented specific traits consistent with the horticultural taxonomy proposed by Pitrat [2]. High morphological variability among botanical groups and among melon landraces was previously reported in many countries [18,35,41].

In the UPGMA dendrogram combining quantitative and qualitative phenotypic traits, the accessions of the non-sweet group (*flexuosus* and *chate*) were separated from the other ones on the basis of fruit shape, weight, and sweetness, in addition to their monoecious sex expression, whereas accessions of the sweet group (*cantalupensis*, *reticulatus*, and *inodorus*) were intermixed. The lack of consistency in the clustering of accessions from the same botanical groups was also observed by Soltani et al. [42] and Aragão et al. [43]. However, Yildiz et al. [35] showed that the accessions of the *flexuosus* group were clustered with the sweet ones. One of the reasons for this variation is the inevitable out-crossing among melon genotypes. Intermediate forms might have been formed among the different groups due to the old farming practices employed by local small-scale melon producers in the different countries.

Compared with morphological traits, molecular analysis is independent of environmental effects and can provide additional and precise information for the assessment of genetic diversity [13,44]. In this study, all SSR markers were polymorphic and displayed a moderate polymorphic information content (PIC = 0.43). The number of alleles ranged from 2–7 with an average of 3.75 alleles per locus, being higher than the 2.54 alleles reported by Henane et al. [23] but lower than the 9.3 alleles reported by Trimech et al. [24]. Aragão et al. [43] and Malik et al. [41] considered that these differences were due to different genetic material and molecular markers used in each experiment.

Despite the allogamous reproductive system of melon and the significant phenotypic variation observed in our collection, the observed heterozygosity was lower (H_0 = 0.1) than the expected heterozygosity (He = 0.12), which revealed an excess in homozygosis further confirmed by the positive value of fixation index or inbreeding coefficient (Fis = 0.04). It is likely that selection, practiced by farmers or CRRHAB's breeders for traits of agronomic interest such as good growth and disease resistance, acted simultaneously upon many loci, controlling a variety of traits under selection. This would greatly reduce diversity throughout the genome and lead to inbreeding depression and higher homozygosity [45].

Two private alleles were detected in the screening collection, allele 206 (in Arbi4, *inodorus* group) and allele 153 (in Horchay, *chate* group). Private alleles were also observed in the *inodorus* 'Yellow Canary' commercial variety and in the *dudaim* 'Chemoum' landrace in Tunisia [24], in a wild and cultivated melon germplasm in China [46], and in heirloom and open-pollinated watermelon commercial cultivars in USA [47]. These alleles may be of interest to conservationists and breeders aiming to preserve and exploit diversity, as they are present only in a single population among a broader collection of populations [48,49].

Both STRUCTURE and SM tree analyses based on SSR markers displayed that landraces and breeding lines were intermixed into five clusters independent of the botanical groups. Insignificant distinctions among sweet and non-sweet genotypes and among *inodorus* and *cantalupensis* genotypes were previously reported by Yildiz et al. [50] and

Sensoy et al. [51]. Intermixed clustering was also observed in accessions from different geographic regions. Malik et al. [41] showed an affinity between Indian and USA melons, and Singh et al. [14] indicated a significant genetic resemblance between Indian modern cultivars/newly derived inbred lines and exotic accessions; this was attributed to a human-mediated transportation of the melon germplasm across geographic locations and/or a great extent of outcrossing among melon genotypes because of the allogamy of the species.

Comparison of the morphological and molecular data in our study demonstrated discordance of grouping the accessions between the two data sets. This was confirmed by the negative correlation ($r = -0.16$) exhibited by the Mantel test, despite being weak and not significant ($p = 0.06$). The reasons might be that (i) the morphogenesis of melon accessions has been shaped not only by genes but also by environmental conditions, with some quantitative traits being especially greatly influenced by the environment [52], or (ii) the polymorphic loci amplified by SSR markers were not linked to the scored agro-morphological traits, such that the basis of classification was different [53]. Moreover, the number of SSR markers used might be one factor for this disparity, as an increased number of SSRs may lead to a greater similarity. Inconsistencies between phenotypic and genotypic data were reported in melon and in other vegetable species [18,35,52–54], and the use of a joint matrix derived from combined phenotypic and molecular matrices was performed in our study in order to increase precision. Indeed, a total discrimination of sweet and non-sweet accessions, as well as a more precise clustering among *inodorus*, *cantalupensis*, and *reticulatus* subgroups, was obtained compared to the separate morphological and molecular data analyses. Both markers are complementary and essential for the efficient conservation of genetic resources and selection of potentially valuable parent lines in breeding programs.

5. Conclusions

This study added further information about the intra and inter variation among local melon genetic resources. Both molecular and morphological features are useful and will facilitate the selection process; the results obtained in the present study can be used for the sustainable conservation and management of melon genetic resources. Landraces with agronomical performance and private alleles can potentially constitute a valuable gene pool for melon breeding, especially in a scenario of rapid climate change. Further phenotypic and molecular studies on national collections, including local varieties, landraces, hybrids, introduced accessions, breeding lines, and wild species, might be necessary for a better understanding of the Tunisian melon gene pool.

Author Contributions: Conceptualization, H.C.-R., N.M. (Najla Mezghani) and A.G.-C.; methodology, H.C.-R., N.M. (Najla Mezghani) and A.G.-C.; validation, H.C.-R., N.M. (Najla Mezghani) and A.G.-C.; formal analysis, H.C.-R., N.M. (Najla Mezghani), S.M. and N.M.; writing—original draft preparation, H.C.-R. and N.M. (Najla Mezghani); writing—review and editing, H.C.-R., N.M. (Najla Mezghani), S.M., N.M. (Neila Mezghani) and A.G.-C.; funding acquisition, A.G.-C. All authors have read and agreed to the published version of the manuscript.

Funding: This research was supported in part by the Spanish Ministerio de Ciencia, Innovación y Universidades, co-funded with FEDER (AGL2017-85563-C2-2-R) and partially financed by Aragon Government grant for Research group, A11-20R-PROVESOS.

Institutional Review Board Statement: Not applicable.

Informed Consent Statement: Not applicable.

Data Availability Statement: The data presented in this study are available on request from the corresponding author.

Acknowledgments: The authors are grateful to Rafika Sta-Baba in CRRHAB-Tunisia for providing seeds of some breeding lines.

Conflicts of Interest: The authors declare no conflict of interest.

References

1. Whitaker, T.W.; Davis, G.N. *Cucurbits: Botany, Cultivation and Utilization*; Interscience Publishers: New York, NY, USA, 1962; 249p.
2. Pitrat, M. Melon Genetic Resources: Phenotypic Diversity and Horticultural Taxonomy. In *Genetics and Genomics of Cucurbitaceae*; Grumet, R., Katzir, N., Garcia-Mas, J., Eds.; Springer: Cham, Switzerland, 2016; Volume 3, pp. 25–60.
3. McCreight, J.D.; Nerson, H.; Grumet, R. Melon (*Cucumis melo* L.). In *Genetic Improvement of Vegetable Crops*, 7th ed.; Kalloo, G., Berch, B.O., Eds.; Pergamon Press: Oxford, UK, 1993.
4. Chikh-Rouhou, H.; Kacem, K.; Letaief, S.; Sta-Baba, R. Evaluation of Tunisian melon genetic resources to biotic stress under greenhouse conditions. In Proceedings of the XIth International Agriculture Symposium (AGROSYM 2020), Jahorina, East Sarajevo, 8–9 October 2020; Book of abstracts. p. 276.
5. Chikh-Rouhou, H.; Garcés-Claver, A.; Sta-Baba, R.; González, V.; Daami-Remadi, M. Screening for resistance to race 1 of *Fusarium oxysporum* f.sp *melonis* in Tunisian melon cultivars using molecular markers. *Comm. Agric. Appl. Biol. Sci.* **2018**, *83*, 87–92.
6. Chikh-Rouhou, H.; Ben Belgacem, A.M.; Sta-Baba, R.; Tarchoun, N.; Gómez-Guillamón, M.L. New source of resistance to *Aphis gossypii* in Tunisian melon accessions using phenotypic and molecular marker approaches. *Phytoparasitica* **2019**, *47*, 405–413. [CrossRef]
7. M'nari, M.; Jebari, H.; Cherif, C. Study of strains of cucumber mosaic virus (CMV) on Tunisian melon and identification of resistant genotypes. *Cah. Agric.* **1993**, *2*, 60–62.
8. Chikh-Rouhou, H.; Tlili, I.; Ilahy, R.; Chaar, H.; Sta-Baba, R. Evaluation de la diversité morphologique de quelques accessions locales de melon (*Cucumis melo* L.). *Annales de l'INRAT* **2020**, *93*, 171–183.
9. Elbekkay, M.; Hamza, H.; Haddad, M.; Ferchichi, A.; Kik, C. Genetic erosion in melon (Cucumis melo): A case study from Tunisia. In *Cucurbitaceae 2008, Proceedings of the IXth EUCARPIA Meeting on Genetics and Breeding of Cucurbitaceae, Avignon, France, 21–24 May 2008*; Pitrat, M., Ed.; INRA: Avignon, France.
10. Escribano, S.; Lazaro, A.; Cuevas, H.E.; Lopez-Sesé, A.I.; Staub, J.E. Spanish melons (*Cucumis melo* L.) of the Madrid provenance: A unique germplasm reservoir. *Genet. Resour. Crop Evol.* **2012**, *59*, 359–373. [CrossRef]
11. Fergany, M.; Kaur, B.; Monforte, A.J.; Pitrat, M.; Lecoq, C.H.; Dhillon, N.P.S.; Dhaliwal, S.S. Variation in melon (*Cucumis melo*) landraces adapted to the humid tropics of southern India. *Genet. Resour Crop. Evol.* **2011**, *55*, 225–243. [CrossRef]
12. Garcia-Mas, J.; Monforte, A.J.; Arus, P. Phylogenetic relationships among *Cucumis* species based on the ribosomal internal transcribed spacer sequence and microsatellite markers. *Plant. Syst. Evol.* **2004**, *248*, 191–203. [CrossRef]
13. Dhillon, N.P.S.; Singh, J.; Fergany, M.; Monforte, A.J.; Sureja, A.K. Phenotypic and molecular diversity among landraces of snapmelon (*Cucumis melo* var. momordica) adapted to the hot and humid tropics of eastern India. *Plant Genet. Resour.* **2009**, *7*, 291–300. [CrossRef]
14. Singh, D.; Leskovar, D.I.; Sharma, S.P.; Sarao, N.K.; Vashisht, V.K. Genetic diversity and inter-relationship among Indian and exotic melons based on fruit morphology, quality components and microsatellite markers. *Physiol. Mol. Biol. Plants.* **2020**, *26*, 985–1002. [CrossRef]
15. Chikh-Rouhou, H.; Tlili, I.; Ilahy, R.; R'him, T.; Sta-Baba, R. Fruit quality assessment and characterization of melon genotypes. *Int. J. Veg. Sci.* **2021**, *27*, 3–19. [CrossRef]
16. Keneni, G.; Jarso, M.; Wolabu, T.; Dino, G. Extent and pattern of genetic diversity for morpho-agronomic traits in Ethiopian highl and pulse landraces II. Faba bean (*Vicia faba* L.). *Genet. Resour. Crop Evol.* **2005**, *52*, 551–561. [CrossRef]
17. Neuhausen, S.L. Evaluation of restriction fragment length polymorphism in *Cucumis melo. Theor. Appl. Genet.* **1992**, *83*, 379–384. [CrossRef]
18. Lopez-Sesé, A.I.; Staub, J.; Katzir, N.; Gómez-Guillamón, M.L. Estimation of between and within accession variation in selected Spanish melon germplasm using RAPD and SSR markers to assess strategies for large collection evaluation. *Euphytica* **2003**, *127*, 41–51. [CrossRef]
19. Garcia-Mas, J.; Seros, M.O.; Gomez-Paniagua, H.; De Vicente, M.C. Comparing AFLP, RAPD and RFLP markers for measuring genetic diversity in melon. *Theor. Appl. Genet.* **2000**, *101*, 860–864. [CrossRef]
20. Merheb, J.; Pawełkowicz, M.; Branca, F.; Bolibok-Bragoszewska, H.; Skarzynska, A.; Plader, A.; Chalak, L. Characterization of Lebanese germplasm of snake melon (*Cucumis melo* subsp. melo var. *flexuosus*) using morphological traits and SSR Markers. *Agronomy* **2020**, *10*, 1293.
21. Lakshmana Reddy, D.C.; Sudurashini, K.V.; Anand, C.R.; Aswath, C.; Avinash, K.N.; Nandini, H.; Sreenivasa Rao, E. Genetic diversity and population structure of Indian melon (*Cucumis melo* L.) landraces with special reference to disease and insect resistance loci. *Plant Breed.* **2016**, *13*, 384–390. [CrossRef]
22. Trimech, R.; Zaouali, Y.; Boulila, A.; Chabchoub, L.; Ghezal, I.; Boussaid, M. Genetic variation in Tunisian melon (*Cucumis melo* L.) germplasm as assessed by morphological traits. *Plant Genet. Resour.* **2013**, *60*, 1621–1628. [CrossRef]
23. Henane, I.; Slimane, R.; Jebari, H. SSR-based genetic diversity analysis of Tunisian varieties of melon (*Cucumis melo* L.) and Fakous (*Cucumis melo* var. *flexuosus*). *Int. J. Adv. Res.* **2015**, *3*, 727–734.
24. Trimech, R.; Afif, M.; Boussaid, M. Genetic diversity of Tunisian melon (*Cucumis melo.* L) landraces and their relationships with introduced varieties as assessed by simple-sequence repeat (SSR) markers. *Afr. J. Biotec.* **2015**, *14*, 86–95.
25. SAS. Statistical Analysis System, version V.9.1. SAS/ STAT Users Guide. SAS Publishing: North Carolina, NY, USA, 1990.
26. Ihaka, R.; Gentleman, R. R: A Language for data analysis and graphics. *J. Comput. Graph Stat.* **1996**, *5*, 299–314.
27. Sokal, R.R.; Michener, C.D. A statistical method for evaluating systematic relationships. *Univ. Kans. Sci. Bull.* **1958**, *38*, 1409–1438.

28. Rodríguez-Maza, M.J.; Garcés-Claver, A.; Park, S. A versatile PCR marker for pungency in *Capsicum* spp. *Mol. Breed.* **2012**, *30*, 889–898. [CrossRef]
29. Fukino, N.; Sakata, Y.; Kunihsa, M.; Matsumoto, S. Characterisation of novel simple sequence repeat (SSR) markers for melon (*Cucumis melo* L.) and their use for genotype identification. *J. Hortic. Sci. Biotec.* **2007**, *82*, 330–334. [CrossRef]
30. Fernandez-Silva, I.; Eduardo, I.; Blanca, J.; Esteras, C.; Pico, B.; Nuez, F.; Arus, P.; Garcia-Mas, J.; Monforte, A.J. Bin mapping of genomic and EST-derived SSRs in melon (*Cucumis melo* L.). *Theor. Appl. Genet.* **2008**, *118*, 139–150. [CrossRef] [PubMed]
31. Diaz, A.; Fergany, M.; Formisano, G. A consensus linkage map for molecular markers and Quantitative Trait Loci associated with economically important traits in melon (*Cucumis melo* L.). *BMC Plant Biol.* **2011**, *11*, 111. [CrossRef] [PubMed]
32. Mallor, C.; Arnedo-Andrés, M.S.; Garcés-Claver, A. Assessing the genetic diversity of Spanish *Allium cepa* landraces for onion breeding using microsatellite markers. *Sci. Hortic.* **2014**, *170*, 24–31. [CrossRef]
33. Peakall, R.; Smouse, P.E. GenAlEx 6: Genetic analysis in Excel. Population genetic software for teaching and research. *Mol. Ecol. Notes* **2006**, *6*, 288–295. [CrossRef]
34. Botstein, D.; White, R.L.; Skalnick, M.; Davis, R.W. Construction of a genetic linkage map in man using restriction fragment length polymorphism. *Am. J. Hum. Genet.* **1980**, *32*, 314–331.
35. Yildiz, M.; Akgul, N.; Sensoy, S. Morphological and Molecular Characterization of Turkish Landraces of *Cucumis melo* L. *Not. Bot. Horti. Agrobo.* **2014**, *42*, 51–58. [CrossRef]
36. Mantel, N.A. The detection of disease clustering and generalized regression approach. *Cancer Res.* **1967**, *27*, 209–220.
37. Pritchard, J.K.; Stephens, M.; Donnelly, P. Inference of population structure using multilocus genotype data. *Genetics.* **2000**, *155*, 945–959. [CrossRef] [PubMed]
38. Evanno, G.G.; Regnaut, S.; Goudet, J. Detecting the number of clusters of individuals using the software structure: A simulation study. *Mol. Ecol.* **2005**, *14*, 2611–2620. [CrossRef] [PubMed]
39. Omari, S.; Kamenir, Y.; Benichou, J.I.C.; Pariente, S.; Sela, H.; Perl-Treves, R. Landraces of snake melon, an ancient Middle Eastern crop, reveal extensive morphological and DNA diversity for potential genetic improvement. *BMC Genet.* **2018**, *19*, 34.
40. Chikh-Rouhou, H.; Gómez-Guillamón, M.L.; González, V.; Sta-Baba, R.; Garcés-Claver, A. *Cucumis melo* L. in Tunisia: Unexploited sources of resistance to Fusarium wilt. *Horticulturae* **2021**. submitted.
41. Malik, A.A.; Vashisht, V.K.; Singh, K.; Sharma, A.; Singh, D.K.; Singh, H.; Monforte, A.J.; McCreight, J.D.; Dhillon, N.P.S. Diversity among melon (*Cucumis melo* L.) landraces from the Indo-Gangetic plains of India and their genetic relationship with USA melon cultivars. *Genet. Resour. Crop Evol.* **2014**, *61*, 1189–1208. [CrossRef]
42. Soltani, F.; Akashi, Y.; Kashi, A.; Zamani, Z.; Mostofi, Y.; Kato, K. Characterization of Iranian melon landraces of *Cucumis melo* L. groups flexuosus and dudaim by analysis of morphological characters and random amplified polymorphic DNA. *Breed. Sci.* **2010**, *60*, 34–45. [CrossRef]
43. Aragão, F.A.S.; Torres Filho, J.; Nunes, G.H.S.; Queiróz, M.A.; Bordallo, P.N.; Buso, G.S.C.; Ferreira, M.A.; Costa, Z.P.; Bezerra Neto, F. Genetic divergence among accessions of melon from traditional agriculture of the Brazilian Northeast. *Genet. Mol. Res.* **2013**, *12*, 6356–6371. [CrossRef]
44. Scarano, D.; Rubio, F.; Ruiz, J.J.; Rao, R.; Corrado, G. Morphological and genetic diversity among and within common bean (*Phaseolus vulgaris* L.) landraces from the Campania region (Southern Italy). *Sci. Hortic.* **2014**, *180*, 72–78. [CrossRef]
45. Tanksley, S.D.; McCouch, S.R. Seed banks and molecular maps: Unlocking genetic potential from the wild. *Science* **1997**, *277*, 1063–1066. [CrossRef]
46. Jianbin, H.; Panqiao, W.; Qiong, L.; Yan, S. Microsatellite analysis of genetic relationships between wild and cultivated melons in Northwest and Central China. *Mol. Biol. Rep.* **2014**, *41*, 7723–7728.
47. Stone, S.; Boyhan, G. Inter- and intra cultivar variation of heirloom and open-pollinated watermelon cultivars. *HortScience* **2019**, *54*, 212–220. [CrossRef]
48. Szpiech, Z.A.; Rosenberg, N.A. On the size distribution of private microsatellite alleles. *Theor. Popul. Biol.* **2011**, *80*, 100–113. [CrossRef] [PubMed]
49. Kalinowski, S.T. Counting alleles with rarefaction: Private alleles and hierarchical sampling designs. *Conserv. Genet.* **2004**, *5*, 539–543. [CrossRef]
50. Yildiz, M.; Ekbicb, E.; Kelesc, D.; Sensoy, S.; Abak, K. Use of ISSR, SRAP, and RAPD markers to assess genetic diversity in Turkish melons. *Sci. Hortic.* **2011**, *130*, 349–353. [CrossRef]
51. Sensoy, S.; Buyukalaca, S.; Abak, K. Evaluation of genetic diversity in Turkish melons (*Cucumis melo* L.) based on phenotypic characters and RAPD markers. *Genet.Resour.Crop Evol.* **2007**, *54*, 1351–1356. [CrossRef]
52. Nkhoma, N.; Shimelis, H.; Laing, M.D.; Shayanowako, A.; Mathew, I. Assessing the genetic diversity of cowpea (*Vignaun guiculata* L.Walp.) germplasm collections using phenotypic traits and SNP markers. *BMC Genet.* **2020**, *21*, 110–126. [CrossRef] [PubMed]
53. Schaal, B.A.; Hayworth, D.A.; Olsen, K.M.; Rauscher, J.T.; Smith, W.A. Phylogeographic studies in plants: Problems and prospects. *Molec. Ecol.* **1998**, *7*, 465–474. [CrossRef]
54. Guidoti, D.T.; Gonela, A.; Vidigal, M.C.; Conrado, T.V.; Romani, I. Interrelationship between morphological, agronomic and molecular characteristics in the analysis of common bean genetic diversity. *Acta Sci. Agron.* **2018**, *40*, 1–9. [CrossRef]

 *agronomy*

Article

Genotype-Dependent Tipburn Severity during Lettuce Hydroponic Culture Is Associated with Altered Nutrient Leaf Content

Virginia Birlanga [1], José Ramón Acosta-Motos [2,3] and José Manuel Pérez-Pérez [4,*]

[1] Monsanto Agricultura España S.L.U., 30319 Miranda, Spain; virginia.birlanga@bayer.com
[2] Group of Fruit Tree Biotechnology, CEBAS-CSIC, 30100 Murcia, Spain; jacosta@cebas.csic.es
[3] Cátedra Emprendimiento en el Ámbito Agroalimentario, Campus de los Jerónimos, Universidad Católica San Antonio de Murcia (UCAM), No. 135 Guadalupe, 30107 Murcia, Spain
[4] Instituto de Bioingeniería, Universidad Miguel Hernández, 03202 Elche, Spain
* Correspondence: jmperez@umh.es; Tel.: +34-966-658-958

Abstract: Cultivated lettuce (*Lactuca sativa* L.) is one of the most important leafy vegetables in the world, and most of the production is concentrated in the Mediterranean Basin. Hydroponics has been successfully utilized for lettuce cultivation, which could contribute to the diversification of production methods and the reduction of water consumption and excessive fertilization. We devised a low-cost procedure for closed hydroponic cultivation and easy phenotyping of root and shoot attributes of lettuce. We studied 12 lettuce genotypes of the crisphead and oak-leaf subtypes, which differed on their tipburn resistance, for three growing seasons (Fall, Winter, and Spring). We found interesting genotype × environment (G × E) interactions for some of the studied traits during early growth. By analyzing tipburn incidence and leaf nutrient content, we were able to identify a number of nutrient traits that were highly correlated with cultivar- and genotype-dependent tipburn. Our experimental setup will allow evaluating different lettuce genotypes in defined nutrient solutions to select for tipburn-tolerant and highly productive genotypes that are suitable for hydroponics.

Keywords: *Lactuca sativa* L.; crisphead; oak-leaf; root system architecture; tipburn; nutritional imbalance

Citation: Birlanga, V.; Acosta-Motos, J.R.; Pérez-Pérez, J.M. Genotype-Dependent Tipburn Severity during Lettuce Hydroponic Culture Is Associated with Altered Nutrient Leaf Content. *Agronomy* **2021**, *11*, 616. https://doi.org/10.3390/agronomy11040616

Academic Editor: Dimitrios Savvas

Received: 4 February 2021
Accepted: 22 March 2021
Published: 24 March 2021

Publisher's Note: MDPI stays neutral with regard to jurisdictional claims in published maps and institutional affiliations.

1. Introduction

Cultivated lettuce (*Lactuca sativa* L.; Asteraceae), which is usually consumed fresh, is one of the most important leafy vegetables in the world. Commercial lettuce varieties are classified based on head and leaf characteristics, and some of the most common horticultural types are romaine, iceberg (also named as crisphead; CHD), oak-leaf (i.e., green oak; GOAK, and red oak; ROAK), and butterhead. Breeding new lettuce cultivars involves manual pollination of genetically stable (i.e., pure) parent lines with agronomic traits of interest, followed by selection based on plant phenotyping and genotyping [1,2]. The availability of detailed genetic maps of cultivated lettuce [3–7] has allowed significant progress for mapping agronomically-important traits and promoted the development of marker-assisted selection (MAS) and candidate gene identification in these species [8]. Several studies have shown that most breeding target traits, such as disease resistance [9], postharvest discoloration [10], thermotolerance in seed germination [7], or water and nitrate capture [6], are complex traits and thus controlled by quantitative trait loci (QTL).

Spain is the third-largest producer of lettuce and chicory in the world after China and the USA, with a total of c.a. 1.1 million tons, with an area of 35,360 hectares dedicated for their cultivation [11]. Most of the production is focused on the southeastern Mediterranean region, with a temperate climate that allows lettuce cultivation throughout the year, making Spain the world's largest lettuce exporting country. However, water scarcity and soil availability are limiting factors for plant cultivation, and inadequate irrigation and fertilization management has increased the environmental impact of agricultural exploitation in this region [12,13]. Therefore, to contribute to sustainable lettuce production, there is a strong requirement for the development of new forms of farming to increase the crop's resource use efficiency and the reduction of production costs. Floating systems or closed hydroponic methods have been successfully engaged for lettuce cultivation [14–16]. This technique allows for the precise control of water and mineral nutrition, saves soil and labor costs, and provides shorter harvest cycles, high product quality, and good consumer acceptance [16].

Tipburn is defined as the localized necrosis found on the distal margins of rapidly expanding leaves. It is a serious problem in controlled lettuce production, as it reduces the quality and shelf life of fresh lettuce, hence resulting in severe economic losses [17–19]. Tipburn is influenced by many environmental factors, such as light intensity, air temperature, and soil conditions, and is considered a calcium deficiency-related physiological disorder, which is usually associated with rapidly growing tissues [17,18,20,21]. In addition, a strong variation for tipburn incidence among different lettuce cultivars has been reported [19,22], allowing for the development of tipburn-resistant varieties [19,23,24]. The use of genomic tools has enabled the identification of QTL for tipburn incidence in several recombinant inbred line (RIL) populations and the development of linked molecular markers [25,26]. However, further research is needed to identify the underlying candidate genes for these QTL and the effect of their introgression into other lettuce cultivars. Also, only a few studies on tipburn incidence have been carried out in lettuce grown in hydroponics [27,28].

Within the framework of a company-based breeding program for lettuce, we devised a low-cost procedure for closed hydroponic cultivation and easy phenotyping of root and shoot attributes during early growth in three growing seasons (Fall, Winter, and Spring). A representative sample of lettuce varieties from different cultivars (CHD, GOAK and ROAK) were selected based on contrasting agronomically relevant traits such as tipburn tolerance. Our results allowed us to define genotype $\times$ environment (G $\times$ E) interactions for some of the studied traits, and to establish a strong correlation between leaf nutrient content and tipburn incidence, which may help to reduce leaf damage through adequate fertilization management.

2. Materials and Methods

2.1. Plant Material and Growth Conditions

We selected 12 lettuce genotypes from the breeding program at Monsanto Agriculture Spain S.L.U. (Murcia, Spain) which differed on their tipburn resistance as visually scored at the company's experimental station (37°41′47.6″ N 1°01′55.2″ W, Murcia, Spain; Table 1). We included four genotypes of the *Lactuca sativa* var. *capitata* L., hereafter referred as crisphead (CHD) or iceberg cultivar; and eight genotypes from *Lactuca sativa* var. *crispa* L., which differed in their leaf color, and which were assigned either to the green oak (GOAK) or to the red oak (ROAK) subtypes (Table 1). Seeds from the cultivars used in this work are available upon request to V.B.

Seedlings were sown in 198-well trays filled with moistened 80% perlite and 20% substrate (FloraGard) and were incubated in darkness for 3 days at 10 $\pm$ 2 °C and 75% relative humidity. Germinated trays were transferred to the nursery chamber set at 20 $\pm$ 2 °C, 65% relative humidity, and under natural photoperiod (Table S1) until the seedlings had 2–3 true leaves (10 mm; Figure 1a). For each cultivar and experiment, eight randomly-selected seedlings were then transferred to 3 L sealed and opaque pots filled with nutrient solution [29] (Table S2); with an eventual air pump (5 $\times$ 2.5 L) for hydroponic growth [30] in a multi-tunnel greenhouse at the company's experimental station and under environmental conditions (0 days after planting, dap; Figure 1b). As previous results indicated that lettuce growth was strongly affected by N application [17], we adjusted the nutrient solution for optimal N supply. To avoid contamination, the nutrient solution was renewed every two weeks.

Table 1. Some details of the lettuce genotypes used in this study.

Cultivar	Genotype	Description	Tipburn Phenotype [18]
CHD	C1	Collected in Summer, crispy leaves	Light
CHD	C3	Collected in late Fall, dark green leaves and ovate leaves, big size head	Severe
CHD	C7	Collected in Winter, dark green leaves and ovate leaves, big size head	Medium
CHD	C8	Collected in late Fall, medium-size head	Medium
GOAK	G1	Collected in Fall, indoor production Voluminous and compact lettuce, strong against bolting	Light
GOAK	G3	Collected in Winter and Spring, indoor and outdoor production. Round shape, dense filling, high weight, strong against bolting	Severe
GOAK	G5	Collected in Spring and Fall, upright and compact leaves, slow bolting	Light
GOAK	G6	Collected in Spring and Summer, indoor production, dark green color, strong against bolting	Medium
ROAK	R2	Collected in Fall and Winter, slight red, small and bit cylindrical heads	Severe
ROAK	R3	Collected in Fall and Spring, dark green color, medium volume	Medium
ROAK	R4	Collected in Fall, good vigor and volume	Light
ROAK	R5	Collected in Fall and Spring, good vigor and volume, strong against bolting	Light

Figure 1. Experimental design for studying growth, tipburn phenotypes, and nutrient concentrations in different lettuce genotypes. (**a**) A representative image of a young seedling from the nursery chamber. (**b**) General view of the hydroponic system used for lettuce cultivation at the experimental station. (**c**) Glass cylinder vase used for image acquisition. (**d**,**e**) A representative image of the roots (**d**) and the shoot (**e**) of a plant grown in hydroponics for 21 days. (**f**) Image segmentation files obtained with Image J software. (**g**) A representative image of leaves collected for nutrient concentration analysis. Scale bars: 50 mm.

2.2. Image Analysis

Five randomly chosen plants were periodically taken for image analyses during the hydroponic culture at 0, 7, 14, 21, 28, 35, and 45 dap. To minimize light variation, a photography box was used with illumination from below. Plants were transferred to a glass cylinder vase (12 × 28 cm) filled with nutrient solution (Figure 1c) and their root and shoot system were respectively imaged (Figure 1d,e) with a still smartphone camera (iPhone 6s, 12 MP f2.2) and saved as an RGB color image in jpeg format (1200 × 2800 pixels). Root area (RA) was measured using the GiA Roots software [31] as described elsewhere [32]. For the shoot area (SA) measurement, the background of the image was removed using Adobe Photoshop CS3, and images were batch-processed using Image J [33] (Figure 1f). Raw measurements were exported to Excel spreadsheets for data analysis.

2.3. Tipburn Evaluation

From 14 dap onwards, tipburn severity (TS) was assessed weekly in individual plants by scoring the presence of necrotic symptoms on the edges of leaves on a scale from 1 to 9, where 1 was no tipburn and 9 was severe tipburn (Figure S1). To obtain these scores, five plants were evaluated per cultivar and season. In addition, tipburn incidence (TBI) was calculated to verify agreement with TS as previously described in [18] with the following formula:

$$\text{TBI} = \frac{(\text{n plants severe tipburn} \times 5 + \text{n plants medium tipburn} \times 3 + \text{n plants light tipburn})}{\text{n plants} \times 5} \times 100$$

2.4. Growth Parameters and Nutrient Content Analysis

At the end of the experiment (45 dap), each plant was separated into shoots and roots to measure their fresh weight (FW). Dry weight (DW) was measured in samples that were oven-dried at 80 °C for 72 h. Root and leaf water content (WC) were determined as: $\frac{(\text{FW}-\text{DW})}{\text{FW}} \times 100$. Stem length and leaf number were also documented.

For nutrient concentration analysis, we randomly selected three 21 dap plants from the Spring season, and mature (M), intermediate (I), and juvenile (J) leaves were collected from each plant and imaged for SA determination (see Section 2.2; Figure 1g). FW, DW, and WC were measured as described above.

The measurement of different macronutrients: potassium (K), calcium (Ca), phosphorus (P), sulfur (S), magnesium (Mg), sodium (Na), and micronutrients: iron (Fe), manganese (Mn), zinc (Zn), copper (Cu), was carried out in a digestion extract containing 100 mg of tissue powder and 50 mL of a mix of HNO_3:$HClO_4$ (2:1 *v/v*) using an inductively coupled plasma optical emission spectrometer (ICP-OES IRIS INTREPID II XDL, Thermo Fisher Scientific Inc., Loughborough, UK) at the Ionomic Services of the CEBAS-CSIC (Murcia, Spain) [34].

2.5. Statistical Analysis

The descriptive statistics (mean, standard error of the mean (SEM), etc.) calculated for samples and different tests described below were performed by using the StatGraphics Centurion XV software (StatPoint Technologies, Inc., Warrenton, VA, USA). The Kolmogorov–Smirnov [35] and Shapiro tests were performed to check the normality of the data by analyzing the goodness-of-fit between the distribution of the data and a given theoretical normal distribution. In addition, to check the homogeneity of the variance, the Bartlett and Levene tests were applied. The data with a normal distribution were analyzed by a one-way ANOVA followed by Fisher's LSD (least significant differences) multiple range Test [36] to separate the treatment means, thereby detecting significant differences (p-value < 0.01). Non-parametric tests were used when necessary. In that case, the median was used instead of the mean, and the data were subjected to the Kruskal–Wallis test (p-value < 0.01). Heatmaps were processed using the pheatmap package in R [37]. Neighbor-joining distance matrixes between genotypes (rows) and between samples (columns) were automatically calculated from average values to build the dendrograms and the heatmap representation. Graphs were drawn with GraphPad Prism 9.0.0 for Windows (GraphPad Software, San Diego, CA, USA).

3. Results

3.1. Quantitative Analysis of Root and Shoot Area during Hydroponic Growth

We followed the growth of the studied lettuce cultivars grown on hydroponic culture by periodically imaging the root and the shoot system between 0 and 35 dap (see Section 2). Estimated root and shoot areas (RA and SA) in the studied CHD genotypes exponentially increased between 0 and 35 dap, following a season-dependent pattern (p-value = 0.002; Figure 2a,b and Table S3). The highest growth rate of the RA occurred during Spring for C3 (Figure S2). Instead, C1 and C7 showed the highest SA growth rate during Fall (Figure S2). In all seasons, C7 and C8 usually showed the lowest RA values, while C3 exhibited the highest RA values at 35 dap (RA_{35}; Figure 2a,c and Figure S3a). In agreement with what was found for RA_{35}, the C8 genotype showed the smallest SA values at 35 dap (SA_{35}) in every season (Figure 2b and Figure S3b), while the SA_{35} values for C7 were much higher in Spring than those in Winter or Fall (Figure 2b,d), despite the RA_{35} in C7 lagging behind in every season (Figure 2a,c). In contrast, the shoot growth and root growth rates of C1 were similar in every season and normally higher than in the other CHD genotypes studied (Figure S2).

Regarding the GOAK genotypes, the estimated RA exponentially increased between 0 and 35 dap. The highest RA growth was observed in Spring (p-value = 0.000), followed by Fall, while in Winter, a slower growth was observed (Figure 3a,c and Figure S4a and Table S3). RA_{35} was similar in all the genotypes in Fall (p-value = 0.624), and Spring (p-value = 0.321), and also slightly significantly different (p-value = 0.046) in Winter (Figure 3a and Figure S4a). RA growth values were quite similar in all GOAK genotypes, with the extreme values shown by G5 in Winter and G3 in Spring (Figure S2a and Table S3). The highest growth rates of the SA were observed during the Fall (Figure S2b and Table S4), and the SA_{35} values significantly differed between GOAK genotypes in every season (Figure 3b and Figure S4b). Overall, G5 showed significantly higher SA_{35} values than the G1 and G6 genotypes, but in Spring, only the SA_{35} values of G5 were significantly higher than the other GOAK genotypes (Figure 3b,d and Figure S4b).

In the ROAK cultivars, we did not find significant differences in the RA_{35} values between ROAK genotypes in Winter (p-value = 0.999) or Spring (p-value = 0.645; Figure 4a and Figure S5a and Table S3). Consistent with the differences observed for RA_{35} values in Fall (Figure 4a and Figure S5a), the lowest growth rate of the RA occurred for the R4 and R5 genotypes in this season (Figure S2a and Table S3).

Conversely, the growth rate of SA was much lower during Spring for all the ROAK genotypes (Figure S2b and Table S3), which also showed similar RA_{35} values in this season (Figure 4b and Figure S5b). The R3 and R4 genotypes showed contrasting RA_{35} and SA_{35} values in Fall (Figure 4a,b), while the R5 genotype showed the smallest RA_{35} and SA_{35} values in this season (Figure 4a,b and Figure S5a,b).

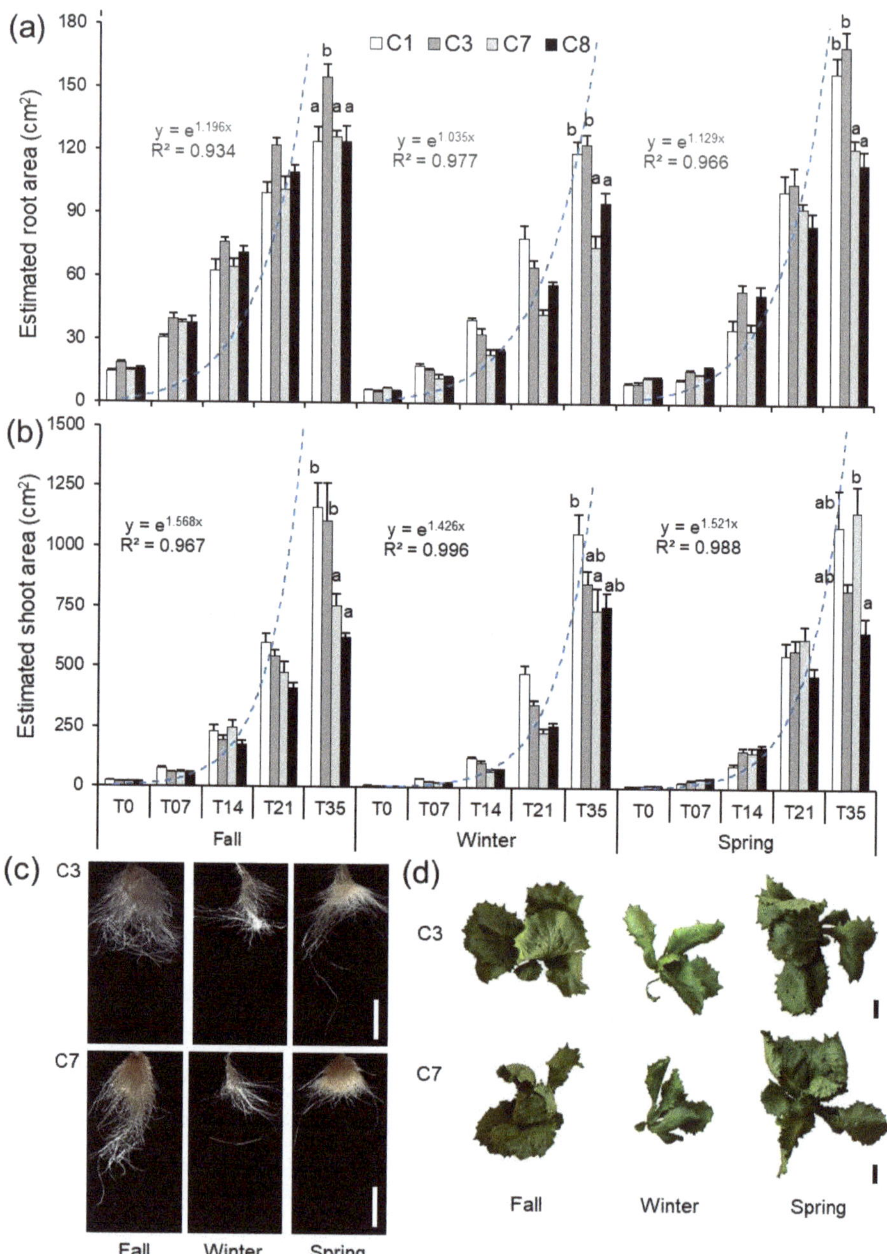

Figure 2. Quantitative analysis of root and shoot area in the studied CHD genotypes during hydroponic growth. (**a**) Average root area (cm^2) and (**b**) average shoot area (cm^2) values in the studied lines (C1, C3, C7, and C8) between 0 (T0) and 35 (T35) days after planting (dap). Theoretical exponential growth curves are depicted in blue. Different letters indicate significant differences at 35 dap (LSD; *p*-value < 0.01). (**c,d**) Representative images of the root (**c**) and shoot (**d**) system of C3 and C7 genotypes at 14 and 21 dap, respectively. Scale bars: 50 mm.

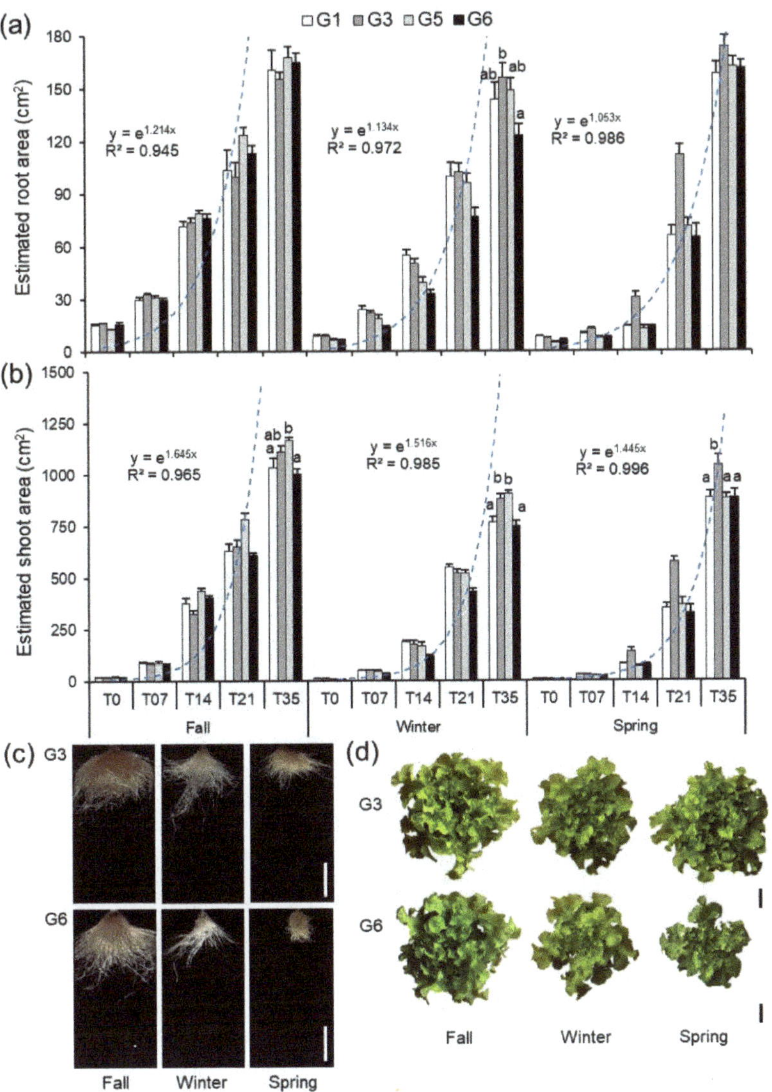

Figure 3. Quantitative analysis of root and shoot area in the studied GOAK genotypes during hydroponic growth. (**a**) Average root area (cm^2) and (**b**) average shoot area (cm^2) values in the studied lines (G1, G3, G5, and G6) between 0 (T0) and 35 (T35) dap. Theoretical exponential growth curves are depicted in blue. Different letters indicate significant differences at 35 dap (LSD; *p*-value < 0.01). (**c**,**d**) Representative images of the root (**c**) and shoot (**d**) system of G3 and G6 genotypes at 14 and 21 dap, respectively. Scale bars: 50 mm.

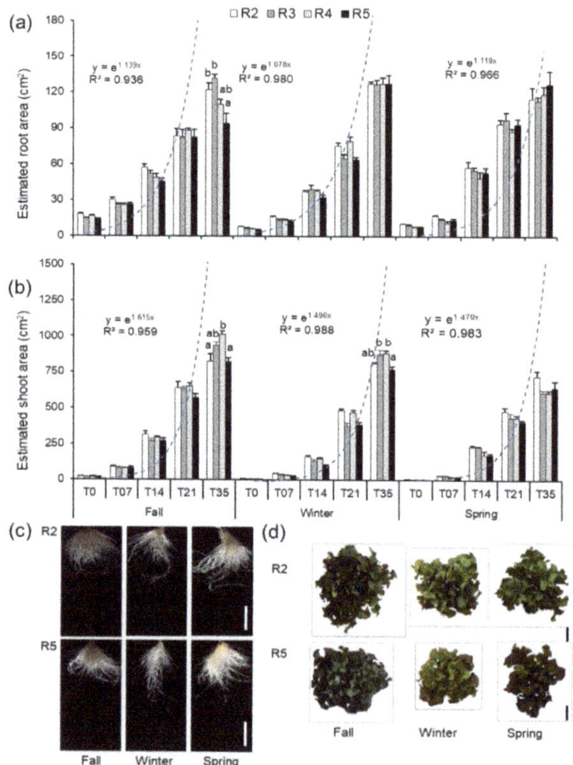

Figure 4. Quantitative analysis of root and shoot area in the studied ROAK genotypes during hydroponic growth. (**a**) Average root area (cm^2) and (**b**) average shoot area (cm^2) values in the studied lines (R2, R3, R4 and R5) between 0 (T0) and 35 (T35) dap. Theoretical exponential growth curves are depicted in blue. Different letters indicate significant differences between genotypes at 35 dap (LSD; *p*-value < 0.01). (**c,d**) Representative images of the root (**c**) and shoot (**d**) system of R2 and R5 genotypes at 14 and 21 dap, respectively. Scale bars: 50 mm.

3.2. Variations in Root and Shoot Weights in the Studied Lettuce Cultivars

We measured several growth-related traits of the root and the shoot system at 45 dap (see Materials and Methods; Table S4 and Figure S6a). Root FW and shoot FW were found to be dependent on the cultivar type (*p*-value = 0.000) and the growing season (*p*-value = 0.002). The GOAK genotypes had significantly heavier root systems (FW = 33.50 ± 0.94 g; DW = 1.39 ± 0.05 g; *n* = 57) than those from ROAK (FW = 24.20 ± 0.58 g; DW = 1.19 ± 0.05 g; *n* = 59) or CHD (FW = 24.60 ± 0.85 g; DW = 1.26 ± 0.06 g; *n* = 60), being the largest in Spring for GOAK and CHD (Figure S6b). As for the CHD genotypes, C1 had significantly (*p*-value < 0.01) heavier root systems (FW = 28.60 ± 1.80 g; DW = 1.63 ± 0.26 g; *n* = 15) than C8 (FW = 21.10 ± 1.02 g; DW = 0.97 ± 0.07 g; *n* = 15; Figure 5a). Among the eight *L. sativa* var. *crispa* genotypes studied, G3 showed the heaviest root system (FW = 41.20 ± 2.03 g; DW = 1.56 ± 0.05 g; *n* = 14), while the R5 root system was the lightest one (FW = 21.00 ± 1.07 g; DW = 0.96 ± 0.08 g; *n* = 14; Figure 5a). Despite the FW and DW values being highly correlated overall (Figure S6b), the root DW values were significantly (*p*-value = 0.000) higher in Winter (1.56 ± 0.06 g; *n* = 59), with the lowest values found in Fall (1.00 ± 0.34 g; *n* = 60), but these were not strongly dependent (*p*-value = 0.014) on the type of cultivar (CHD, GOAK or ROAK).

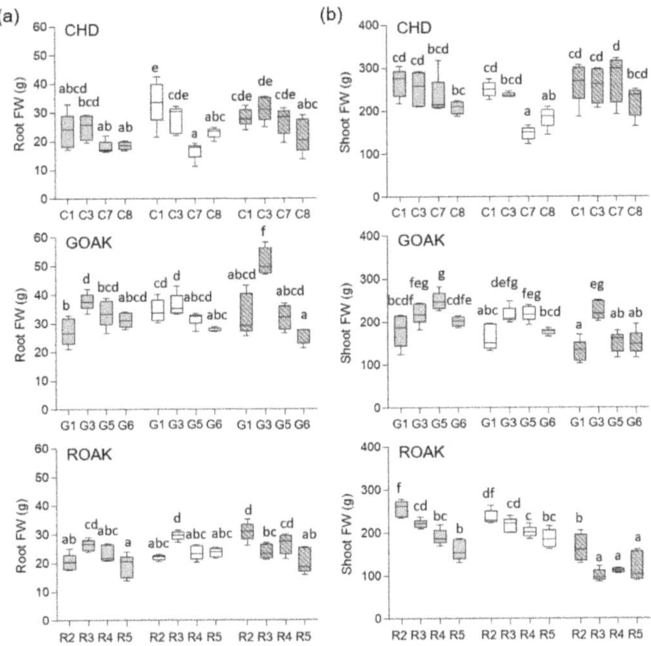

Figure 5. Fresh weight of the studied genotypes at the end of the experiment. (**a**) Root FW and (**b**) shoot FW (Fall, grey-filled bars; Winter, white-filled bars; Spring, lined-filled bars). Different letters indicate significant (p-value < 0.01) differences between CHD, GOAK and ROAK samples.

Despite their small root systems, shoot weights were significantly (p-value = 0.000) higher in the CHD genotypes (FW = 233.30 ± 6.18 g; DW = 8.84 ± 0.30 g; n = 59) than those in GOAK (FW = 190.90 ± 5.36 g; DW = 7.81 ± 0.25 g; n = 60) or ROAK (FW = 182.10 ± 6.72 g; DW = 7.71 ± 0.19 g; n = 59), even though the CHD shoots had significantly (p-value = 0.000) less leaves (15.90 ± 0.50; n = 20) than GOAK shoots (29.80 ± 0.64; n = 60) or ROAK (28.40 ± 0.42; n = 60) shoots (Table S4). Remarkably, a statistically significant (p-value = 0.000) G × E interaction affected shoot FW in GOAK and ROAK genotypes. The shoot FW values from G3 and R2 were much higher than other GOAK or ROAK genotypes only in Spring (Figure 5b). The shoot DW values were also significantly (p-value = 0.000) higher in Winter (9.27 ± 0.26 g; n = 59), but surprisingly, the lowest shoot DW values were found in Spring (7.00 ± 0.21 g; n = 59). The root-to-shoot ratio (R:S ratio) steadily increased from Fall to Spring in the CHD and ROAK cultivars, while non-significant differences were found for the R:S ratio of GOAK in Winter and Spring (Figure S6c).

Root water content (RWC) varied from 94.50 ± 0.36% in C1 to 96.10 ± 0.20% in G3, with a clear effect of the growing season, with lower RWC values in Winter and higher RWC values in Fall (Table S4). Also, shoot water content (SWC) was significantly (p-value = 0.001) lower in ROAK cultivars (95.50 ± 0.15%; n = 57), with the highest values found in C8 (96.50 ± 0.21%; n = 15). Intriguingly, water content (both in the shoot and in the root) was negatively and significantly (p-value = 0.000) correlated with root DW and most considerably in Winter (Figure S6d).

3.3. Tipburn Severity during Hydroponic Growth

Tipburn was scored weekly on cultivars grown in hydroponic culture by means of a visual scaling rate (Figure S1), and it was found that the symptoms steadily increased from 3–4 weeks after planting onwards (Table S5). We found a significant G × E interaction for tipburn severity in the CHD genotypes (p-value = 0.000), with a higher contribution of the Fall and Spring seasons on the scores at 45 dap (Figure 6a and Table S5). While C3 showed higher scores in every season (6.3 ± 0.5; n = 18), others only showed tipburn symptoms during Spring (C8) or Fall (C7 and C1). Additionally, tipburn phenotypes were highly variable within individual plants in C1, C3, and C8 (Table S5), as estimated by their variance values at 45 dap (7.22; n = 54) compared with those of C7 (2.24; n = 18). The tipburn severity in the ROAK genotypes (measured at 45 dap) was not dependent on the growing season (p-value = 0.296), but a significant dependency on the genotype was observed (p-value = 0.000), with R2 leaves showing similar tipburn symptoms and much higher scores (8.0 ± 0.9; n = 18) than those of the other ROAK genotypes studied (2.1 ± 1.5; n = 54). On the other hand, tipburn symptoms of the GOAK leaves were dependent on the growing season and the genotype (p-value = 0.041), with higher scores in Spring for G3 and G6 (5.0 ± 2.4; n = 16) as compared with the other GOAK genotypes studied (1.8 ± 1.2; n = 56).

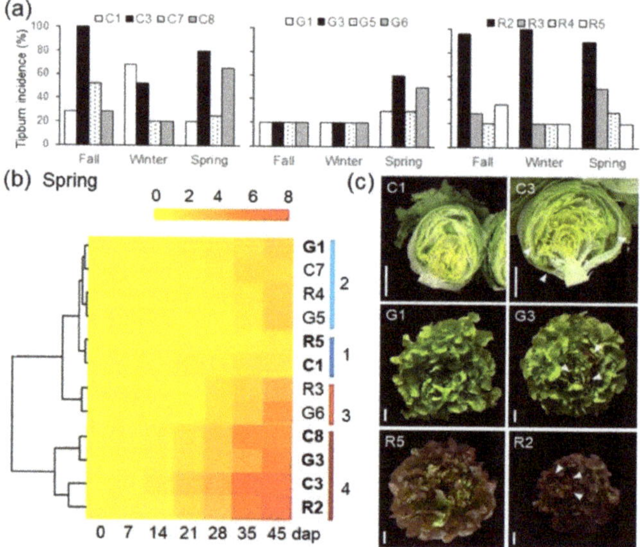

Figure 6. Tip-burn assessment in the studied genotypes. (**a**) Tipburn incidence of the studied genotypes at 45 days after planting (dap) (**b**) Heatmap of tipburn scoring values during the Spring season. Colored bars indicate the severity of the tipburn phenotypes in the studied genotypes, from highly tolerant (0, yellow) to highly sensitive (8, red). Genotypes were grouped into four groups (tolerant: 1,2; sensitive: 3,4). (**c**) Representative images of rosettes of genotypes with extreme tipburn phenotypes (tolerant, left panels; sensitive: right panels) at 45 dap. White arrowheads point to regions where tipburn lesions are present. Scale bars: 50 mm.

Next, we classified the studied genotypes based on their tipburn severity symptoms during the Spring season into four groups: (i) highly tolerant (R5, C1), (ii) intermediate (C7, R4, G1, G5), (iii) sensitive (R3, G6), and (iv) highly sensitive (C8, G3, R2, C3) (Figure 6b). Some highly sensitive genotypes, such as R2 or C3, obtained higher scores in all the seasons, while others, such as C8, G3, or G6 only showed tipburn symptoms during the Spring season (Table S5). Representative pictures of some of the most tolerant and most sensitive genotypes for tipburn symptoms are shown in Figure 6c.

3.4. Leaf Nutrient Variation in the Spring Season in the Studied Lettuce Cultivars

We determined nutrient concentration in mature, intermediate, and juvenile leaves of the studied genotypes in the Spring season at the end of the experiment (Table S6). For the studied nutrients, most of the variation was found associated with leaf type (*p*-value between 0.0000 (Ca, P, Mg, Na, Fe, Mn, and Cu) and 0.0078 (K)) or cultivar type (p-value between 0.0000 (K, P, S, Mg, Na, Fe, Mn, and Cu) and 0.0029 (Zn)) (Tables 2 and 3). In the CHD cultivar, we found the lowest nutrient levels, while in the GOAK cultivars, their nutrient levels were significantly higher. Nutrient concentrations in the ROAK cultivar were intermediate and similar to GOAK, except for P, where the highest values were observed (Table 2). In regard to the studied macronutrients (K, Ca, P, S, Mg, and Na), all cultivars showed higher P levels in the juvenile leaves, while the Ca, Mg, and Na levels were higher in the mature leaves, albeit not significant for Mg or Na in CHD (Table 2). K levels were significantly higher in mature leaves in the GOAK and ROAK cultivars than in intermediate or juvenile leaves; S showed similar behavior in the CHD and ROAK cultivars with higher concentrations in the juvenile leaves than the GOAK cultivar but without significant differences (Table 2). For the studied micronutrients (Fe, Mn, Zn, and Cu), we found significantly higher concentrations of Mn and Cu in mature leaves of all the cultivars. On the other hand, Fe was significantly higher in mature leaves in GOAK and ROAK, and Zn showed a contrasting behavior in these two cultivars, with significantly higher levels in juvenile leaves in ROAK (Table 2). We next analyzed the mature-to-juvenile (M/J) ratio and found higher values of Ca, Na, and Mn in mature leaves irrespectively of the cultivar. Interestingly, the CHD cultivar showed low ($\leq$1) M/J ratios of K, S, Mg, Fe, Zn, and Cu as compared to those in the GOAK and ROAK cultivars (Tables 2 and 3).

Considering the nutrient concentration of all leaves, we found significant differences between the studied genotypes (Table 4). Overall, the CHD genotypes contained a lower nutrient concentration than the ROAK and GOAK genotypes (Table 4). G3 and R4 had the highest nutrient concentration, which almost doubled those found in C1 and C7 (Table 4). We did not find a clear association between total nutrient concentration and tipburn scores (Figure 6a,b).

After analyzing each nutrient individually, we found that all three cultivars showed similar trends for macronutrient (K > Ca > P > S > Mg > Na) and micronutrient (Fe > Mn > Zn > Cu) concentrations (see percentage in italics in Tables 5 and 6). However, some genotypes displayed substantial differences in the amounts of specific nutrients compared with those in other genotypes of the same cultivar. As regards the CHD genotypes, C1 and C7 had higher concentrations or percentages for all the nutrients analyzed except for Mn, Zn, and Cu, where C1 showed the highest percentages. Comparing the most differentiated genotypes (C1 and C7 vs. C3 and C8), we observed significant differences in K (*p*-value = 0.0003), Ca (*p*-value = 0.0047), S (*p*-value = 0.0026), Mg (*p*-value = 0.0070), Na (*p*-value = 0.0024), and Mn (*p*-value = 0.0025). We classified the CHD genotypes based on their statistically significant nutrient levels, as follows: C1 > C7 > C8 = C3 (Tables 5 and 6). In relation to the GOAK genotypes, G3 contained higher levels of most nutrients compared to those found in G5, G1, and G6 (Tables 5 and 6). Conversely, G6 showed lower levels of some macronutrients (K, P, S, Mg, and Na), being the most malnourished GOAK genotype but without showing significant differences with respect to G1 to G5, which were nutritionally more balanced (Table 5). We found a significantly (*p*-value = 0.0004) higher nutrient concentration in G3 as compared to those in G5, G1, and G6. We ordered the GOAK genotypes as regards to their statistically significant nutrient levels, as follows: G3 > G1 > G5 > G6 Table 6). We did not find significant differences between the ROAK genotypes in regards to total macronutrient and micronutrient content (*p*-value = 0.0650). However, we observed significant differences between R4 and the other ROAK genotypes for P (*p*-value = 0.0021) and S (*p*-value = 0.0040), which resulted in sorting the ROAK genotypes based on significant P and S nutrient concentrations from R4 > R5 > R2 > R3 (Tables 5 and 6). Tipburn incidence and tipburn scores (Figure 6a,b) associated with the R2 and R3 genotypes containing lower P and S content (Table 5).

Table 2. Analysis at the cultivar level (CHD, GOAK, and ROAK) and by leaf type of the macronutrients studied in this trial (ppm). Percentages to the proportion of each nutrient in each type of leaf. Data are means of 12 replicates with different letters for each column indicating significant differences (p-value < 0.01) as determined by LSD multiple comparisons test in different types of leaves analyzed.

Leaf Type	CHD						GOAK						ROAK					
	K	Ca	P	S	Mg	Na	K	Ca	P	S	Mg	Na	K	Ca	P	S	Mg	Na
Mature	1349.00 a 67.52%	340.68 b 17.05%	92.26 a 4.62%	101.45 b 5.08%	65.91 a 3.30%	38.02 b 1.90%	3290.63 b 66.49%	794.50 b 16.05%	228.85 a 4.62%	309.97 b 6.26%	185.36 b 3.75%	109.80 b 2.22%	3204.74 b 69.85%	608.37 b 13.26%	272.64 a 5.94%	179.97 ab 3.92%	184.19 b 4.01%	115.74 c 2.52%
Intermediate	1400.84 a 72.72%	201.41 a 10.46%	137.12 b 7.12%	92.88 a 4.82%	56.78 a 2.95%	30.11 ab 1.56%	2228.40 a 72.06%	316.88 a 10.25%	208.25 a 6.73%	177.64 a 5.74%	105.75 a 3.42%	41.41 a 1.34%	2328.67 a 73.58%	243.09 a 7.68%	283.54 b 8.96%	139.78 a 4.42%	95.44 a 3.02%	64.21 b 2.03%
Juvenile	1657.43 a 73.72%	155.36 a 6.91%	218.00 c 9.70%	121.04 b 5.38%	64.44 a 2.87%	22.90 a 1.02%	2415.40 ab 72.69%	227.05 a 6.83%	326.71 b 9.83%	199.86 a 6.01%	111.22 a 3.35%	28.52 a 0.86%	2424.82 a 72.62%	141.51 a 4.24%	440.74 b 13.20%	192.42 b 5.76%	104.57 a 3.13%	23.21 a 0.69%
p-value	0.1711	0.0001	0.0000	0.1196	0.4857	0.0211	0.0609	0.0001	0.0159	0.0164	0.0133	0.0001	0.0373	0.0000	0.0012	0.0332	0.0000	0.0000
M/J ratio	0.81	2.19	0.42	0.84	1.02	1.66	1.36	3.50	0.70	1.55	1.67	3.85	1.32	4.30	0.62	1.02	1.76	4.99

Table 3. Analysis at the cultivar level (CHD, GOAK, and ROAK) and by leaf type of the micronutrients studied in this trial (ppm). Percentages to the proportion of each nutrient in each type of leaf. Data are means of 12 replicates with different letters for each column indicating significant differences (p-value < 0.01) as determined by LSD multiple comparisons test in different types of leaves analyzed.

Leaf Type	CHD				GOAK				ROAK			
	Fe	Mn	Zn	Cu	Fe	Mn	Zn	Cu	Fe	Mn	Zn	Cu
Mature	4.92 a 0.25%	3.90 b 0.20%	1.67 ab 0.08%	0.23 b 0.01%	16.72 b 0.34%	8.89 b 0.18%	3.35 b 0.07%	0.61 b 0.01%	13.57 b 0.30%	6.75 b 0.15%	1.54 a 0.03%	0.37 b 0.01%
Intermediate	4.12 a 0.21%	1.99 a 0.10%	1.17 a 0.06%	0.13 a 0.01%	8.37 a 0.27%	3.61 a 0.12%	1.67 a 0.05%	0.28 a 0.01%	6.01 a 0.19%	2.45 a 0.08%	1.28 a 0.04%	0.12 a 0.00%
Juvenile	5.23 a 0.23%	1.85 a 0.08%	1.93 b 0.09%	0.21 b 0.01%	8.58 a 0.26%	3.00 a 0.12%	2.35 ab 0.09%	0.30 a 0.01%	6.92 a 0.21%	2.04 a 0.06%	2.50 b 0.07%	0.19 a 0.01%
p-value	0.4398	0.0015	0.0505	0.0029	0.0096	0.0001	0.0431	0.0066	0.0002	0.0000	0.0006	0.0000
M/J ratio	0.94	2.11	0.87	1.09	1.95	2.96	1.43	2.03	1.96	3.31	0.62	1.95

Table 4. Global analysis using all nutrients (ppm) at the cultivar and genotype level. Data are means ± SEM of 9 replicates with different letters for each column (genotypes) or rows (cultivars) indicating significant differences (p-value < 0.01) as determined by LSD multiple comparisons test.

C1	7670.09 ± 634.81 b	G1	10,310.58 ± 338.97 a	R2	10,446.25 ± 804.42 a
C3	5222.15 ± 182.30 a	G3	15,708.86 ± 1099.30 b	R3	10,015.28 ± 1445.66 a
C7	6833.17 ± 294.16 b	G5	10,011.790 ± 990.28 a	R4	13,405.84 ± 1389.01 a
C8	4966.47 ± 392.13 a	G6	9424.43 ± 1482.28 a	R5	9924.48 ± 661.77 a
p-value	0.0004		0.0004		0.0650
Average	6172.97 ± 381.09 a		11,363.91 ± 886.52 b		11,091.40 ± 691.06 b

Table 5. Individual analysis of each macronutrient studied in this trial (ppm) at the genotype level for each cultivar. Percentages refer to the proportion of each nutrient in each type of genotype. Data are means of 9 replicates with different letters for each column indicating significant differences (p-value < 0.01) as determined by LSD multiple comparisons test in the four genotypes analyzed per cultivar (p-value).

CHD	K	Ca	P	S	Mg	Na	GOAK	K	Ca	P	S	Mg	Na	ROAK	K	Ca	P	S	Mg	Na
C1	5493.20 b 71.62%	918.54 b 11.98%	468.06 b 6.10%	390.30 b 5.09%	237.80 b 3.10%	128.51 c 1.68%	G1	7171.16 a 69.55%	1310.47 a 13.05%	655.84 b 6.53%	614.46 a 6.12%	339.65 a 3.38%	170.41 b 1.70%	R2	7649.10 a 73.22%	917.81 ab 8.79%	726.56 a 6.96%	559.32 b 5.35%	396.68 a 3.80%	145.85 a 1.40%
C3	3806.33 a 72.89%	542.29 a 10.38%	361.88 a 6.93%	268.90 a 5.15%	151.76 a 2.91%	69.21 ab 1.33%	G3	10721.92 b 68.25%	2184.11 b 13.90%	886.81 b 5.65%	944.94 b 6.01%	598.40 b 3.81%	287.79 b 1.83%	R3	7129.90 a 71.19%	990.55 ab 9.89%	883.95 b 8.83%	400.07 a 3.99%	377.61 a 3.77%	190.05 ab 1.90%
C7	4849.21 b 70.97%	755.94 b 11.06%	549.96 b 8.05%	355.06 b 5.20%	193.96 b 2.84%	99.52 bc 1.46%	G5	7094.33 a 70.86%	902.42 a 9.01%	858.71 b 8.58%	628.49 a 6.28%	339.99 a 3.40%	140.02 a 1.40%	R4	9485.96 a 70.76%	1253.07b 9.35%	1267.85 c 9.46%	622.57 b 4.64%	441.41 a 3.29%	289.24 b 2.16%
C8	3480.33 a 70.08%	573.01 a 11.54%	409.62 a 8.25%	247.21 a 4.98%	165.00 a 3.32%	66.90 a 1.35%	G6	6750.30 a 71.63%	956.47 a 10.15%	653.87 b 6.94%	561.99 a 5.96%	331.25 a 3.51%	120.72 a 1.28%	R5	7188.79 a 72.43%	784.60 a 7.91%	981.54 b 9.89%	445.11 a 4.48%	323.07 a 3.26%	164.03 a 1.65%
Total	4407.27	697.45	447.38	315.37	187.13	91.03	Total	7934.43	1338.42	763.81	687.47	402.32	179.73	Total	7958.22	992.97	996.92	512.18	384.20	203.16
p-value	0.0003	0.0047	0.0251	0.0026	0.007	0.0024	p-value	0.0012	0.0002	0.2007	0.0009	0.0001	0.0002	p-value	0.1335	0.0879	0.0021	0.004	0.1973	0.0152

Table 6. Individual analysis of each macronutrient studied in this trial (ppm) at the genotype level for each cultivar. Percentages refer to the proportion of each nutrient in each type of genotype. Data are means of 9 replicates with different letters for each column indicating significant differences (*p*-value < 0.01) as determined by LSD multiple comparisons test in the four genotypes analyzed per cultivar (*p*-value).

CHD	Fe	Mn	Zn	Cu	GOAK	Fe	Mn	Zn	Cu	ROAK	Fe	Mn	Zn	Cu
C1	15.43 a 0.20%	11.34 b 0.15%	6.15 b 0.08%	0.75 b 0.01%	G1	28.40 a 0.28%	13.17 a 0.13%	6.09 a 0.06%	0.92 a 0.01%	R2	31.74 a 0.30%	14.83 b 0.14%	3.83 a 0.04%	0.53 a 0.01%
C3	11.73 a 0.22%	5.54 a 0.10%	4.01 a 0.08%	0.50 a 0.01%	G3	49.52 b 0.32%	23.19 b 0.15%	10.37 b 0.07%	1.59 b 0.01%	R3	25.85 a 0.26%	11.85 ab 0.12%	4.81 a 0.05%	0.64 a 0.01%
C7	17.34 a 0.25%	7.34 a 0.11%	4.32 a 0.06%	0.52 a 0.01%	G5	27.97 a 0.28%	12.07 a 0.12%	6.67 a 0.07%	1.12 ab 0.01%	R4	27.05 a 0.20%	11.47 ab 0.09%	6.44 a 0.05%	0.79 a 0.01%
C8	12.57 a 0.25%	6.72 a 0.14%	4.59 a 0.09%	0.50 a 0.01%	G6	28.80 a 0.31%	13.61 a 0.14%	6.31 a 0.07%	1.12 ab 0.01%	R5	22.87 a 0.23%	8.24 a 0.08%	5.53 a 0.06%	0.69 a 0.01%
Total	14.27	7.74	4.77	0.57	Total	33.67	15.51	7.36	1.19	Total	26.50	11.25	5.32	0.68
p-value	0.051	0.0025	0.0235	0.0147	*p*-value	0.0016	0.0003	0.0026	0.0214	*p*-value	0.3675	0.1888	0.1744	0.2709

We wondered whether differences in nutrient levels between mature and juvenile leaves could account for the observed differences in tipburn scores and tipburn incidence in the GOAK and ROAK genotypes (Figure 6a,b). We found that G1 and R5, with the lowest tipburn scores, showed a mild decrease in K, Ca, Mg, Fe, and Mn concentrations between mature and juvenile leaves (Figure 7). The genotypes with the highest tipburn scores, G3 and R2, showed higher differences in K, Ca, Mg, Fe, and Mn concentrations between mature and juvenile leaves (Figure 7). A similar trend was found for S and Cu, although the R2 genotype showed higher S and Cu levels in juvenile leaves than the other three genotypes (Figure 7). Na concentration was also higher in mature leaves, and their levels were similarly reduced in juvenile leaves in the four genotypes (Figure 7). On the other hand, P and Zn showed the highest levels in juvenile leaves except for the G3 genotype (Figure 7).

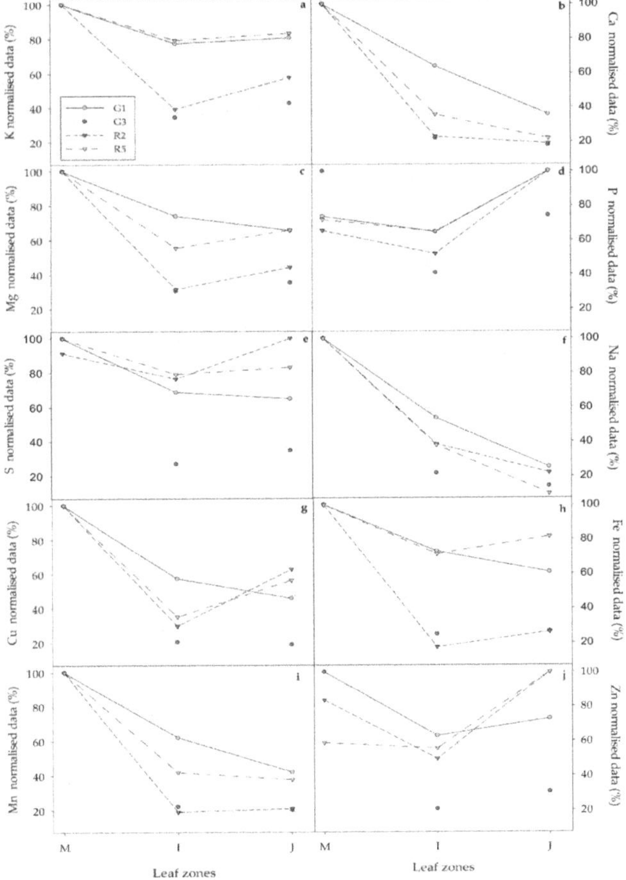

Figure 7. Nutrient analysis of the studied macronutrients in different leaves (M, mature; I, intermediate; J, juvenile). Only the most divergent genotypes for tipburn scoring and tipburn incidence for GOAK (G1 and G3) and ROAK (R2 and R5) cultivars are shown. The nutrients analyzed were (**a**) Potassium (K), (**b**) Calcium (Ca), (**c**) Magnesium (Mg), (**d**) Phosphorus (P), (**e**) Sulphur (S), (**f**) Sodium (Na), (**g**) Copper (Cu), (**h**) Iron (Fe), (**i**) Manganese (Mn) and (**j**) Zinc (Zn). Data are normalized as regards the leaf with the highest amount for a given nutrient.

4. Discussion

There is an increased demand for fresh, locally grown, and safe vegetables among the EU consumers [38]. However, intensive agricultural exploitation might lead to water shortage and soil salinization, among other environmental damages [39]. The greenhouse production of vegetables in closed hydroponic systems is a resource-efficient technique for the production of high-quality and high-yield crops [40]. Here, we devised a low-cost hydroponic system (i.e., floating rafts) for lettuce cultivation, which was used to evaluate the early growth and quality parameters of 12 genotypes from different lettuce cultivars (CHD, GOAK, and ROAK) in three growing seasons (Fall, Winter, and Spring). These genotypes were selected based on agronomically-relevant traits.

Research on the role of root system architecture (RSA) traits that could enhance nutrient and water use efficiency has not received broad attention in lettuce breeding programs until quite recently [6]. We found striking differences among the studied lettuce cultivars in regard to their root system (Figure 2a,c, Figure 3a,c, and Figure 4a,c). The CHD genotypes showed deeper roots as compared to those from GOAK and ROAK. As it is known that deeper roots are crucial for improving drought resistance in plants [41,42], CHD cultivars may be more drought tolerant than GOAK and ROAK genotypes, although this hypothesis could not be directly tested in our hydroponics system. On the other hand, GOAK and ROAK root systems were heavier and more superficial than those in the CHD cultivar. Indeed, GOAK and ROAK are oak-leaf cultivars located on the same genetic clade [43], which are mainly differentiated by their leaf anthocyanin content [1]. The differences in the RSA of the CHD and GOAK/ROAK cultivars may thus account for the genotype-dependent behavior of cultivated lettuce in saline soils [44] or in response to water and nutrient deficiency [6]. Our experimental setup will allow evaluating growth responses under different soil stresses through the adjustment of the nutrient solution and/or the experimental conditions (i.e., temperature, aeration, etc.). In addition, the contrasting RSA phenotypes of the G3 and R5 genotypes (Figures 3c and 4c), with a three-fold difference in their root fresh-weight during the Spring season (Figure 5), may be used for the identification of the genetic determinants involved in RSA variation in the oak-leaf lettuce clade through the implementation of QTL mapping. We estimated the shoot growth rates of lettuce in hydroponic culture through dedicated image analysis (Figure S2). Overall, shoot growth was much lower during Winter than in Spring or Fall, which is in agreement with previous studies where higher temperatures and high irradiance were found to be key factors, which affected growth product quality in these species [16,22]. However, we found an interesting G × E interaction for the estimated SA in some of the studied genotypes. On the one hand, the ROAK genotypes showed lower SA values in Spring than in Winter (Figure 4b). On the other hand, SA values in C7 were highly affected by the growing season, as higher growth occurred during Spring for this genotype (Figure 2b). However, the SA and RA values estimated from images were inaccurate descriptors of yield, as confirmed by the low correlations found between FW and DW values of the shoot system and the root system at the end of the experiment (Figure S6). We found that FW was highly correlated with DW (both for root or shoot) for all the studied cultivars and during the different growing seasons (Figure S6), and that their WC variation ranged from 92% to 98% (Table S4). Interestingly, we found that WC (either in roots or shoots) was negatively correlated with root DW but not with shoot DW, which suggest that thinner roots may be more efficient in water uptake in lettuce plants grown in hydroponics, as compared to those plants grown in soil where root diameter may be directly related to the ability to penetrate the drying soil [41]. In addition, the R:S ratio allowed us to identify genotypes with contrasting yield genotypes, such as C7 and R3 (Table S4). While C7 had the lowest R:S values (and hence higher yield) in Fall, R3 showed the highest R:S ratio (thus lower yield) during Spring, indicating a G × E interaction for this trait, as well.

In Spain and Italy, two of the fifth largest lettuce and chicory producers in the world [11], greenhouse lettuce production is often limited by the extent of tipburn and premature bolting. Tipburn is a physiological disorder characterized by necrotic lesions at the margins of the developing leaves, resulting from a localized Ca deficiency [45]. Tipburn development in lettuce depends on environmental factors that promote growth [46]. Ca translocation from the roots to the shoots occurs through the xylem due to transpiration, and Ca cannot be mobilized from older leaves to younger ones [47]. Some of the climatic factors that characterize the Mediterranean region, specifically high temperature, high radiation, and long photoperiod, lead to the rapid shoot growth of lettuce, which cannot match Ca translocation from the roots. The lettuce genotypes studied in this work were selected based on their contrasting tipburn incidence when grown in soil. A previous study using a small number of lettuce cultivars grown in hydroponics showed that tipburn was not observed in the late Winter season, whereas it was severe during Spring [48]. We found that tipburn incidence was higher during Spring but lower in Winter for most of the studied cultivars grown in hydroponics (Figure 6a). And we also observed that the CHD cultivars showed a higher variation for tipburn incidence as compared to the studied oak-leaf types (GOAK and ROAK). These results were consistent with the greater genetic variability for tipburn responses found in the CHD cultivar as a result of earlier breeding efforts for tipburn tolerance in this cultivar [19,22,24,25]. In a recent study [28], early bolting and tipburn behavior were studied on 18 genotypes from different lettuce cultivars grown in hydroponics at high temperature and extensive differences were also observed among them. Hence, the combined effect of high growth rates and high temperatures during Spring may lead to the reduced nutrient supply to the developing leaves, resulting in the observed enhancement of the tipburn severity during Spring. Only two of the studied genotypes, C3 and R2 showed severe tipburn symptoms in every season (Figure 6a,c). Another two genotypes, C8 and G3 showed intermediate-to-severe tipburn symptoms only during Spring (Figure 6a,c). On the other hand, C1, G1, and R5 showed tipburn tolerance when grown in hydroponics (Figure 6a,c). These results perfectly matched the tipburn severity symptoms found in the studied genotypes when grown in soil (V.B., unpublished), which validates our experimental setup for the fast and high-throughput evaluation of tipburn responses in lettuce germplasm collections grown in hydroponics.

To investigate the nutritional causes of tipburn incidence during Spring in the studied genotypes, we measured the levels of several macro and micronutrients in leaves of different ages at the end of the experiment (45 dap; Table S6). Ca and Na levels showed the highest M/J ratio, irrespectively of the cultivar type (Table 2), which is consistent with the low Ca mobilization from mature tissues [45] and the higher Na accumulation in older leaves [49]. On the other hand, P displayed the lowest M/J values within the studied macronutrients (Table 2), with lower P levels in the CHD than in the GOAK/ROAK genotypes. These latter results could be explained by the differences in RSA between the studied cultivars, as root responses to low phosphate favor the exploration of the shallower part of the soil, where phosphate tends to be more abundant [50]. We noticed that the CHD genotypes contained a lower nutrient concentration than the GOAK and ROAK ones (Table 4). Nutritional differences between lettuce cultivars have been described previously [51]. The tipburn incidence and tipburn scores of the CHD genotypes perfectly matched their total nutrient content, hence, the genotypes with the lowest nutrient levels (C3 and C8) showed severe tipburn symptoms (Figure 6a,b and Tables 5 and 6). Our results suggest that tipburn in the studied CHD genotypes may be related to some nutrient imbalance, as has been proposed earlier in lettuce [52]. The high growth rates observed during Spring for C3 and C8, combined with their contrasting R:S ratios, may result in decreased Ca concentrations in leaves and thus increased tipburn, as has been previously reported in other lettuce genotypes [21,45]. Because of the restricted Ca transport within the head-enclosed leaves of the CHD genotypes, Ca levels are much lower in the leaf margins, where tipburn symptoms arise early; high K levels in this region might also contribute to enhanced tipburn in CHD genotypes, as suggested previously [45]. Overall, the studied ROAK and GOAK genotypes were less sensitive to tipburn, which was consistent with previous results which suggested a narrow genetic variation for this trait in oak-leaf type cultivars [19]. R2 and G3, which displayed severe tipburn during the Spring season, were characterized by a strong decrease in K levels between mature and juvenile leaves, as compared with the tipburn tolerant G1 and R5 genotypes (Figure 7). To assess whether tipburn in the studied lettuce genotypes is caused by an altered Ca/K homeostasis, we plan to evaluate tipburn susceptibility using an in vitro evaluation system [24], with some modifications.

We also found striking differences in regard to the studied micronutrients (Fe, Mn, Zn, and Cu) depending on cultivar type and genotypes (Tables 3 and 6). Alterations in micronutrient homeostasis (such as Fe and Mn) have commonly been associated with the appearance of shoot tip necrosis during pistachio in vitro culture [53], which very much resembles the tipburn symptoms found in the ROAK and GOAK genotypes. Fe and Mn levels in leaves, as well as their M/J ratios, were much higher in the GOAK and ROAK genotypes than in the CHD ones (Table 3). Consistent with previous results on K levels, Fe and Mn levels strongly decreased in the R2 and G3 genotypes from mature to juvenile leaves (Figure 7), suggesting that a nutritional unbalance of some micronutrients (Fe and Mn) could explain tipburn in oak-leaf susceptible genotypes. Further experiments with additional ROAK and GOAK genotypes will allow us to confirm this hypothesis.

5. Conclusions

We devised a multi-factorial approach for the study of several growth and quality traits of lettuce (*Lactuca sativa* L.) using a low-cost and high-throughput scalable hydroponic system. By analyzing tipburn incidence and leaf nutrient content, we were able to identify a number of nutrient traits that were highly correlated with cultivar- and genotype-dependent tipburn, suggesting that tipburn is a complex trait in this species. Indeed, the genetic dissection of tipburn resistance in lettuce has recently gained from a detailed study using seven RIL populations in multiple environments and years that allowed the identification of two major QTL affecting this trait [26]. The forthcoming availability of linked molecular markers will allow the evaluation of our germplasm collection.

Supplementary Materials: The following are available online at https://www.mdpi.com/2073-4395/11/4/616/s1, Figure S1: A representation of the scale used for tipburn severity assessment; Figure S2: The growth rate of the studied cultivars between 0 and 35 dap; Figure S3: Representative images of the studied CHD genotypes; Figure S4: Representative images of the studied GOAK genotypes; Figure S5: Representative images of the studied ROAK genotypes; Figure S6: Growth quantification of the studied genotypes at 45 dap; Table S1: Details of the experimental design used; Table S2: Nutrient solution composition; Table S3:, Raw data of root and shoot area; Table S4: raw data of root and shoot weights; Table S5, Raw data of tipburn phenotype assessment, Table S6: Raw data of the nutrient analysis.

Author Contributions: Conceptualization, V.B., J.R.A.-M., and J.M.P.-P.; data curation, V.B. and J.R.A.-M.; funding acquisition, J.M.P.-P.; investigation, V.B.; methodology, V.B. and J.R.A.-M.; supervision, J.M.P.-P.; validation, V.B., J.R.A.-M., and J.M.P.-P.; writing—original draft, J.R.A.-M. and J.M.P.-P.; writing—review and editing, V.B., J.R.A.-M., and J.M.P.-P. All authors have read and agreed to the published version of the manuscript.

Funding: This research received no external funding.

Institutional Review Board Statement: Not applicable.

Informed Consent Statement: Not applicable.

Data Availability Statement: The data presented in this study are available online at www.mdpi.com/xxx/s1.

Acknowledgments: We are grateful to María José Ñíguez Gómez (Universidad Miguel Hernández) for her expert technical assistance, Jorge Benítez Vega (Bayer CropScience) for his support in the experimental designs and genotype information, José Antonio Hernández for his help with nutrient analyses at the CEBAS-CSIC Ionomics laboratory and Mario Fon for his help with English editing.

Conflicts of Interest: The authors declare no conflict of interest.

References

1. Hayes, R.J.; Simko, I. Breeding lettuce for improved fresh-cut processing. *Acta Hortic.* **2016**, 65–76. [CrossRef]
2. Patella, A.; Palumbo, F.; Galla, G.; Barcaccia, G. The Molecular Determination of Hybridity and Homozygosity Estimates in Breeding Populations of Lettuce (*Lactuca sativa* L.). *Genes* **2019**, *10*, 916. [CrossRef]
3. Truco, M.J.; Antonise, R.; Lavelle, D.; Ochoa, O.; Kozik, A.; Witsenboer, H.; Fort, S.B.; Jeuken, M.J.W.; Kesseli, R.V.; Lindhout, P.; et al. A high-density, integrated genetic linkage map of lettuce (*Lactuca* spp.). *Theor. Appl. Genet.* **2007**, *115*, 735–746. [CrossRef]
4. Truco, M.J.; Ashrafi, H.; Kozik, A.; van Leeuwen, H.; Bowers, J.; Wo, S.R.C.; Stoffel, K.; Xu, H.; Hill, T.; Van Deynze, A.; et al. An Ultra-High-Density, Transcript-Based, Genetic Map of Lettuce. *G3 Genes Genomes Genet.* **2013**, *3*, 617–631. [CrossRef]
5. Rauscher, G.; Simko, I. Development of genomic SSR markers for fingerprinting lettuce (*Lactuca sativa* L.) cultivars and mapping genes. *BMC Plant Biol.* **2013**, *13*, 11. [CrossRef]
6. Kerbiriou, P.J.; Maliepaard, C.A.; Stomph, T.J.; Koper, M.; Froissart, D.; Roobeek, I.; Lammerts Van Bueren, E.T.; Struik, P.C. Genetic Control of Water and Nitrate Capture and Their Use Efficiency in Lettuce (*Lactuca sativa* L.). *Front. Plant Sci.* **2016**, *7*, 343. [CrossRef]
7. Yoong, F.-Y.; O'Brien, L.K.; Truco, M.J.; Huo, H.; Sideman, R.; Hayes, R.; Michelmore, R.W.; Bradford, K.J. Genetic Variation for Thermotolerance in Lettuce Seed Germination Is Associated with Temperature-Sensitive Regulation of ethylene response factor1 (ERF1). *Plant Physiol.* **2016**, *170*, 472–488. [CrossRef]
8. Damerum, A.; Selmes, S.L.; Biggi, G.F.; Clarkson, G.J.; Rothwell, S.D.; Truco, M.J.; Michelmore, R.W.; Hancock, R.D.; Shellcock, C.; Chapman, M.A.; et al. Elucidating the genetic basis of antioxidant status in lettuce (*Lactuca sativa*). *Hortic. Res.* **2015**, *2*, 15055. [CrossRef]
9. Simko, I.; Hayes, R.J.; Mou, B.; McCreight, J.D. Lettuce and Spinach. In *CSSA Special Publications*; Smith, S., Diers, B., Specht, J., Carver, B., Eds.; American Society of Agronomy and Soil Science Society of America: Madison, WI, USA, 2015; pp. 53–85. ISBN 978-0-89118-619-9. [CrossRef]
10. Atkinson, L.D.; McHale, L.K.; Truco, M.J.; Hilton, H.W.; Lynn, J.; Schut, J.W.; Michelmore, R.W.; Hand, P.; Pink, D.A.C. An intra-specific linkage map of lettuce (*Lactuca sativa*) and genetic analysis of postharvest discolouration traits. *Theor. Appl. Genet.* **2013**, *126*, 2737–2752. [CrossRef]
11. Food and Agriculture Organization of the United Nations. *FAOSTAT Statistical Database*; FAO: Rome, Italy, 2019; Available online: http://www.fao.org/faostat/es/#data/QC/metadata (accessed on 15 January 2021).
12. Romero-Gámez, M.; Audsley, E.; Suárez-Rey, E.M. Life cycle assessment of cultivating lettuce and escarole in Spain. *J. Clean. Prod.* **2014**, *73*, 193–203. [CrossRef]
13. Guaita-García, N.; Martínez-Fernández, J.; Barrera-Causil, C.J.; Esteve-Selma, M.Á.; Fitz, H.C. Local perceptions regarding a social–ecological system of the mediterranean coast: The Mar Menor (Región de Murcia, Spain). *Environ. Dev. Sustain.* **2020**, *23*, 2882–2909. [CrossRef]
14. Coronel, G.; Chang, M.; Rodríguez-Delfín, A. Nitrate reductase activity and chlorophyll content in lettuce plants grown hydroponically and organically. *Acta Hortic.* **2009**, 137–144. [CrossRef]
15. Kotsiras, A.; Vlachodimitropoulou, A.; Gerakaris, A.; Bakas, N.; Darras, A.I. Innovative harvest practices of Butterhead, Lollo rosso and Batavia green lettuce (*Lactuca sativa* L.) types grown in floating hydroponic system to maintain the quality and improve storability. *Sci. Hortic.* **2016**, *201*, 1–9. [CrossRef]
16. Petropoulos, S.A.; Chatzieustratiou, E.; Constantopoulou, E.; Kapotis, G. Yield and Quality of Lettuce and Rocket Grown in Floating Culture System. *Not. Bot. Hortic. Agrobot.* **2016**, *44*, 603–612. [CrossRef]
17. Saure, M.C. Causes of the tipburn disorder in leaves of vegetables. *Sci. Hortic.* **1998**, *76*, 131–147. [CrossRef]
18. Frantz, J.M.; Ritchie, G.; Cometti, N.N.; Robinson, J.; Bugbee, B. Exploring the Limits of Crop Productivity: Beyond the Limits of Tipburn in Lettuce. *J. Am. Soc. Hortic. Sci.* **2004**, *129*, 331–338. [CrossRef]
19. Jenni, S.; Hayes, R.J. Genetic variation, genotype × environment interaction, and selection for tipburn resistance in lettuce in multi-environments. *Euphytica* **2010**, *171*, 427–439. [CrossRef]
20. Lee, J.G.; Choi, C.S.; Jang, Y.A.; Jang, S.W.; Lee, S.G.; Um, Y.C. Effects of air temperature and air flow rate control on the tipburn occurrence of leaf lettuce in a closed-type plant factory system. *Hortic. Environ. Biotechnol.* **2013**, *54*, 303–310. [CrossRef]
21. Sago, Y. Effects of Light Intensity and Growth Rate on Tipburn Development and Leaf Calcium Concentration in Butterhead Lettuce. *HortScience* **2016**, *51*, 1087–1091. [CrossRef]
22. Jenni, S.; Yan, W. Genotype by environment interactions of heat stress disorder resistance in crisphead lettuce. *Plant Breed.* **2009**, *128*, 374–380. [CrossRef]
23. Ryder, E.J.; Waycott, W. Crisphead lettuce resistant to tipburn: Cultivar tiber and eight breeding lines. *HortScience* **1998**, *33*, 903–904. [CrossRef]
24. Koyama, R.; Sanada, M.; Itoh, H.; Kanechi, M.; Inagaki, N.; Uno, Y. In vitro evaluation of tipburn resistance in lettuce (*Lactuca sativa*. L). *Plant Cell Tissue Organ. Cult.* **2012**, *108*, 221–227. [CrossRef]
25. Jenni, S.; Truco, M.J.; Michelmore, R.W. Quantitative trait loci associated with tipburn, heat stress-induced physiological disorders, and maturity traits in crisphead lettuce. *Theor. Appl. Genet.* **2013**, *126*, 3065–3079. [CrossRef]
26. Macias-González, M.; Truco, M.J.; Bertier, L.D.; Jenni, S.; Simko, I.; Hayes, R.J.; Michelmore, R.W. Genetic architecture of tipburn resistance in lettuce. *Theor. Appl. Genet.* **2019**, *132*, 2209–2222. [CrossRef]

27. Carassay, L.R.; Bustos, D.A.; Golberg, A.D.; Taleisnik, E. Tipburn in salt-affected lettuce (*Lactuca sativa* L.) plants results from local oxidative stress. *J. Plant Physiol.* **2012**, *169*, 285–293. [CrossRef]
28. Holmes, S.C.; Wells, D.E.; Pickens, J.M.; Kemble, J.M. Selection of Heat Tolerant Lettuce (*Lactuca sativa* L.) Cultivars Grown in Deep Water Culture and Their Marketability. *Horticulturae* **2019**, *5*, 50. [CrossRef]
29. Incrocci, L.; Moolhuyzen, M.; Guzmán, M. *Nutrient Solution Calculator ES*; IFAPA: Almería, Spain, 2020. [CrossRef]
30. Berry, W.L.; Knight, S. *Plant Culture in Hydroponics*; Iowa State University: Ames, IA, USA, 1997; pp. 119–131.
31. Galkovskyi, T.; Mileyko, Y.; Bucksch, A.; Moore, B.; Symonova, O.; Price, C.A.; Topp, C.N.; IyerPascuzzi, A.S.; Zurek, P.R.; Fang, S.; et al. GiA Roots: Software for the high throughput analysis of plant root system architecture. *BMC Plant Biol.* **2012**, *12*, 116. [CrossRef]
32. Birlanga, V.; Villanova, J.; Cano, A.; Cano, E.A.; Acosta, M. Quantitative Analysis of Adventitious Root Growth Phenotypes in Carnation Stem Cuttings. *PLoS ONE* **2015**, *10*, e0133123. [CrossRef]
33. Schneider, C.A.; Rasband, W.S.; Eliceiri, K.W. NIH Image to ImageJ: 25 years of image analysis. *Nat. Methods* **2012**, *9*, 671–675. [CrossRef] [PubMed]
34. Acosta-Motos, J.R.; Hernández, J.A.; Álvarez, S.; BarbaEspín, G.; SánchezBlanco, M.J. The long term resistance mechanisms, critical irrigation threshold and relief capacity shown by Eugenia myrtifolia plants in response to saline reclaimed water. *Plant Physiol. Biochem.* **2017**, *111*, 244–256. [CrossRef] [PubMed]
35. Massey, F. The Kolmogorov Smirnov test for goodness of fit. *J. Am. Stat. Assoc.* **1951**, *46*, 253. [CrossRef]
36. Fisher, R. The correlation between relatives on the supposition of Mendelian inheritance. *Trans. R. Soc. Edinb.* **1918**, *52*, 399–433. [CrossRef]
37. The R Project for Statistical Computing. Available online: https://www.r-project.org/ (accessed on 2 February 2021).
38. Baselice, A.; Colantuoni, F.; Lass, D.A.; Nardone, G.; Stasi, A. Trends in EU consumers' attitude towards fresh-cut fruit and vegetables. *Food Qual. Prefer.* **2017**, *59*, 87–96. [CrossRef]
39. Gomiero, T. Soil Degradation, Land Scarcity and Food Security: Reviewing a Complex Challenge. *Sustainability* **2016**, *8*, 281. [CrossRef]
40. Hosseinzadeh, S.; Bonarrigo, G.; Verheust, Y.; Roccaro, P.; Van Hulle, S. Water reuse in closed hydroponic systems: Comparison of GAC adsorption, ion exchange and ozonation processes to treat recycled nutrient solution. *Aquac. Eng.* **2017**, *78*, 190–195. [CrossRef]
41. Lynch, J.P.; Chimungu, J.G.; Brown, K.M. Root anatomical phenes associated with water acquisition from drying soil: Targets for crop improvement. *J. Exp. Bot.* **2014**, *65*, 6155–6166. [CrossRef] [PubMed]
42. Acosta-Motos, J.R.; Rothwell, S.A.; Massam, M.J.; Albacete, A.; Zhang, H.; Dodd, I.C. Alternate wetting and drying irrigation increases water and phosphorus use efficiency independent of substrate phosphorus status of vegetative rice plants. *Plant Physiol. Biochem.* **2020**, *155*, 914–926. [CrossRef] [PubMed]
43. Simko, I. Development of EST-SSR Markers for the Study of Population Structure in Lettuce (*Lactuca sativa* L.). *J. Hered.* **2009**, *100*, 256–262. [CrossRef] [PubMed]
44. Xu, C.; Mou, B. Evaluation of Lettuce Genotypes for Salinity Tolerance. *HortScience* **2015**, *50*, 1441–1446. [CrossRef]
45. Barta, D.J.; Tibbitts, T.W. Calcium Localization and Tipburn Development in Lettuce Leaves during Early Enlargement. *J. Am. Soc. Hortic. Sci.* **2000**, *125*, 294–298. [CrossRef] [PubMed]
46. Wissemeier, A.H.; Zühlke, G. Relation between climatic variables, growth and the incidence of tipburn in field-grown lettuce as evaluated by simple, partial and multiple regression analysis. *Sci. Hortic.* **2002**, *93*, 193–204. [CrossRef]
47. Gilliham, M.; Dayod, M.; Hocking, B.J.; Xu, B.; Conn, S.J.; Kaiser, B.N.; Leigh, R.A.; Tyerman, S.D. Calcium delivery and storage in plant leaves: Exploring the link with water flow. *J. Exp. Bot.* **2011**, *62*, 2233–2250. [CrossRef]
48. Assimakopoulou, A.; Kotsiras, A.; Nifakos, K. Incidence of lettuce tipburn as related to hydroponic system and cultivaR. *J. Plant Nutr.* **2013**, *36*, 1383–1400. [CrossRef]
49. Almeida, D.M.; Oliveira, M.M.; Saibo, N.J.M. Regulation of Na+ and K+ homeostasis in plants: Towards improved salt stress tolerance in crop plants. *Genet. Mol. Biol.* **2017**, *40*, 326–345. [CrossRef]
50. Péret, B.; Desnos, T.; Jost, R.; Kanno, S.; Berkowitz, O.; Nussaume, L. Root Architecture Responses: In Search of Phosphate. *Plant Physiol.* **2014**, *166*, 1713–1723. [CrossRef]
51. Mou, B. Genetic Variation of Beta-carotene and Lutein Contents in Lettuce. *J. Am. Soc. Hortic. Sci.* **2005**, *130*, 870–876. [CrossRef]
52. Ashkar, S.A.; Ries, S.K. Lettuce tipburn as related to nutrient imbalance and nitrogen composition. *J. Am. Soc. Hortic. Sci.* **1971**, *96*, 448–452.
53. Nezami-Alanagh, E.; Garoosi, G.-A.; Landín, M.; Gallego, P.P. Computer-based tools provide new insight into the key factors that cause physiological disorders of pistachio rootstocks cultured in vitro. *Sci. Rep.* **2019**, *9*, 9740. [CrossRef] [PubMed]

Article

Genetic Diversity of Soybeans (*Glycine max* (L.) Merr.) with Black Seed Coats and Green Cotyledons in Korean Germplasm

Hyun Jo [1,†], Ji Yun Lee [2,†], Hyeontae Cho [1], Hong Jib Choi [2], Chang Ki Son [2], Jeong Suk Bae [2], Kristin Bilyeu [3], Jong Tae Song [1] and Jeong-Dong Lee [1,*]

1. Department of Applied Biosciences, Kyungpook National University, Daegu 41566, Korea; johyun@knu.ac.kr (H.J.); best94@knu.ac.kr (H.C.); jtsong68@knu.ac.kr (J.T.S.)
2. Gyeongsangbuk-do Provincial Agricultural Research & Extension Service, Daegu 41404, Korea; dock0409@korea.kr (J.Y.L.); chj1217@korea.kr (H.J.C.); sck3058@korea.kr (C.K.S.); jsbae24@korea.kr (J.S.B.)
3. USDA-ARS, Plant Genetics Research Unit, 110 Waters Hall, University of Missouri-Columbia, Columbia, MO 65211, USA; kristin.bilyeu@usda.gov
* Correspondence: jdlee@knu.ac.kr
† H. Jo and J.Y. Lee contributed equally for this work.

Abstract: Soybeans (*Glycine max* (L.) Merr.) with black seed coats and green cotyledons are rich in anthocyanins and chlorophylls known as functional nutrients, antioxidants and compounds with anticarcinogenic properties. Understanding the genetic diversity of germplasm is important to determine effective strategies for improving the economic traits of these soybeans. We aimed to analyze the genetic diversity of 470 soybean accessions by 6K single nucleotide polymorphic loci to determine genetic architecture of the soybeans with black seed coats and green cotyledons. We found soybeans with black seed coats and green cotyledons showed narrow genetic variability in South Korea. The genotypic frequency of the *d1d2* and *psbM* variants for green cotyledon indicated that soybean collections from Korea were intermingled with soybean accessions from Japan and China. Regarding the chlorophyll content, the nuclear gene variant pair *d1d2* produced significantly higher chlorophyll *a* content than that of chloroplast genome *psbM* variants. Among the soybean accessions in this study, flower color plays an important role in the anthocyanin composition of seed coats. We provide 36 accessions as a core collection representing 99.5% of the genetic diversity from the total accessions used in this study to show potential as useful breeding materials for cultivars with black seed coats and green cotyledons.

Keywords: genetic diversity; black soybean; green cotyledon; anthocyanin; chlorophyll

Citation: Jo, H.; Lee, J.Y.; Cho, H.; Choi, H.J.; Son, C.K.; Bae, J.S.; Bilyeu, K.; Song, J.T.; Lee, J.-D. Genetic Diversity of Soybeans (*Glycine max* (L.) Merr.) with Black Seed Coats and Green Cotyledons in Korean Germplasm. *Agronomy* **2021**, *11*, 581. https://doi.org/10.3390/agronomy11030581

Academic Editors: Gregorio Barba-Espín and Jose Ramon Acosta-Motos

Received: 23 February 2021
Accepted: 17 March 2021
Published: 19 March 2021

Publisher's Note: MDPI stays neutral with regard to jurisdictional claims in published maps and institutional affiliations.

1. Introduction

The composition of soybean (*Glycine max* [L.] Merr.) seeds is 40% protein, 20% oil, and 15% soluble carbohydrates, making it one of the most economically important crops in the world. Soybean production in western countries primarily focuses on producing high-protein meals for livestock and vegetable oils, whereas in Asian countries soybeans have traditionally been used as a staple food and consumed as soymilk, tofu, soy sprouts, fermented soy foods, and soy sauce [1,2]. Soybeans with black seed coats have been attracting interest as a soybean food [3]. Soybeans with black seed coats can be classified into two groups based on their cotyledon colors, which are either green or yellow. Soybean with black seed coats and green cotyledons (BLG) have been used as traditional ingredients in medicinal treatments in China, Japan, and Korea, unlike yellow commodity soybeans [4]. Several studies have reported that daily consumption of black soybeans is associated with a reduced risk of breast cancer and cardiovascular diseases due to their content of potentially active phytochemicals, such as isoflavones, sterols, phytic acid, saponins, and anthocyanins [5–8]. In addition, BLG soybeans are preferred by consumers for health benefits, and are often cooked with rice and other side dishes in Korea.

Soybeans with black seed coats do not differ from the yellow commodity soybeans in seed composition and plant morphology, and the black seed coat is caused by the accumulation of flavonoids and anthocyanins in the epidermal layer of the soybean seed coat [9–11]. Anthocyanins are important functional nutrients and antioxidants that benefit human health by having a positive effect on obesity, diabetes, and cardiovascular diseases [12,13]. Black seed coats are controlled by multiple classic genetic loci, such as *I*, *T*, *W1*, *R*, and *O*. The epistatic interaction of those loci contributes to soybean seed coloration. The black seed coat is controlled by the expression level of the chalcone synthase gene, which plays an important role in the anthocyanin biosynthesis pathway [14,15]. In the seed coat of black soybeans, there are eight anthocyanins: delphinidin 3-galactoside, delphinidin 3-O-glucoside, cyanidin 3-O-galactoside, cyanidin 3-O-glucoside, petunidin 3-O-glucoside, pelargonidin 3-O-glucoside, peonidin 3-O-glucoside, and cyanidin chloride [16]. Among them, three anthocyanins, cyanidin 3-O-glucoside, delphinidin 3-O-glucoside, and petunidin 3-O-glucoside are the primary anthocyanins in black seed coat [17].

In pea genetics, green cotyledon was one of earliest traits studied by Gregor Mendel [18]. The dominant allele at the *I* locus determined the yellow cotyledon color in the pea, whereas the *i* mutant controlled the green cotyledon color during seed maturation and maintained the green color of leaves during senescence compared with a wild type [19,20]. Delayed leaf senescence in the pea *i* mutant is known as the "stay-green" phenomenon, and it is caused by the impairment of chlorophyll degradation. Studies confirmed that Mendel's *I* locus encoded *SGR* in pea, which encodes a Mg-dechelatase for catalyzing the first step of the chlorophyll *a* (Chl *a*) degradation pathway [21,22]. Chlorophyll plays the primary role in harvesting the light energy during photosynthesis [23]. There are two kind of chlorophylls, Chl *a* and chlorophyll *b* (Chl *b*). Chl *a* and Chl *b* are present in light-harvesting complexes I (LHCI) and II (LHCII), including photosystem I and photosystem II. The senescence process in the plant is caused by chlorophyll degradation and a breakdown of LHCI and LHCII, which results in delayed leaf yellowing and is known as the stay-green phenotype [24–26]. Degradation of chlorophyll is associated with delayed green leaves during senescence, as well as the green cotyledon in plants.

Due to its advantage regarding the ease and simplicity of observation, inheritance studies on the morphological colors of soybean seed have been conducted since the early 1900s to evaluate seed coat and cotyledon colors [27,28]. Terao [28] reported that in soybeans there are two kinds of inheritance for the green cotyledon phenotype, such as the nuclear inheritance and cytoplasmic inheritance. Genes for the green cotyledon strain were recently cloned for both inheritances. First, *D1* (*Glyma.01g214600* in W82.a2.v1 assembly) and *D2* (*Glyma.11g027400*) are two paralogous nuclear genes in soybeans and are homologs of *SGR* genes from other plant species [29]. Double variants of the *D1* and *D2* gene result in the stay-green phenotype, including delayed yellowing of leaves during senescence, with green seed cotyledons [29,30]. Second, Terao [28] reported the maternal inheritance for the green cotyledon phenotype in soybeans, which are present in the chloroplast genome [30]. The 5-bp insertion in the soybean chloroplast genome results in a frameshift in *PsbM*, which encodes one of the small subunits of photosystem II [31]. The genotyping of *D1*, *D2*, and *PsbM* from 212 soybeans with green cotyledons revealed that all lines carry either *d1d2* or *psbM* with the known mutations, which suggests that naturally occurring *d1*, *d2*, and *psbM* mutations are rare [31].

Diverse germplasm accessions increase genetic diversity in soybean breeding programs and preserve the rare alleles that contribute to unique germplasm collections. Understanding the genetic diversity of germplasm sets is important for determining effective strategies that improve economic traits for crop development. In soybeans, studies to assess genetic diversity have been conducted on accessions from three major gene pools in China, Korea, and Japan by utilizing molecular markers [32–37].

Frankel et al. [38] first proposed the construction of a germplasm core subset, which is representative of the entire population, by maximizing the genetic variation and minimizing repetitiveness. Utilization of the core subset increased the efficiency to overcome the size, cost, and labor issues during evaluating an entire population by focusing on a limited but representative subset instead. Therefore, core subsets have been constructed for various crop species based on morphological and phenotypic observations [39–47]. Because quantitative phenotypic traits are often affected by environmental factors, a core subset that covers the genetic diversity of the entire population cannot be perfectly constructed by phenotype [48]. Molecular markers directly reflect genetic diversity rather than phenotypic assessments [49].

Due to increasing consumer awareness regarding the BLG soybean, it has become a preferred food ingredient soybean in South Korea. However, little information on the genetic diversity of BLG accessions has been obtained thus far. Korea is one of the centers of origin for domesticated soybeans, with a long history of cultivation [50]. There are ~17,000 improved and landrace cultivars (*Glycine max*) in the National Agrobiodiversity Center of Rural Development Administration in Korea [51]. Among them, we used 405 randomly selected collections in this study. The objective of this study was to analyze the genetic diversity of BLG accessions by 6K single nucleotide polymorphism (SNP) loci to understand their genetic architecture. In addition, a core subset of accessions was selected to represent the BLG accessions.

2. Materials and Methods

2.1. Growth Conditions of BLG Germplasms

To understand the genetic diversity, a collection of 405 BLG accessions was distributed from the National Agrobiodiversity Center in Jeonju, Republic of Korea. The collection comprised 385 landraces, eight breeding lines, five unknown, and seven plant inductions from the U.S. National Genetic Resources Program (Table S1). Among the 405 BLG accessions, 397 were collected from Korea, two from Japan, one from China, one from the U.S., and four were unknown. However, 47 accessions from the National Agrobiodiversity Center showed different seed characteristics, plant appearance, flowering and maturity. Therefore, we conducted a pure line selection and added 47 accessions as independent individuals (Table S2). Fifteen additional BLG accessions were locally collected from Gyeongsanbuk-do, Republic of Korea. Two BLG cultivars with green cotyledons, Cheongja [52], Cheongja 3 [53], and soybean cultivar with the yellow seed coat and yellow cotyledon, Uram [53], were used as check cultivars for this study. The 470 accessions including three check cultivars formed the total population and were used for further analyses. The entire set of 470 accessions was grown at Gyeongsanbuk-do Agricultural Research and Extension, Daegu, Republic of Korea over three years and planting dates (14 June 2013, 29 May 2014, and 15 June 2015). Each soybean accession was planted in single rows that were 1.5 m long and spaced 80 cm apart, with two replications. Seeds were planted by hand on hills in rows spaced 15 cm apart and thinned to a final stand of two seedlings per hill. Each plot grown from each year was harvested in bulk at the plant's full maturity (R8 stage) [54] for further seed analysis.

2.2. DNA Extraction and Determination of Genotyping for Soybean Accessions

Young trifoliate leaves were collected from soybean accessions with three check cultivars in the summer of 2015. Before DNA extraction, the leaves of each line were frozen in liquid nitrogen and ground into a fine powder. Next, 20 mg of leaf tissue from each sample was placed into tubes, and each DNA sample was isolated using the cetyltrimethylammonium bromide method with a minor modification [55]. Quantification and qualification of the genomic DNA of each accession was determined by electrophoresis running on 1.5% agarose gel. Next, 30 µL of genomic DNA at 100 ng/µL from 470 accessions, including three check cultivars, Cheongja, Cheongja 3, and Uram, were sent to the National Instrumentation Center for Environmental Management (NICEM; Seoul, Korea) at Seoul National University to genotype the soybean accessions using BARCsoySNP6K BeadChip [56]. The NICEM staff performed the assay procedures encompassing a series of approaches, such as incubation, DNA amplification, preparation of the bead assay, hybridization of samples of the bead assay, extension, staining of the samples, and imaging of the bead assay [57]. The SNP alleles were called using the Genome Studio Genotyping Module (Illumina, Inc. San Diego, CA, USA) [57]. Total 4459 SNPs were used for further analyses after filtering through the TASSEL software to exclude those with >20% missing data and rare SNPs.

2.3. Basic Population Genetic Parameters, Population Structure, and Construction of a Core Subset Accession

Minor allele frequency, genetic diversity index, polymorphism information content (PIC), and heterozygosity were evaluated using 469 BLG accessions with Uram and 4459 SNP markers using PowerMarker 3.25 software [58]. Principal component analysis (PCA) was conducted with 470 soybean accessions using the R package SNPRelate tool. This plot showed the first principal component (PC1) against the second principal component (PC2). For the phylogenetic tree in the present study, an unweighted pair group method with arithmetic mean (UPGMA) tree was constructed with entire accessions using the calculation of the distance based on a modified Euclidean distance between each pair of accessions by TASSEL [59]. The admixture model was used to analyze the genotypic data using STRUCTURE software [60], which is one of the most widely used genotypic clustering software. Three runs of STRUCTURE were executed for each number of the population (*K*) from 1 to 10. The burn-in time and replication number were set to 100,000 in each run. For the construction of a core collection, the 4,459 SNP genotypic data of the total BLG accessions was analyzed using GenoCore [61]. In this study, 99.5% of the coverage and 0.001% of the delta value were applied to select accessions for a core subset representing the entire collection.

2.4. Genotyping Assays for D1, D2, and PsbM

To genotype *D1* for the presence of the 1-bp deletion [29], a polymerase chain reaction (PCR) was conducted using a forward primer 5′-CGTTGTTGGGTTTGTCTGATGG-3′, and two reverse primers, 5′-GCGGGCTCGTCCACTCCTAAGAATAAAACC-3′ and 5′-GCGGGCAGGGCGGCTCGTCCACTCCTAAGAATAAACC-3′. The total volume of the PCR reaction was 20 µL, containing 1× real-time PCR smart mixtures (2.5 U/µL h-Taq DNA polymerase, 1× h-Taq reaction buffer, and 200 µM dNTP) with EvaGreen (SolGent Co. Ltd. Daejeon, Korea), 0.5 µM primers, and 5–50 ng of genomic DNA template. The PCR conditions for the D1 assay were used for the Gene Touch PCR thermal cycler (Hangzhou Bioer Technology Co. Ltd., Hangzou, China) as follows: 95 °C for 15 min, followed by 35 cycles of 95 °C for 20 s, 60 °C for 20 s, and 72 °C for 20 s. Melting curve analysis was conducted using a Roche LightCycler 480 II (Roche Applied Sciences, Indianapolis, IN, USA) by increasing the temperature from 65 °C to 90 °C and reading every 0.1 °C. The homozygous wild type allele of the *D1* gene was detected with a peak at 86 °C, and the homozygous mutant allele containing the *D1* gene showed a peak at 87 °C.

The *D2* and *PsbM* genotype assays were conducted as previously described [29,31]. In the *D2* assay, DNA was amplified with two forward primers, 5′-TGATACGAAACACCCACTACGA-3′ and 5′-GACTATCTCATCTCATCTCTGAATGC-3′, and a reverse primer, 5′-TTGCTACTGCTATTTCGTTATTTAA-3′. In the *PsbM* assay, DNA was amplified with the dCAPS forward primer 5′-GCACTGTTTATTCTAGTTCCTACTGCT TTTTTAGATAT-3′ and the dCAPS reverse primer 5′-TATCTGGATTACGGTGATTGTAGTCCG-3′, and then digested with EcoR V (Enzynomics, Daejeon, Korea). The PCR and digested products were detected using agarose gel electrophoresis at 120 V.

2.5. Phenotype Determination by High Performance Liquid Chromatography (HPLC)

The anthocyanin and chlorophyll contents in the BLG accessions were determined using Thermo Scientific Dionex UltiMate 3000 HPLC (Thermo Scientific Dionex, Waltham, MA, USA). To estimate the genetic and environmental variations, the anthocyanin and chlorophyll contents were measured for seeds from each replicated plot during the three years.

For anthocyanin, a hand-peeled seed coat (0.1 g) from each accession, including three check cultivars, were ground into a powder. Next, 100 mg of powder was extracted with 10 mL of 1% HCl and 20% MeOH for 24 h at 4 °C in a shaking incubator at 110 rpm/min in darkness. The extracted solutions were filtered through Whatman No. 2 filter paper and a syringe filter (0.2 μm). Six extracted anthocyanins (delphinidin 3-O-glucoside, cyanidin 3-O-glucoside, petunidin 3-O-glucoside, pelargonidin 3-O-glucoside, peonidin 3-O-glucoside, and malvidin 3-O-glucosie) were separated in a YMC-pack Pro C18 RS analytical column (250 mm × 4.6 mm, 5 μm). Next, 10 μL of extracted anthocyanins was injected into the column at 1.0 mL/min rates at a temperature of 30 °C. The mobile phases were composed of H_2O/HCOOH (90/10, v/v) (mobile phase A) and CH_3CN/CH_3OH/H_2O/HCOOH (22.5/22.5/40/10, v/v) (mobile phase B). The gradient conditions were as follows: 0 min 7% B, 35 min 35% B, 45 min 65% B, 46 min 100% B. Each anthocyanin was detected using a VIS detector at 520 nm and was quantified based on the standard curves generated for each anthocyanin. The anthocyanin content was converted to milligram per gram (mg/g).

For the chlorophyll content, green cotyledons without a seed coat from each accession, including three check cultivars, were ground into powder. One gram of powder was extracted with 10 mL of 85% $(CH_3)_2CO$ for three hours at 40 °C in a shaking incubator at 110 rpm/min in darkness. The extracted solutions were filtered through Whatman No. 2 filter paper and a syringe filter (0.2 μm). Two extracted chlorophylls (Chl *a*, and Chl *b*) were separated in a YMC-pack ODS-A analytical column (150 mm × 6.0 mm, 5 μm). Next, 20 μL of extracted chlorophylls was injected into the column with 1.0 mL/min rates at a temperature of 30 °C. The mobile phases were composed of 75% MeOH in water (mobile phase C) and 100% EtOAc (mobile phase D). The gradient conditions were as follows: 0 min, 30% D; 15 min, 90% D; 20 min, 30% D. Each chlorophyll was detected using a UV-vis detector at 430 nm and was quantified based on the standard curves generated for each chlorophyll content. Each chlorophyll content was converted to microgram per gram (μg/g).

2.6. Genome Wide Association Study (GWAS) for Anthocyanin

GWAS was conducted using TASSEL software and the GAPIT R package. PCA was constructed with 4459 SNPs [62]. The kinship coefficient matrix was used to provide an estimate of additive genetic variance [62,63]. In the present study, we used models wherein a mixed linear model produced p values to populate Manhattan plots [63,64]. The significance of associations between SNPs and traits was based on Bonferroni's correction and false discovery rate analyses.

2.7. Statistical Data Analysis

All statistical analyses in this study were conducted in SAS v9.4 (SAS Institute, Cary, NC, USA, 2013). Analysis of variance (ANOVA) was conducted to evaluate differences over the three years using PROC GLM in SAS. A comparison of the measured chlorophyll and anthocyanin between the two groups was determined using genotyping, and a Student's *t*-test analysis ($p \leq 0.05$) was conducted using PROC TTEST in SAS.

3. Results

3.1. Genetic Diversity and Population Structure

To estimate the genetic diversity index of BLG accessions for Korean germplasm, 6K SNP markers were used. The genetic diversity index per SNP marker varied from 0.02 to 0.59, with an average of 0.26 (Figure 1A). The PIC value revealed allelic diversity and frequency of the SNP markers (Figure 1B). PIC values of 0.512 and 0.017 were revealed as the maximum and minimum, respectively, with an average of 0.220. The SNP (Gm11_7661182_T_C) at 7,661,182 on chromosome 11 was the marker that showed the most diversity in this study based on the genetic diversity index and the PIC value. Minor allele frequencies ranged from 0.01 to 0.50, with an average of 0.18 (Figure 1C).

Heterozygotes were rare for most of the analyzed SNP markers, and 15 SNP markers showed >0.5 heterozygote frequency with the BLG accessions (Figure 1D).

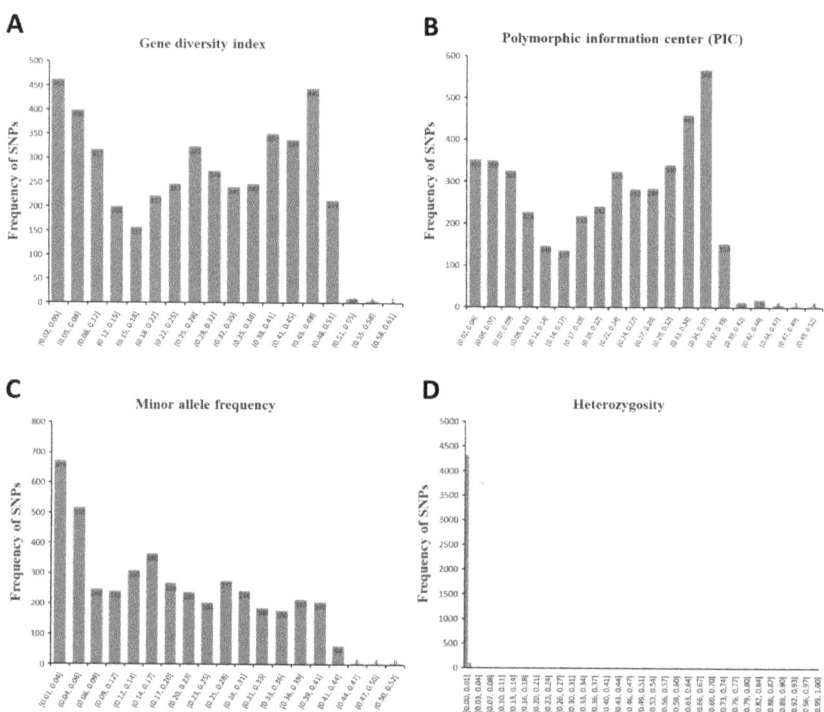

Figure 1. Distribution of the genetic diversity of black soybean accessions with green cotyledon. Genetic diversity index (**A**), polymorphic information center (**B**), minor allele frequency (**C**), and heterozygosity (**D**) of 4459 SNPs across 467 black soybean collections with green cotyledon with three check cultivars.

Population clustering was performed using STRUCTURE software to identify a possible population structure without introducing any prior information. The number of subpopulations could not be identified from the plot of Ln (likelihood probability) for K. However, the ΔK value identified the choice of $K = 2$ as the highest structural level. Admixture plots of K from 2 to 5 are shown in Figure 2A. Both PCA and UPGMA phylogenetic tree were analyzed to validate the number of clusters from the STRUCTURE result and the genetic diversity of entire BLG accessions (Figure 2B,C). Most accessions in cluster 1 showed no genetic differences based on PCA (Figure 2B) and the phylogenetic tree (Figure 2C). The second highest value of ΔK was at $K = 3$ as the final number of subpopulations in the present study. Therefore, these soybean germplasms could be divided into three clusters, which were colored gray for cluster 1, orange for cluster 2, and blue for cluster 3 (Figure 2A–C). PCA and phylogenetic tree analyses showed good agreement with $K = 3$ from the STRUTURE result (Figure 2B,C). Cluster 1 comprised 231 accessions, whereas the other two clusters comprised 107 in cluster 2 and 132 accessions in cluster 3 (Table S1). Two BLG cultivars, Cheongja and Cheongja 3, and the yellow soybean cultivar, Uram, as a check, belonged to cluster 2 (Figure 2B,C).

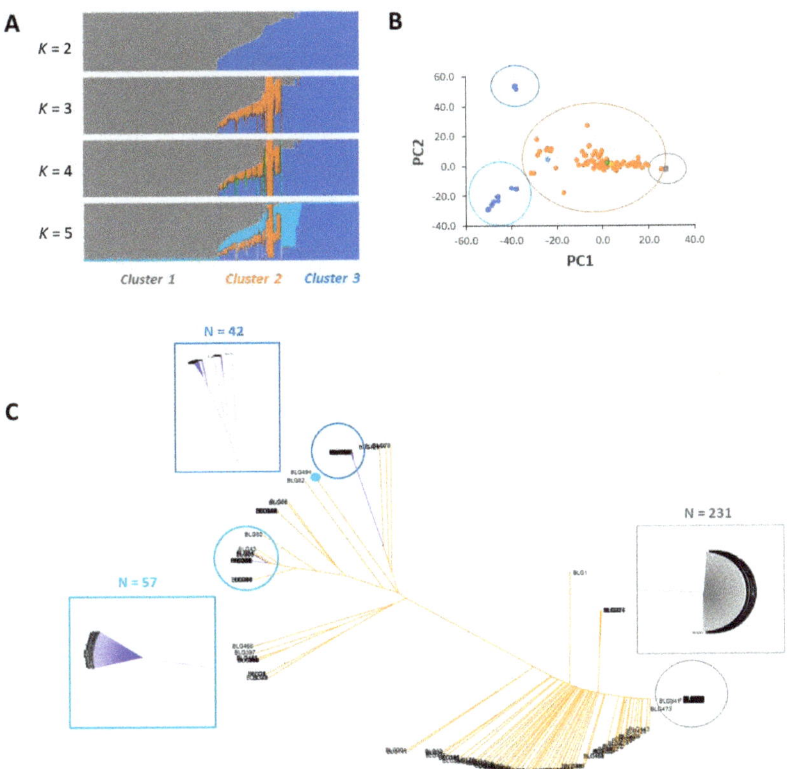

Figure 2. Cluster analyses and a phylogenetic tree of 469 black soybean accessions with green cotyledon and a yellow cultivar. (**A**) ADMIXUTRE plot. Clustering from 2 to 5 of *K* value for the entire set of black soybean accessions with green cotyledon with yellow soybean. Each accession is showed to a vertical bar representing the proportion of the accession's genome from clusters. (**B**) Principal components of SNP variation. Each PC1 and PC2 explained 26.0% and 14.4% of variance in the data. Cluster 1, cluster 2 and cluster 3 are shown by grey, orange, and blue color, respectively. Yellow, green, and light blue dot represented as Cheongja 3 (black soybean cultivar with green cotyledon), Cheongja (black soybean cultivar with green cotyledon) and Uram (yellow soybean), respectively (**C**) UPGMA (unweighted pair group method with arithmetic mean) phylogenetic tree of entire set of 470 accessions. Circle with light blue is yellow soybean, Uram and red circle consists of Cheongja and Cheongja 3.

3.2. Construction of Core Collection

To construct the core collection, 4459 SNPs from the 467 BLG accessions were used. The number of the core collection included thirty-six accessions accounting for 7.7% of the total population (Table S1). The core collection explained 99.5% of the genetic diversity of the total population. Among the core subset, 32 accessions belonged to cluster 2, accounting for 89.0% of the core collections (Figure 3A). PCA was performed to reveal the core collection representing the genetic diversity of the total population. The result showed that the selected accessions in the core collection by GenoCore evenly covered the total populations of PC1 and PC2 (Figure 3B).

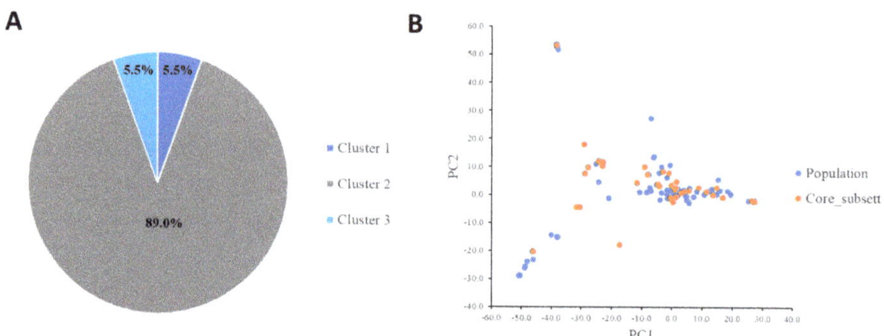

Figure 3. Principal component and a phylogenetic tree of core subset containing 36 black soybean accessions with green cotyledon. (**A**) Distribution of three clusters in core subset (**B**) Principal components of SNP variation. Entire population and core subset are shown by blue and orange dots, respectively.

3.3. Frequency of the D1, D2, and PsbM Alleles and Their Relationship with the Chlorophyll Content in Cotyledon

Genotyping of *D1*, *D2*, and *PsbM* for the total accessions was conducted to determine the frequency of genes in the Korean soybean germplasm. The result revealed that the 467 BLG accessions and the two check BLG cultivars contained either the double recessive mutant alleles of the *D1* and *D2* nuclear genes or the 5-bp insertion in the chloroplast gene, *PsbM*. However, accession BLG397 (IT263333) and BLG467 (IT263849) had heterozygous *D1* and *D2*, whereas BLG397 contained a variant of the *PsbM* gene (Table S1). Seventy-six percent of the analyzed total accessions carried *psbM*, whereas 24% contained double mutant alleles of *D1* and *D2* (Figure 4A). The genotype frequency of *d1d2* (9 of 35, 25.7%) and *psbM* (26 of 35, 74.3%) in the core collection were similar to the total population (Figure 4B). The genotype frequencies of *d1d2* and *psbM* were different based on the three clusters; all 231 accessions in cluster 1 and 83 of 104 accessions (80%) in the group 2 cluster carried *psbM*, whereas 90 of 132 accessions (68%) in group 3 cluster had the *d1d2* genotype (Figure 4C–E).

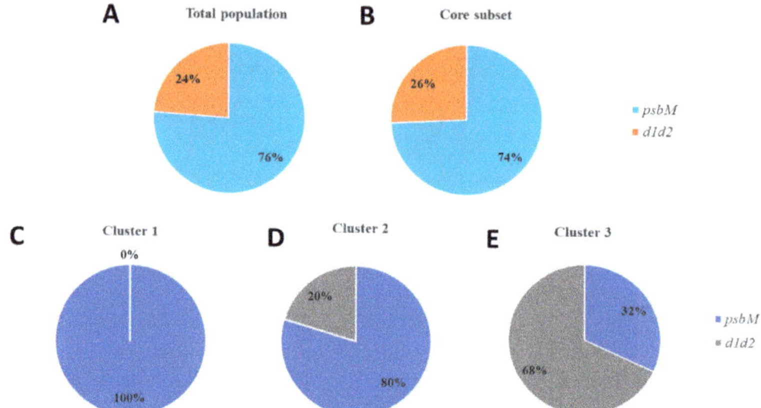

Figure 4. Distribution of *d1d2* and *psbM* among black soybean accessions with green cotyledon in South Korea. Percentage of *d1d2* genotype and *psbM* genotype are provided in each graph. (**A**) Total population. (**B**) Core set. (**C**) Cluster 1. (**D**) Cluster 2. (**E**) Cluster 3.

To identify the relationship of the cloned genes for stay-green and seed phenotype, the seed chlorophyll content was measured for the total BLG accessions using HPLC. The Chl *a*, Chl *b*, and total chlorophyll content of the total collection ranged from 11.5 µg/g to 88.4 µg/g with the mean of 33.9 µg/g, from 7.7 µg/g to 40.8 µg/g with the mean of 21.9 µg/g, and from 22.6 µg/g to 120.0 µg/g with the mean of 55.8 µg, respectively (Figure S1). The chlorophyll *a*/*b* ratio ranged from 0.8 to 5.3, with the mean of 1.8. Mean values of the measured chlorophyll content of individual accessions in two stay-green genotypic groups, *d1d2* and *psbM*, are shown in Figure 5. The Chl *a* and total chlorophyll contents in the *d1d2* genotypic group were significantly higher than those of the *psbM* genotypic group, whereas Chl *b* in the *d1d2* genotypic group was statistically lower than that of the *psbM* genotypes (Figure 5A). The chlorophyll *a*/*b* ratio in the *d1d2* genotypes was ~4-fold higher than that of the *psbM* genotypic group in the total population (Figure 5C). Each measured chlorophyll content was significant for the genotype, environment, and genotype by environment interaction during the three years (Table S3).

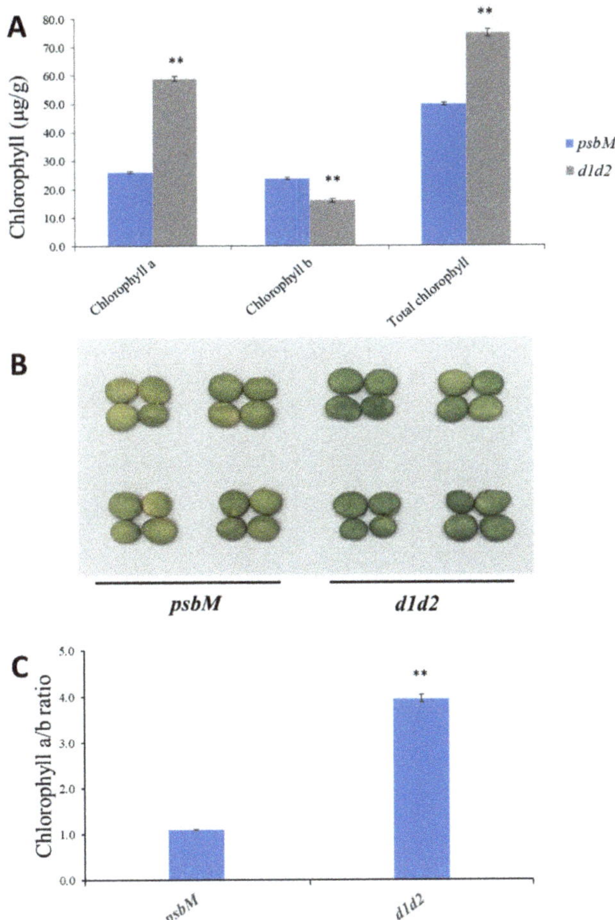

Figure 5. Chlorophyll content of *d1d2* and *psbM* genotype. (**A**) Chlorophyll *a*, *b* and total content of *d1d2* and *psbM* genotype. (**B**) Green cotyledon color without black seed coat for *psbM* and *d1d2* genotype (**C**) Chlorophyll *a*/*b* ratio of *d1d2* and *psbM* genotype. Statistical analysis was conducted using the Student's *t* test (** *p* < 0.01). Bars indicate standard error.

3.4. GWAS for Anthocyanin

To understand the association between phenotype and genes or loci involved in the anthocyanin pathway, a biochemical characterization was performed using HPLC to measure the content of each anthocyanin with 469 BLG accessions. The mean of delphinidin 3-O-glucoside, cyanidin 3-O-glucoside, petunidin 3-O-glucoside, pelargonidin 3-O-glucoside, peonidin 3-O-glucoside, and malvidin 3-O-glucosie over the three years was 2.53 ± 0.04 mg/g, 9.76 ± 0.19 mg/g, 0.57 ± 0.01 mg/g, 0.16 ± 0.01 mg/g, 0.06 ± 0.001 mg/g, and 0.04 ± 0.002 mg/g, respectively (Figure S2). The ranges of delphinidin 3-O-glucoside, cyanidin 3-O-glucoside, petunidin 3-O-glucoside, pelargonidin 3-O-glucoside, peonidin 3-O-glucoside, and malvidin 3-O-glucosie were 0.61–4.78 mg/g, 1.65–19.83 mg/g, 0.04–1.61 mg/g, 0.00–0.71 mg/g, 0.00–0.14 mg/g, and 0.00–0.25 mg/g, respectively. The total anthocyanin content ranged from 2.6 mg/g to 24.3 mg/g, with the mean of 13.11 mg/g (Figure S2). Each measured anthocyanin composition was significant for the genotype, environment, and genotype by environment interaction over the three years (Table S4).

The seed coat color (*I* locus), pubescence color (*T* locus), and flower color (*W1* locus) were involved in the anthocyanin pathway (Figure 6A). All accessions in this study, except Uram, were black seed coat (*ii*) and tawny pubescence color (*TT*). Regarding the flower color, 2.6% of the total accessions (12/469 accessions) except the yellow soybean check showed a white flower (*w1w1*). For delphinidin 3-O-glucoside, petunidin 3-O-glucodsied, and malvidin 3-O-glucoside, ones with the *w1w1* allele (white flower) were significantly lower than ones with the *W1W1* allele (purple flower) (Figure 6B). There was no significant difference between the white and purple flower regarding cyanidin 3-O-glucoside, peonidin 3-O-glucoside, and pelargonidin 3-O-glucoside. The total measured anthocyanins content was 13.1 mg/g and 8.1 mg/g for the purple and white flower, respectively. To identify the genes controlling the anthocyanin content, GWAS was performed with 4459 SNPs (Figure 6C). For cyanidin 3-O-glucoside and the total anthocyanin content, the most significant SNPs were colocalized at 4,873,149 (Wm82.a2.v1 assembly) on chromosome 8. This SNP was near to the *O* locus, which corresponded with an anthocyanidin reductase gene. For delphinidin 3-O-glucoside and pelargonidin 3-O-glucoside, SNPs were located on chromosome 9, which were near the R locus.

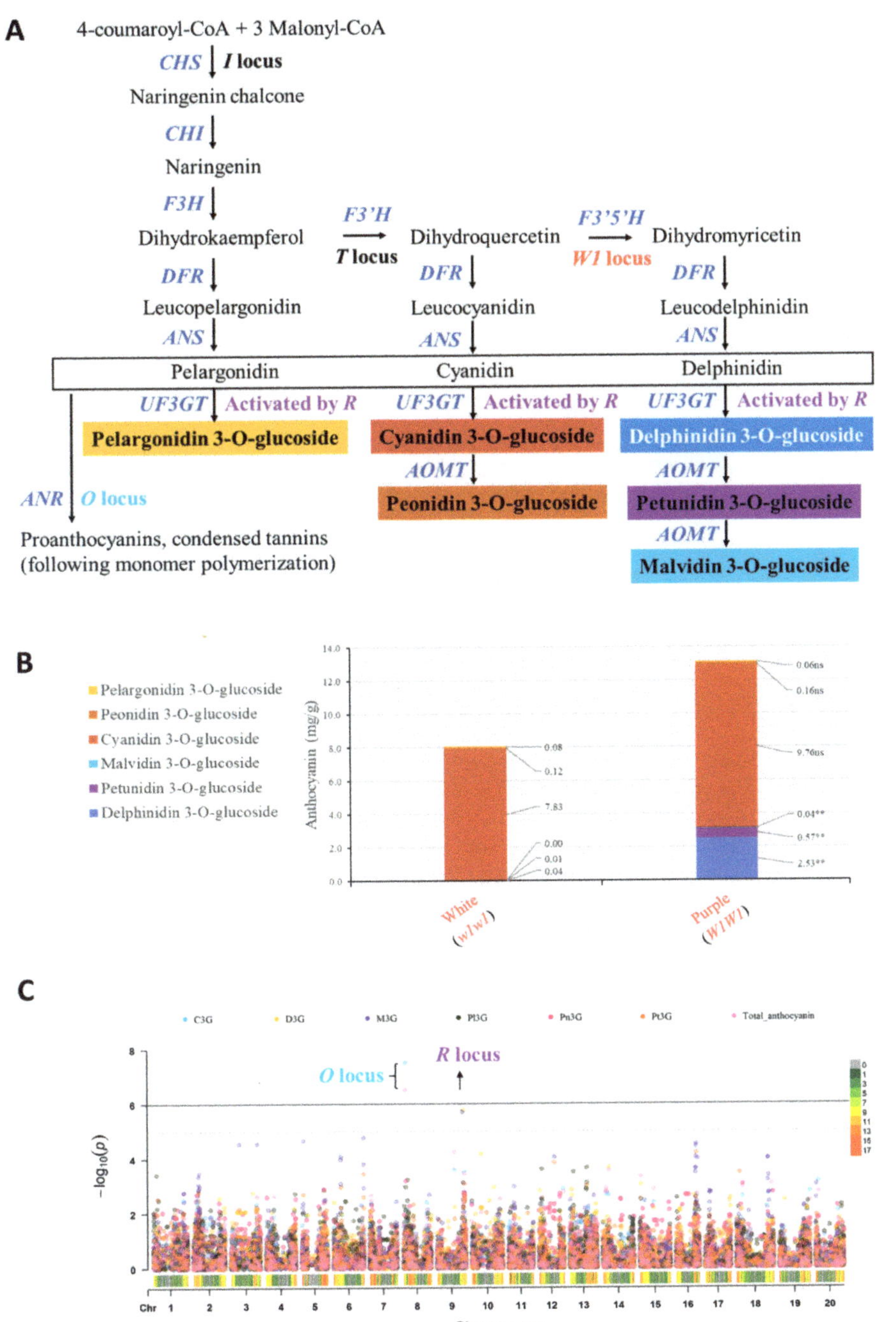

Figure 6. Schematic for genes in the anthocyanin biosynthetic pathway with comparison of measured anthocyanin compositions and GWAS anlaysis. (**A**) the biosynthetic pathway of anthocyanins. Genes or corresponding enzymes are denoted with the capital letters of abbreviated names, as follows: Chalcone synthase (*CHS*, *I* locus); Flavonoid 3′-hydroxylase (*F3′H*, *T* locus); Flavonoid 3′,5′-hydroxylase (*F3′5′H*, *W1* locus); Chalcone isomerase (*CHI*); Dihydroflavonol-4-reductase

(*DFR*); Anthocyanidin synthase (*ANS*); Anthocyanidin reductase (*ANR, O* locus); Flavonoid 3-O-glucosyltransferase (*UF3GT*); Anthocyanin O-methyltransferase (*AOMT*). (**B**) Comparison of measured anthocyanin content of *W1W1* and *w1w1* genotype. Statistical analysis was conducted using the Student's t test (** $p < 0.01$, ns; no significant). (**C**) GWAS anlaysis on mean of each anthocyanin and total content over three years. Delphinidin 3-O-glucoside (D3G); Cyanidin 3-O-glucoside (C3G); Petunidin 3-O-glucoside (P3G); Pelargonidin 3-O-glucoside (Pl3G); Peonidin 3-O-glucoside (Pn3G); Malvidin 3-O-glucosie (M3G).

4. Discussion

Molecular markers can directly reflect genetic diversity, in contrast to phenotypic assessments, because quantitative traits are often influenced by environmental factors [65]. The average genetic diversity index and PIC value of SNPs for the total BLG accessions were 0.26 and 0.22, respectively (Figure 1). However, Liu et al. [36] reported a genetic diversity study on soybean cultivars and advanced breeding lines from the U.S. and China, for which the genetic diversity index and PIC values were 0.3489 and 0.2769, respectively. Hao et al. [66] primarily used Chinese soybean landraces for a genetic diversity study, which resulted in 0.391 for the genetic diversity index and 0.313 for the PIC value, respectively. In addition, Lee et al. [34] and Yoon et al. [67] reported similar genetic diversity indexes for Korean soybean accessions, with an average of 0.70 and 0.71, respectively. The genetic diversity of a population is affected by the number of markers, diversity index of markers, number of accessions, and geographical distribution of accessions used in study. Compared with previously reported results for genetic diversity studies of soybean, the level of genetic diversity of the BLG collections was relatively low because we used only soybean accessions having the BLG phenotype, which were intensively collected from Korea [34,36,66–69]. The population structure analysis used in the present study revealed that there were three clusters in the 469 BLG accessions (Figure 2A). PCA and the UPGMA-phylogenetic tree showed that cluster 1 and cluster 3 had relatively lower genetic diversity than cluster 2 (Figure 2B,C). The result revealed that BLG collections in cluster 1 and cluster 3 may spread over a wide range of geographical areas by farmer's distribution due to a better performance and yield for a long history of soybean cultivation in South Korea. Those accessions have been collected from different areas by the National Agrobiodiversity Center of Rural Development Administration in Korea [51]. Considered together, this may be one of the primary reasons that the BLG collections exhibited narrow genetic variability in this study.

Studies have shown that 5%–20% of the total population was generally constructed as a core collection, thereby representing 99.0% of the genetic diversity of the total population [65,70]. The size of the core collections could be determined by the genetic diversity of the total population [65]. In this study with 4,459 SNPs, 36 accessions were selected as the core collection to improve the utilization of useful breeding material for BLG phenotypes. Although the number of the core collection accounted for 7.7% of the total population, the core collection represented 99.5% of the genetic diversity of the entire population (Figure 3). The core collection was evenly distributed for the total population. Therefore, 36 accessions can be used as breeding materials for the BLG trait in the soybean breeding program in South Korea.

In this study, we found that all 469 BLG accessions contained the same mutation of the *d1d2* or *psbM* allele (Figure 4). Kohzuma et al. [31] revealed that most of the green cotyledon soybean strains from Japan carried *psbM* (99.5%), whereas all Chinese green cotyledon soybeans consisted of *d1d2* (100%). However, the Korean accessions showed both *d1d2* (24.1%) and *psbM* (75.9%) (Figure 4A). In addition, we determined the frequencies of *d1d2* and *psbM* in three clusters (Figure 4C–E), which were different by three clusters: 100.0% of cluster 1, 80.0% of cluster 2, and 32.0% of cluster 3 had the *psbM* allele (Figure 4). Cluster 3 was divided into two subgroups based on the PCA analysis (Figure 2B). Note that all accessions in the light blue of cluster 3 contained the *d1d2* genotype, whereas the ones in the dark blue of cluster 3 had green cotyledons due to the *psbM* allele (Figure 2B,C). Several studies mentioned that Korean soybeans were closely related to Japanese soybeans and were genetically distinct from the Chinese soybean population [34,35,71]. The genotypic frequency of the *d1d2* and *psbM* alleles in this study supported that soybean accessions from Korea were a mixture of genetic contexts related to soybean from both Japan and China [72].

Because chlorophyll has anticarcinogenic properties [73,74], increasing the chlorophyll content is one of the targeted traits in breeding programs for soybean-based foods. The compositions of chlorophyll in green cotyledon were different based on the *d1d2* or *psbM* genotypes. The amount of Chl *a* in *d1d2* was ~2-fold higher than Chl *a* in the *psbM* genotype (Figure 5). The Chl *a* content was 4-fold higher than the Chl *b* content in the *d1d2* genotypes, which resulted in a higher chlorophyll *a/b* ratio than that of the *psbM* genotype. Our result supported that *d1d2* genotypes contain higher chlorophyll content [75]. In the soybean breeding program, *d1d2* genotypes may be a good genetic resource for increasing the total chlorophyll contents in cotyledon seeds.

The soybean seed coat is controlled by multiple loci, such as *I, T, W1, R,* and *O* loci involved in the anthocyanin biosynthetic pathway [76–78]. The *I* locus encodes chalcone synthase, which is the key enzyme of the flavonoid pathway (Figure 6A) [79]. A black seed coat forms primarily due to the accumulation of anthocyanins and is controlled by the *i* allele. In most soybean with black seed coats, the three major anthocyanins are cyanidin 3-O-glucoside, delphinidin 3-O-glucoside, and petunidin 3-O-glucoside [17]. Lee et al. [16] reported six additional anthocyanin compounds from black seed coats, including a major amount of pelargonidin 3-O-glucosied and a minor amount of peonidin 3-O-glucoside. In the present study, we analyzed six anthocyanins and found that cyanidin 3-O-glucoside, delphinidin 3-O-glucoside, and petunidin 3-O-glucoside were major components of anthocyanin compositions in total BLG accessions (Figure 6B). The *W1* locus, flavonoid 3′,5′-hydroxylase (*F3′5′H*), plays an important role in the accumulation of delphinidin 3-O-glucoside, petunidin 3-O-glucoside, and malvidin 3-O-glucoside under the genetic background of the *iiTT* genotype [80]. Recently, Kim et al. [81] reported that the *iiRRTTw1w1* genotype would prohibit the accumulation of delphinidin 3-O-glucoside and petunidin 3-O-glucoside compounds, whereas the *T* locus, flavonoid 3′-hydroxylase (*F3′H*), can control the production of cyanidin-derived and delphinidin-derived anthocyanin compounds. The result of this study was that the purple flower (*W1W1*) produced ~5.0 mg/g more total anthocyanin (Figure 6B). Flower color can be used as a possible selection trait to increase the total anthocyanin composition in the black soybean breeding program.

Although 97.4% of the total BLG accessions showed a black seed coat (*ii* allele), tawny pubescence (*TT* allele), and purple flower (*W1W1* allele) in the present study, each anthocyanin composition still showed phenotypic variations. The GWAS result showed that SNPs on chromosomes 8 and 9 were associated with the production of delphinidin 3-O-glucoside, pelargonidin 3-O-glucoside, cyanidin 3-O-glucoside, and total anthocyanins. These SNPs were near the *R* locus and *O* locus on chromosomes 8 and 9, respectively (Figure 6C). The *R* locus is the R2R3 MYB transcription factor for upregulating UDP-glycose: flavonoid 3-O-glycosyltranferase (*UF3GT*) in black soybeans [76]. For the *O* locus, the expression level of anthocyanidin reductase (*ANR*) in the soybean genome is associated with the red-brown and black color of the soybean seed coat [82]. The GWAS result concluded that the extent of the expression levels of *UF3GT* and *ANR* may be associated with the amount of anthocyanin in BLG soybean.

5. Conclusions

In conclusion, BLG soybeans exhibited narrow genetic variability due to artificial selection, and a core collection representing the genetic diversity of the total BLG soybeans was constructed based on the 6K SNP genotypes. This core collection can provide useful genetic resources for the development of new cultivars for BLG seeds with increased anthocyanin and chlorophyll contents as part of a breeding program for soybean food.

Supplementary Materials: The following are available online at https://www.mdpi.com/2073-439 5/11/3/581/s1, Figure S1: Frequency distribution of an average of two chlorophyll content and sum of total chlorophyll contents in 469 accessions over three years, Figure S2: Frequency distribution of an average of two six anthocyanin content and sum of total anthocyanin contents in 469 accessions over three year, Table S1: List of 470 soybean accessions, subpopulation, core collection and genotype of *D1*, *D2* and *PsbM* allele, Table S2: List of pure line selection from 47 accessions, Table S3: Mean squares from analysis of variances of each measured chlorophyll for soybean accessions over 3 years, Table S4: Mean squares from analysis of variances of each measured anthocyanin for soybean accessions over 3 years.

Author Contributions: H.J. and J.Y.L. contributed equally for this work. Conceptualization, J.-D.L.; Methodology, H.J., J.Y.L. and H.C.; Formal analysis, H.J. and J.Y.L.; Visualization, H.J.; Investigation, J.Y.L., H.J.C., C.K.S. and J.S.B.; Writing—Original Draft Preparation, H.J.; Writing—Review & Editing, J.Y.L., K.B., J.T.S. and J.-D.L. All authors have read and agreed to the published version of the manuscript.

Funding: This research was funded by a grant from the Republic of Korea Rural Development Administration New Breeding Technologies Development Program (Project No. PJ01488301).

Institutional Review Board Statement: Not applicable.

Informed Consent Statement: Not applicable.

Data Availability Statement: The datasets generated during this study are available from the corresponding author on reasonable request.

Acknowledgments: The authors would like to acknowledge the personnel from the Plant Genetics and Breeding lab at the Kyungpook National University and Gyeongsangbuk-do Provincial Agricultural Research & Extension Service for their time and work on the field experiments.

Conflicts of Interest: The authors declare no conflict of interest.

References

1. Liu, K. Chemistry and nutritional value of soybean components. In *Soybeans*; Springer: Boston, MA, USA, 1997; pp. 25–113. [CrossRef]
2. Watanabe, T. *Science of Tofu, Chapter 2, Manufacture of Tofu*; Food Journal Co. Ltd.: Kyoto, Japan, 1997.
3. Willcox, J.K.; Ash, S.L.; Catignani, G.L. Antioxidants and prevention of chronic disease. *Crit. Rev. Food Sci. Nutr.* **2004**, *44*, 275–295. [CrossRef]
4. Xu, B.; Chang, S.K. Antioxidant capacity of seed coat, dehulled bean, and whole black soybeans in relation to their distributions of total phenolics, phenolic acids, anthocyanins, and isoflavones. *J. Agric. Food Chem.* **2008**, *56*, 8365–8373. [CrossRef]
5. Do, M.H.; Lee, S.S.; Jung, P.J.; Lee, M.H. Intake of fruits, vegetables, and soy foods in relation to breast cancer risk in Korean women: A case-control study. *Nutr. Cancer* **2007**, *57*, 20–27. [CrossRef]
6. Takahashi, R.; Ohmori, R.; Kiyose, C.; Momiyama, Y.; Ohsuzu, F.; Kondo, K. Antioxidant activities of black and yellow soybeans against low density lipoprotein oxidation. *J. Agric. Food Chem.* **2005**, *53*, 4578–4582. [CrossRef] [PubMed]
7. Ganesan, K.; Xu, B. A critical review on polyphenols and health benefits of black soybeans. *Nutrients* **2017**, *9*, 455. [CrossRef]
8. Jhan, J.K.; Chung, Y.C.; Chen, G.H.; Chang, C.H.; Lu, Y.C.; Hsu, C.K. Anthocyanin contents in the seed coat of black soya bean and their anti-human tyrosinase activity and antioxidative activity. *Int. J. Cosmet. Sci.* **2016**, *38*, 319–324. [CrossRef]
9. Hymowitz, T.; Newell, C. Taxonomy, speciation, domestication, dissemination, germplasm resources, and variation in the genus Glycine. In *Advances in Legume Science*; Summerfield, R.J.B.A., Ed.; Royal Botanical Gardens: Richmond, UK, 1980; pp. 251–264.
10. Gonzali, S.; Perata, P. Anthocyanins from Purple Tomatoes as Novel Antioxidants to Promote Human Health. *Antioxidants* **2020**, *9*, 1017. [CrossRef]
11. De Masi, L.; Bontempo, P.; Rigano, D.; Stiuso, P.; Carafa, V.; Nebbioso, A.; Piacente, S.; Montoro, P.; Aversano, R.; D'Amelia, V.; et al. Comparative phytochemical characterization, genetic profile, and antiproliferative activity of polyphenol-rich extracts from pigmented tubers of different Solanum tuberosum varieties. *Molecules* **2020**, *25*, 233. [CrossRef] [PubMed]
12. Norberto, S.; Silva, S.; Meireles, M.; Faria, A.; Pintado, M.; Calhau, C. Blueberry anthocyanins in health promotion: A metabolic overview. *J. Funct. Foods* **2013**, *5*, 1518–1528. [CrossRef]
13. Tsuda, T. Dietary anthocyanin-rich plants: Biochemical basis and recent progress in health benefits studies. *Mol. Nutr. Food Res.* **2012**, *56*, 159–170. [CrossRef]
14. Clough, S.J.; Tuteja, J.H.; Li, M.; Marek, L.F.; Shoemaker, R.C.; Vodkin, L.O. Features of a 103-kb gene-rich region in soybean include an inverted perfect repeat cluster of *CHS* genes comprising the *I* locus. *Genome* **2004**, *47*, 819–831. [CrossRef]
15. Todd, J.J.; Vodkin, L.O. Duplications that suppress and deletions that restore expression from a chalcone synthase multigene family. *Plant Cell* **1996**, *8*, 687–699. [CrossRef]

16. Lee, J.H.; Kang, N.S.; Shin, S.O.; Shin, S.H.; Lim, S.G.; Suh, D.Y.; Baek, I.Y.; Park, K.Y.; Ha, T.J. Characterisation of anthocyanins in the black soybean (*Glycine max* L.) by HPLC-DAD-ESI/MS analysis. *Food Chem.* **2009**, *112*, 226–231. [CrossRef]

17. Choung, M.G.; Baek, I.Y.; Kang, S.T.; Han, W.Y.; Shin, D.C.; Moon, H.P.; Kang, K.H. Isolation and determination of anthocyanins in seed coats of black soybean (*Glycine max* (L.) Merr.). *J. Agric. Food Chem.* **2001**, *49*, 5848–5851. [CrossRef]

18. Mendel, G. Versuche uber pflanzen-hybriden. *Verh. Nat. Verein Brunn* **1866**, *4*, 3–47. [CrossRef]

19. Thomas, H.; Schellenberg, M.; Vicentini, F.; Matile, P. Gregor Mendel's green and yellow pea seeds. *Bot. Acta* **1996**, *109*, 3–4. [CrossRef]

20. White, O.E. Studies of inheritance in Pisum. II. The present state of knowledge of heredity and variation in peas. *Proc. Am. Philos. Soc.* **1917**, *56*, 487–588.

21. Armstead, I.; Donnison, I.; Aubry, S.; Harper, J.; Hörtensteiner, S.; James, C.; Mani, J.; Moffet, M.; Ougham, H.; Roberts, L.; et al. Cross-species identification of Mendel's I locus. *Science* **2007**, *315*, 73. [CrossRef]

22. Sato, Y.; Morita, R.; Nishimura, M.; Yamaguchi, H.; Kusaba, M. Mendel's green cotyledon gene encodes a positive regulator of the chlorophyll degrading pathway. *Proc. Natl. Acad. Sci. USA* **2007**, *104*, 14169–14174. [CrossRef] [PubMed]

23. Croce, R.; Van Amerongen, H. Natural strategies for photosynthetic light harvesting. *Nat. Chem. Biol.* **2014**, *10*, 492–501. [CrossRef]

24. Eckhardt, U.; Grimm, B.; Hortensteiner, S. Recent advances in chlorophyll biosynthesis and breakdown in higher plants. *Plant Mol. Biol.* **2004**, *56*, 1–14. [CrossRef]

25. Hortensteiner, S.; Krautler, B. Chlorophyll breakdown in higher plants. *Biochim. Biophys. Acta* **2011**, *1807*, 977–988. [CrossRef] [PubMed]

26. Thomas, H.; Howarth, C.J. Five ways to stay green. *J. Exp. Bot.* **2000**, *51*, 329–337. [CrossRef] [PubMed]

27. Owen, F.V. Inheritance studies in soybeans. II. Glabrousness, color of pubescence, time of maturity, and linkage relations. *Genetics* **1927**, *12*, 519. [CrossRef]

28. Terao, H. Maternal inheritance in the soybean. *Am. Nat.* **1918**, *52*, 51–56. [CrossRef]

29. Fang, C.; Li, C.; Li, W.; Wang, Z.; Zhou, Z.; Shen, Y.; Wu, M.; Wu, Y.; Li, G.; Kong, L.A.; et al. Concerted evolution of *D1* and *D2* to regulate chlorophyll degradation in soybean. *Plant J.* **2014**, *77*, 700–712. [CrossRef]

30. Guiamet, J.J.; Schwartz, E.; Pichersky, E.; Nooden, L.D. Characterization of cytoplasmic and nuclear mutations affecting chlorophyll and chlorophyll-binding proteins during senescence in soybean. *Plant Physiol.* **1991**, *96*, 227–231. [CrossRef]

31. Kohzuma, K.; Sato, Y.; Ito, H.; Okuzaki, A.; Watanabe, M.; Kobayashi, H.; Nakano, M.; Yamatani, H.; Masuda, Y.; Nagashima, Y.; et al. The non-Mendelian green cotyledon gene in soybean encodes a small subunit of photosystem II. *Plant Physiol.* **2017**, *173*, 2138–2147. [CrossRef]

32. Hirata, T.; Abe, J.; Shimamoto, Y. Genetic structure of the Japanese soybean population. *Genet. Resour. Crop Evol.* **1999**, *46*, 441–453. [CrossRef]

33. Lee, J.D.; Yu, J.K.; Hwang, Y.H.; Blake, S.; So, Y.S.; Lee, G.J.; Nguyen, H.T.; Shannon, J.G. Genetic diversity of wild soybean (*Glycine soja* Sieb. and Zucc.) accessions from South Korea and other countries. *Crop Sci.* **2008**, *48*, 606–616. [CrossRef]

34. Lee, J.D.; Vuong, T.D.; Moon, H.; Yu, J.K.; Nelson, R.L.; Nguyen, H.T.; Shannon, J.G. Genetic diversity and population structure of Korean and Chinese soybean [*Glycine max* (L.) Merr.] accessions. *Crop Sci.* **2011**, *51*, 1080–1088. [CrossRef]

35. Li, Z.; Nelson, R.L. Genetic diversity among soybean accessions from three countries measured by RAPDs. *Crop Sci* **2001**, *41*, 1337–1347. [CrossRef]

36. Liu, Z.; Li, H.; Wen, Z.; Fan, X.; Li, Y.; Guan, R.; Guo, Y.; Wang, S.; Wang, D.; Qiu, L. Comparison of genetic diversity between Chinese and American soybean (*Glycine max* (L.)) accessions revealed by high-density SNPs. *Front. Plant Sci.* **2017**, *8*, 2014. [CrossRef]

37. Xu, D.H.; Abe, J.; Gai, J.Y.; Shimamoto, Y. Diversity of chloroplast DNA SSRs in wild and cultivated soybeans: Evidence for multiple origins of cultivated soybean. *Theor. Appl. Genet.* **2002**, *105*, 645–653. [CrossRef]

38. Frankel, O.H.; Brown, A.H. Plant genetic resources today: A critical appraisal. In *Crop Genetic Resources*; Holden, J.H.W., Williams, J.T., Eds.; Allen and Unwin: London, UK, 1984; pp. 249–257.

39. Dong, Y.S.; Cao, Y.S.; Zhang, X.Y.; Liu, S.C.; Wang, L.F.; You, G.X.; Pang, B.S.; Li, L.H.; Jia, J.Z. Establishment of candidate core collections in Chinese common wheat germplasm. *J. Plant Genet. Resour.* **2003**, *4*, 1–8.

40. Diwan, N.; McIntosh, M.S.; Bauchan, G.R. Methods of developing a core collection of annual Medicago species. *Theor. Appl. Genet.* **1995**, *90*, 755–761. [CrossRef]

41. Holbrook, C.C.; Anderson, W.F. Evaluation of a core collection to identify resistance to late leafspot in peanut. *Crop Sci.* **1995**, *35*, 1700–1702. [CrossRef]

42. Hu, J.; Wang, P.; Su, Y.; Wang, R.; Li, Q.; Sun, K. Microsatellite diversity, population structure, and core collection formation in melon germplasm. *Plant Mol. Biol. Rep.* **2015**, *33*, 439–447. [CrossRef]

43. Lee, H.Y.; Ro, N.Y.; Jeong, H.J.; Kwon, J.K.; Jo, J.; Ha, Y.; Jung, A.; Han, J.W.; Venkatesh, J.; Kang, B.C. Genetic diversity and population structure analysis to construct a core collection from a large Capsicum germplasm. *BMC Genet.* **2016**, *17*, 142. [CrossRef]

44. Li, Z.C.; Zhang, H.I.; Cao, Y.S.; Qiu, Z.E.; Wei, X.H.; Tang, S.X.; Yu, P.; Wang, X. Studies on the sampling strategy for primary core collection of Chinese ingenious rice. *Acta Agron. Sin.* **2003**, *29*, 20–24.

45. Wang, L.; Guan, Y.; Guan, R.; Li, Y.; Ma, Y.; Dong, Z.; Liu, X.; Zhang, H.; Zhang, Y.; Liu, Z.; et al. Establishment of Chinese soybean *Glycine max* core collections with agronomic traits and SSR markers. *Euphytica* **2006**, *151*, 215–223. [CrossRef]

46. Xu, C.; Gao, J.; Du, Z.; Li, D.; Wang, Z.; Li, Y.; Pang, X. Identifying the genetic diversity, genetic structure and a core collection of *Ziziphus jujuba* Mill. var. jujuba accessions using microsatellite markers. *Sci. Rep.* **2016**, *6*, 31503. [CrossRef]

47. Xu, H.; Mei, Y.; Hu, J.; Zhu, J.; Gong, P. Sampling a core collection of island cotton (*Gossypium barbadense* L.) based on the genotypic values of fiber traits. *Genet. Resour. Crop Evol.* **2006**, *53*, 515–521. [CrossRef]

48. Hu, J.; Zhu, J.; Xu, H.M. Methods of constructing core collections by stepwise clustering with three sampling strategies based on the genotypic values of crops. *Theor. Appl. Genet.* **2000**, *101*, 264–268. [CrossRef]

49. Liu, X.B.; Li, J.; Yang, Z.L. Genetic diversity and structure of core collection of winter mushroom (*Flammulina velutipes*) developed by genomic SSR markers. *Hereditas* **2018**, *155*, 3. [CrossRef] [PubMed]

50. Hymowitz, T. The history of the soybean. In *Soybeans: Chemistry, Production, Processing, and Utilization*; Johnson, L.A., White, P.J., Galloway, R., Eds.; AOCS Press: Urbana, IL, USA, 2008; pp. 1–31. [CrossRef]

51. National Agrobiodiversity Center. Available online: http://genebank.rda.go.kr/ (accessed on 15 December 2020).

52. Baek, I.Y.; Kang, S.T.; Shin, D.C.; Choung, M.G.; Han, W.Y.; Kwack, Y.H.; Moon, H.P. A new black soybean variety with green cotyledon, early maturity and large seed size "Cheongjakong". *Korean J. Breed Sci.* **2001**, *33*, 240–241.

53. Lee, C.; Choi, M.S.; Kim, H.T.; Yun, H.T.; Lee, B.; Chung, Y.S.; Kim, R.W.; Choi, H.K. Soybean [*Glycine max* (L.) Merrill]: Importance as a crop and pedigree reconstruction of Korean varieties. *Plant Breed Biotech.* **2015**, *3*, 179–196. [CrossRef]

54. Fehr, W.R.; Caviness, C.E.; Burmood, D.T.; Pennington, J.S. Stage of development descriptions for soybeans, *Glycine Max* (L.) Merrill[1]. *Crop Sci.* **1971**, *11*, 929–931. [CrossRef]

55. Doyle, J.; Doyle, J.L. Genomic plant DNA preparation from fresh tissue-CTAB method. *Phytochem. Bull.* **1987**, *19*, 11–15.

56. Akond, M.; Liu, S.; Schoener, L.; Anderson, J.A.; Kantartzi, S.K.; Meksem, K.; Song, Q.; Wang, D.; Wen, Z.; Lightfoot, D.A.; et al. A SNP-based genetic linkage map of soybean using the SoySNP6K Illumina Infinium BeadChip genotyping array. *J. Plant Genome Sci.* **2013**, *1*, 80–89. [CrossRef]

57. Song, Q.; Hyten, D.L.; Jia, G.; Quigley, C.V.; Fickus, E.W.; Nelson, R.L.; Cregan, P.B. Development and evaluation of SoySNP50K, a high-density genotyping array for soybean. *PLoS ONE* **2013**, *8*, e54985. [CrossRef]

58. Liu, K.; Muse, S.V. PowerMarker: An integrated analysis environment for genetic marker analysis. *Bioinform* **2005**, *21*, 2128–2129. [CrossRef] [PubMed]

59. Kumar, A.; Sharma, D.; Tiwari, A.; Jaiswal, J.P.; Singh, N.K.; Sood, S. Genotyping-by-sequencing analysis for determining population structure of finger millet germplasm of diverse origins. *Plant Genome* **2016**, *9*, 1–5. [CrossRef] [PubMed]

60. Pritchard, J.K.; Wen, W.; Daniel, D. *Documentation for Structure Software*, version 2; University of Chicago: Chicago, IL, USA, 2010.

61. Jeong, S.; Kim, J.Y.; Jeong, S.C.; Kang, S.T.; Moon, J.K.; Kim, N. GenoCore: A simple and fast algorithm for core subset selection from large genotype datasets. *PLoS ONE* **2018**, *12*, e0181420. [CrossRef] [PubMed]

62. Bradbury, P.J.; Zhang, Z.; Kroon, D.E.; Casstevens, T.M.; Ramdoss, Y.; Buckler, E.S. TASSEL: Software for association mapping of complex traits in diverse samples. *Bioinform* **2007**, *23*, 2633–2635. [CrossRef]

63. Zhang, Z.; Ersoz, E.; Lai, C.Q.; Todhunter, R.J.; Tiwari, H.K.; Gore, M.A.; Bradbury, P.J.; Yu, J.; Arnett, D.K.; Ordovas, J.M.; et al. Mixed linear model approach adapted for genome-wide association studies. *Nat. Genet.* **2010**, *42*, 355–360. [CrossRef]

64. Yu, J.; Pressoir, G.; Briggs, W.H.; Bi, I.V.; Yamasaki, M.; Doebley, J.F.; McMullen, M.D.; Gaut, B.S.; Nielsen, D.M.; Holland, J.B.; et al. A unified mixed-model method for association mapping that accounts for multiple levels of relatedness. *Nat. Genet.* **2006**, *38*, 203–208. [CrossRef] [PubMed]

65. Jeong, N.; Kim, K.S.; Jeong, S.; Kim, J.Y.; Park, S.K.; Lee, J.S.; Jeong, S.C.; Kang, S.T.; Ha, B.K.; Kim, D.Y.; et al. Korean soybean core collection: Genotypic and phenotypic diversity population structure and genome-wide association study. *PLoS ONE* **2019**, *14*, e0224074. [CrossRef]

66. Hao, D.; Cheng, H.; Yin, Z.; Cui, S.; Zhang, D.; Wang, H.; Yu, D. Identification of single nucleotide polymorphisms and haplotypes associated with yield and yield components in soybean (*Glycine max*) landraces across multiple environments. *Theor. Appl. Genet.* **2012**, *124*, 447–458. [CrossRef] [PubMed]

67. Yoon, M.S.; Lee, J.; Kim, C.Y.; Kang, J.H.; Cho, E.G.; Baek, H.J. DNA profiling and genetic diversity of Korean soybean (*Glycine max* (L.) Merrill) landraces by SSR markers. *Euphytica* **2009**, *165*, 69–77. [CrossRef]

68. Li, Y.H.; Smulders, M.J.; Chang, R.Z.; Qiu, L.J. Genetic diversity and association mapping in a collection of selected Chinese soybean accessions based on SSR marker analysis. *Conserv. Genet.* **2011**, *12*, 1145–1157. [CrossRef]

69. Zhou, Z.; Jiang, Y.; Wang, Z.; Gou, Z.; Lyu, J.; Li, W.; Yu, Y.; Shu, L.; Zhao, Y.; Ma, Y.; et al. Resequencing 302 wild and cultivated accessions identifies genes related to domestication and improvement in soybean. *Nat. Biotechnol.* **2015**, *33*, 408–414. [CrossRef] [PubMed]

70. Van Hintum, T.J.; Brown, A.H.; Spillane, C. *Core Collections of Plant Genetic Resources*; IPGRI Technical Bulletin No. 3; International Plant Genetic Resources Institute: Rome, Italy, 2020.

71. Griffin, J.D.; Palmer, R.G. Variability of thirteen isozyme loci in the USDA soybean germplasm collections. *Crop Sci.* **1995**, *35*, 897–904. [CrossRef]

72. Abe, J.; Xu, D.; Suzuki, Y.; Kanazawa, A.; Shimamoto, Y. Soybean germplasm pools in Asia revealed by nuclear SSRs. *Theor. Appl. Genet.* **2003**, *106*, 445–453. [CrossRef] [PubMed]

73. Donaldson, M.S. Nutrition and cancer: A review of the evidence for an anti-cancer diet. *Nutr. J.* **2004**, *3*, 19. [CrossRef] [PubMed]

74. Li, W.T.; Tsao, H.W.; Chen, Y.Y.; Cheng, S.W.; Hsu, Y.C. A study on the photodynamic properties of chlorophyll derivatives using human hepatocellular carcinoma cells. *Photochem. Photobiol. Sci.* **2007**, *6*, 1341–1348. [CrossRef] [PubMed]

75. Kang, S.; Seo, M.; Moon, J.; Yun, H.; Lee, Y.; Kim, S.; Hwang, Y.; Lee, S.; Choung, M. Introduction of stay green mutant for the development of black seed coat and green cotyledon soybean variety. *Korean J. Crop Sci.* **2010**, *55*, 187–194.

76. Gillman, J.D.; Tetlow, A.; Lee, J.D.; Shannon, J.G.; Bilyeu, K. Loss-of-function mutations affecting a specific *Glycine max* R2R3 MYB transcription factor result in brown hilum and brown seed coats. *BMC Plant Biol.* **2011**, *11*, 1–12. [CrossRef]

77. Song, J.; Liu, Z.; Hong, H.; Ma, Y.; Tian, L.; Li, X.; Li, Y.H.; Guan, R.; Guo, Y.; Qiu, L.J. Identification and validation of loci governing seed coat color by combining association mapping and bulk segregation analysis in soybean. *PLoS ONE* **2016**, *11*, e0159064. [CrossRef]

78. Toda, K.; Yang, D.; Yamanaka, N.; Watanabe, S.; Harada, K.; Takahashi, R. A single-base deletion in soybean flavonoid 3′-hydroxylase gene is associated with gray pubescence color. *Plant Mol. Biol.* **2002**, *50*, 187–196. [CrossRef]

79. Tuteja, J.H.; Clough, S.J.; Chan, W.C.; Vodkin, L.O. Tissue-specific gene silencing mediated by a naturally occurring chalcone synthase gene cluster in *Glycine max*. *Plant Cell* **2004**, *16*, 819–835. [CrossRef] [PubMed]

80. Zabala, G.; Vodkin, L.O. A rearrangement resulting in small tandem repeats in the F3′5′ H gene of white flower genotypes is associated with the soybean *W1* locus. *Crop Sci.* **2007**, *47*, S113–S124. [CrossRef]

81. Kim, J.H.; Park, J.S.; Lee, C.Y.; Jeong, M.G.; Xu, J.L.; Choi, Y.; Jung, H.W.; Choi, H.K. Dissecting seed pigmentation-associated genomic loci and genes by employing dual approaches of reference-based and k-mer-based GWAS with 438 Glycine accessions. *PLoS ONE* **2020**, *15*, e0243085. [CrossRef]

82. Kovinich, N.; Saleem, A.; Arnason, J.T.; Miki, B. Combined analysis of transcriptome and metabolite data reveals extensive differences between black and brown nearly-isogenic soybean (*Glycine max*) seed coats enabling the identification of pigment isogenes. *BMC Genom.* **2011**, *12*, 381. [CrossRef] [PubMed]

Article

The Genetic Diversity and Structure of Tomato Landraces from the Campania Region (Southern Italy) Uncovers a Distinct Population Identity

Martina Caramante, Youssef Rouphael and Giandomenico Corrado *

Department of Agricultural Sciences, University of Naples Federico II, 80055 Portici, Italy;
martina.caramante@gmail.com (M.C.); youssef.rouphael@unina.it (Y.R.)
* Correspondence: giandomenico.corrado@unina.it; Tel.: +39-081-2539297

Abstract: Italy is one of the main producers and processors of tomato and it is considered a secondary center of diversity. In some areas, such as the Campania region (Southern Italy), a range of traditional tomato landraces is still cultivated. The distinction of this heritage germplasm is often based only on folk taxonomy and a more comprehensive definition and understanding of its genetic identity is needed. In this work, we compared a set of 15 local landraces (representative of traditional fruit types) to 15 widely used contemporary varieties, using 14 fluorescent Simple Sequence Repeat (SSR) markers. Each of the accessions possessed a unique molecular profile and overall landraces had a genetic diversity comparable to that of the contemporary varieties. The genetic diversity, multivariate, and population structure analysis separated all the genotypes according to the pre-defined groups, indicating a very reduced admixture and the presence of a differentiated (regional) population of landraces. Our work provides solid evidence for implementing conservation actions and paves the way for the creation of a premium regional brand that goes beyond the individual landrace names of the Campania region known throughout the world.

Keywords: *Solanum lycopersicum*; agro-biodiversity; crops; breeding; DNA markers

Citation: Caramante, M.; Rouphael, Y.; Corrado, G. The Genetic Diversity and Structure of Tomato Landraces from the Campania Region (Southern Italy) Uncovers a Distinct Population Identity. *Agronomy* **2021**, *11*, 564. https://doi.org/10.3390/agronomy11030564

Academic Editors: Gregorio Barba Espin and Jose Ramon Acosta-Motos

Received: 9 February 2021
Accepted: 12 March 2021
Published: 17 March 2021

Publisher's Note: MDPI stays neutral with regard to jurisdictional claims in published maps and institutional affiliations.

1. Introduction

Italy is the leading European nation for tomatoes, considering the cultivated area, the total yield, and the output of the processing industry, which is second in the world only to that of the USA (https://tinyurl.com/y2n6pg5n; accessed 10 March 2021). Moreover, the tomato is a dominant element of the Italian gastronomy and a symbol of a culinary identity that was globally spread by Italian immigrants in the XX century [1].

The tomato was domesticated in the Americas and was introduced to Italy soon after the Columbian exchange [2]. However, its culinary use started in Italy since the second half of the 17th century when, despite the inability to contribute significantly to the caloric intake of the farmers, it was recognized the gustatory value of tomato as (cooked) sauce, condiment and in combination with other foods [3]. In that period, a regional diversification in the use of tomato has also begun, influenced by the political, cultural, and linguistic division of Italy at the time. The success of tomatoes in Italy was not uniform and even today, there are areas where relatively few tomatoes are consumed and cultivated. In the 18th century, the tomato was widespread in the Italian gastronomy especially in the Kingdom of Naples and Sicily (current Southern Italy), at that time under the Spanish rule of the House of Bourbon [4]. Tomato was consumed by commoners both fresh and as a slow cooked sauce, also because of the influence of the Spanish gastronomy [5]. It is especially in this century that varieties with different fruit shapes started to have different destinations of use, a phenomenon that has led to the selection of specific tomato types (e.g., the San Marzano) later employed for industrial canning [3]. After WWII, Italian production reached its peak (in terms of commercial product and dedicated area) during the 1970s, after the widespread diffusion of contemporary varieties [6]. This expansion was accompanied by a strong decline in landrace cultivation, which until recently are predominantly grown in amateur family gardens and in small-size farms for local markets.

In tomato, as well as in other crops, landraces are being recognized as an important genetic material for their yield stability, cultural value, adaptability to low-input and organic farming, resilience to stress, and fruit quality traits [7–10]. Tomato landraces are of increasing scientific interest also as a source of adaptive traits in the face of climate change and for sustainable fruit quality [11,12]. Moreover, landraces are of commercial importance following the rediscovery of local food systems and the promotion of short food supply chains [13]. These values led to the introduction of EU quality schemes, which promoted the cultivation of some traditional landraces (i.e., San Marzano and Pomodorino Vesuviano, for the Campania region), also because of the possibility of limiting the unauthorized production or marketing of goods using such a name. Regrettably, besides some globally recognized names, most of this germplasm is neglected and at risk of extinction.

In Italy the loss of tomato landrace has been significant especially in the regions where tomato was most diffused [14]. Until the 1950s, tomato breeding was predominantly based on mass and pedigree selection using local germplasm, while hybridization was later exploited [6]. Therefore, in germplasm collections, traditional varietal names can be associated with both landraces and old selected lines [15]. Another problem is the presence of so-called contaminant genotypes, deriving from the fortuitous or voluntary introduction in cultivation of morphologically related varieties, deriving also from crosses and/or on-farm seed propagation of hybrid cultivars [16–18]. Moreover, the intrinsic lack of uniformity and unequivocal descriptors of traditional landraces creates the opportunity for spurious, uncontrolled seed marketing, especially from on-line and unprofessional retailers [19–21]. Landraces are mainly distinguished by their geographical location and folk taxonomy. Often, tomato landraces or landrace groups are recognized by farmers in terms of fruit type [22]. Nonetheless, the presence of a clear genetic distinction to support landraces' authenticity is not often available, increasing the need for a molecular discrimination [15].

Southern Italy is considered a secondary center of diversity for tomato [23,24], and in view of the above-mentioned reasons, it is not surprising that the Campania region, and more generally, Southern Italy is a rich reservoir of crop landraces [25]. During the transition to intensive modern agriculture, the seed industry in Southern Italy was in its infancy and it is possible to presume that most of the local germplasm has not been involved in formal breeding programs, for instance, by being engaged in the selection of (pre-)breeding material. Therefore, still today the pool of tomato landraces that are locally cultivated may be a genetically different group from contemporary varieties. In this work, to test this hypothesis, we analyzed and compared a population of local landraces (e.g., collected from farms of the Campania region) with an equivalent set of contemporary cultivars, picked among the most diffuse in cultivation. Among the different DNA molecular markers available for tomato [26], we used Simple Sequence Repeats (SSRs) because these highly variable multiallelic DNA markers are considered most suitable to reveal recent demographic events [27,28]. Previous works were dedicated to specific (fruit) types, single or national collections of tomato landraces [29–34]. Our specific aims were: (i) to evaluate the level and the distribution of the molecular diversity among and within the collected landraces and a set of common contemporary varieties, going beyond the single tomato landrace or landrace group evaluation; (ii) to quantify the genetic differentiation and highlight locus-specific differences that may contribute to the maintenance of a genetic diversity. Our work provided a solid and experimentally validated basis for the uniqueness of tomato landraces of a specific region, with the prospect of strengthening market opportunities and adding value for cultivation.

2. Materials and Methods

2.1. Plant Material and DNA Isolation

This work was carried out on a selection of 15 tomato (*Solanum lycopersicum* L.) landraces of the Campania region, mainly collected in the area to produce the "Pomodoro San Marzano dell'Agro Sarnese-nocerino DOP" and of the "Pomodorino del Piennolo del Vesuvio DOP", two Protected Denomination of Origin (PDO) EU label schemes for the tomatoes of the Campania regions. Seeds were multiplied before analysis at the Department of Agricultural Sciences, University of Naples Federico II. This population was compared to an equal number of cultivars, from different seed companies, selected among the most diffused in the Campania and Puglia regions in the 2009–2014 period according to the cultivated area. As possible output for cladistic classification, we also analyzed two tomato wild relatives, namely *S. habrochaites* and *S. neorickii*. The list, code, and main characteristics of the fruit are reported in the Supplementary Table S1. Seeds were germinated on moist filter paper in Petri dishes and then plantlets were transferred to polystyrene seed trays in a growing chamber (24 °C). For DNA isolation, the first two true leaves from two different plants per accession were collected and immediately frozen in liquid nitrogen. Total DNA isolation and quantification were performed as previously described, starting from approximately 150 mg of finely ground frozen tissue [35].

2.2. SSR Analysis

To genotype the plant material, we used 14 Simple Sequence Repeat (SSR) loci that represent different repeat classes [36,37]. The loci and their main characteristics are presented in the Supplementary Table S2. Polymerase Chain Reaction (PCR) amplifications were performed independently on two plants per accession. Each reaction was assembled using 20 ng genomic DNA template, 1.5 mM $MgCl_2$, 100 µM deoxyribonucleotide-triphosphates, 0.2 µM fluorescently labeled forward primer and unlabeled reverse primer (Supplementary Table S2), and 0.5 U GoTaq DNA polymerase (Promega, Milan, Italy), for a final volume of 25 µL. PCR reactions were performed in a Verity Pro thermal cycler (Thermo Fisher, Milan, Italy). The PCR cycle was as previously described using the annealing temperature indicated in the Supplementary Table S2 [38]. For allelic discrimination, amplicons were analyzed by fluorescent capillary electrophoresis (dye-labels are reported in Supplementary Table S2), with the POP6 polymer (Thermo Fisher) in an ABI Prism 3100 Genetic Analyzer (Thermo Fisher). Signal peak height and allele sizes were calculated using the ABI Prism GeneMapper software v. 4.1 (Thermo Fisher, Milan, Italy) calibrated on the GeneScan 500Liz dye size standard (Thermo Fisher). Values were rounded to integer, and if necessary scaled, based on the SSR core size (Supplementary Table S2), minimizing for each locus the average offset of the scaled alleles within the instrumental resolution of the DNA separation (±1 bp) (Supplementary Figure S1).

2.3. Data Analysis

For each SSR locus, we calculated the number of different alleles (for the whole collection and for *S. lycopersicum*), the observed heterozygosity (Ho; number of heterozygotes/N), the number of effective alleles (Ne; = $1/(Sum\ p_i^2)$), the Polymorphic Information Content (PIC; $1 - Sum\ (p_i^2)$, equivalent to the expected heterozygosity), the Shannon's Information Index (I; $-1 \times Sum\ (p_i \times Ln\ (p_i))$), and the Wright's fixation index, where N is number of individuals, p_i is the frequency of the i-th allele of a locus, and $Sum\ (p_i^2)$ represents the sum of the squared p_i. These calculations were performed with Genalex 6.5 [39]. Evenness of alleles was obtained with poppr [40]. This R-library was also employed to calculate pairwise genetic distances between genotypes according to the Prevosti coefficient [41]. Hierarchical clustering based on the unweighted pair-group method with arithmetic averages (UPGMA) algorithm, cophenetic correlation and tree visualization were performed in R using the factoextra package [42]. The Agglomerative Cluster (AC) value was computed as reported [43]. The genotype accumulation curve was built by boxplotting the number of multilocus genotypes obtained for an increasing number of SSRs (from 1 to plateau, 10 for cultivars and 12 for landraces) for each predefined population. The data distribution for each number of SSRs was obtained by random sampling loci (n = 10,000) [40]. As model-free ordination technique we used the Principal Coordinate Analysis (PCoA), carried out via covariance matrix with distance data standardization. The Analysis of Molecular Variance (AMOVA) was performed using the pairwise distance matrix with 9999 permutations to test for significance. These calculations were carried out using the Genalex 6.5 software [39]. To validate the population structure, the Fst and Gst between *S. lycopersicum* populations were calculated per locus and globally and statistically evaluated against a null distribution obtained from 10,000 permutations on alleles, using the MSA 4.05 software [44].

3. Results

3.1. Genetic Diversity of the Germplasm under Investigation

The analysis of the genetic diversity of the entire set of genotypes (n = 32) was carried out with 14 SSRs. Differences within each accession were not detected. All loci were polymorphic yet, the locus LEcaa001 was polymorphic only between wild species and *S. lycopersicum* and therefore, it was excluded for comparisons between landraces and contemporary cultivars. Main genetic parameters of the tomato collection under investigation are presented in Table 1. We detected 75 different alleles for an average of 5.35 alleles per locus. Large differences were present in the number of alleles per locus, which ranged from 2 (LELE25 and LEcaa001) to 11 (LEEF1Aa). The latter was the most diverse locus also considering the number of effective alleles, a measure of diversity weighted for allele frequencies. There was not a significant difference (p > 0.05; Student's *t*-test) between the allelic richness in *S. lycopersicum* subgroups, with an average of 3.0 (respectively, 2.8) alleles per locus in the landraces (resp., contemporary varieties). The observed heterozygosity (Ho) was low (on average, 0.17) but, as expected, highly affected by the tomato groups, because heterozygote genotypes were present at one locus for the open-pollinated landraces (Legaa003) and at 12 SSRs for the contemporary (mostly hybrid) varieties. Nonetheless, two polymorphic loci (LELE25 and LEct001) were fixed in the whole germplasm collection and lacked heterozygotes. A significant correlation between the number of alleles per locus and its observed heterozygosity was present (p = 0.005; Spearman's Rho). Considerable differences among SSRs were also present in the PIC, which ranged from 0.03 (LEcaa001) to 0.81 (LEEF1Aa), the locus with the highest number of alleles. However, the most diverse loci were LE20592 and LEtat002 considering the evenness of distribution of alleles for each SSR.

To evaluate the relationship between the genotypes, pairwise genetic distances were used to build a dendrogram (Figure 1). In this analysis, the two wild species served as possible outgroup. Genetic distances were summarized in an heatmap to simultaneously visualize how the genetic relatedness varies between genotypes and clusters of genotypes (Supplementary Figure S2). The cophenetic correlation was high (Pearson's product-moment correlation: 0.80) and significant (p < 0.001). The Agglomerative Cluster value was 0.65. The dendrogram illustrated that each genotype had a distinct SSR profile and duplicate accessions were not present in our landrace collection. Moreover, the dendrogram indicated that the genotypes could be clearly separated in biologically meaningful groups at k = 4. The most taxonomically distant wild species had a distinctive position at the highest hierarchical node (k = 2). Within the *S. lycopersicum* germplasm, the landraces and the contemporary varieties agglomerated in two separate clusters, suggesting the presence of two genetically distinct populations (Figure 1).

Table 1. Main genetic indices estimated with the SSR.

SSR	Na	NaSl	Ne	Ho	PIC	I	F	Ev	ALC
LE20592	5	3	3.35	0.25	0.70	1.30	0.64	0.88	8
LELE25	2	2	1.07	0.00	0.06	0.15	1.00	0.44	2
LE21085	5	3	2.34	0.19	0.57	1.11	0.67	0.66	8
LEEF1Aa	11	11	5.22	0.16	0.81	1.99	0.81	0.66	13
LEta003	8	5	3.39	0.25	0.71	1.45	0.65	0.75	9
LEat002	4	2	1.95	0.13	0.49	0.85	0.74	0.71	6
LEga003	6	4	1.78	0.28	0.44	0.94	0.36	0.50	7
LEtat002	5	3	3.35	0.25	0.70	1.30	0.64	0.88	9
LEaat002	5	3	2.42	0.13	0.59	1.04	0.79	0.78	7
LEcaa001	2	1	1.03	0.03	0.03	0.08	−0.02	0.38	2
LEaat007	5	3	2.42	0.13	0.59	1.04	0.79	0.77	7
LEct001	4	3	2.32	0.00	0.57	0.98	1.00	0.78	5
LEctt001	6	5	2.05	0.38	0.51	1.05	0.27	0.56	7
LEta015	7	6	3.80	0.19	0.74	1.54	0.74	0.77	9

Legend: Na: total number of alleles; NaSl: total number of alleles in *S. lycopersicum*; Ne: number of effective alleles; Ho: observed heterozygosity; PIC: polymorphic information content; I = Shannon's Information Index; F: fixation index; Ev: evenness; ALC: allelic combinations.

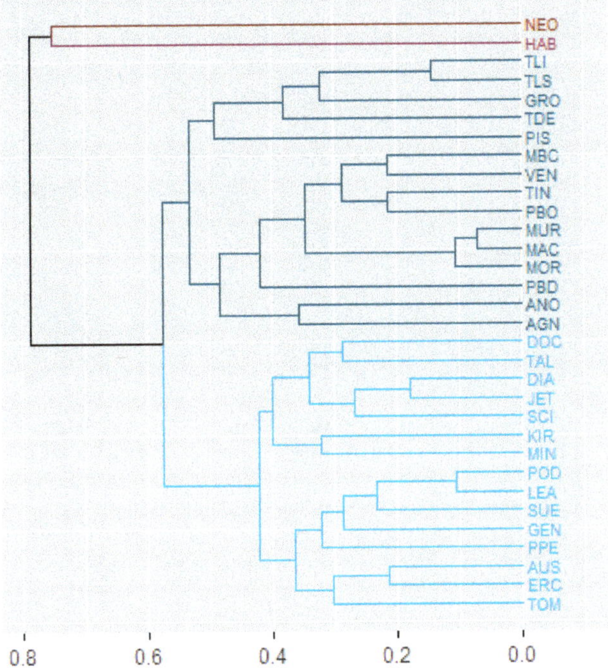

Figure 1. Dendrogram of the tomato genotypes based on the UPGMA cluster using the Prevosti (1975) genetic distance (the scale is reported at the bottom). Different colors denote the subclusters at *k* = 4 namely, red garnet: *S. neorickii*; purple: *S. habrochaites*; blue navy: *S. lycopersicum* landraces; azure: *S. lycopersicum* contemporary varieties. The codes of the genotypes are available in Table S1.

Within the landraces, some small groups of genotypes with a similar fruit shape (Supplementary Table S1) could be also identified, such as the MUR-MAC-MOR subcluster (with "San Marzano"- type fruit and collected in the related PDO area), the ANO-AGN ("pomodorino" type, both collected in the valley of the Sarno river), and TLI-TLS ("tondo" type, but originating from different cultivation areas).

Finally, before testing for a possible genetic structure, we verified if the number of employed SSRs is suitable to capture the diversity in our germplasm. To this aim, we built genotype accumulation curves, considering the landrace or the contemporary varieties group as defined by our a priori classification. The result indicated that, also taking into account the possible specific polymorphisms within each group, the employed number of SSR loci is above the minimum number needed to fully capture the diversity in our sub-groups (Figure 2). For every number of randomly sampled SSRs, there was not a statistical difference in the average number of identified multilocus genotypes considering landraces or cultivars ($p > 0.05$, Student's *t*-test).

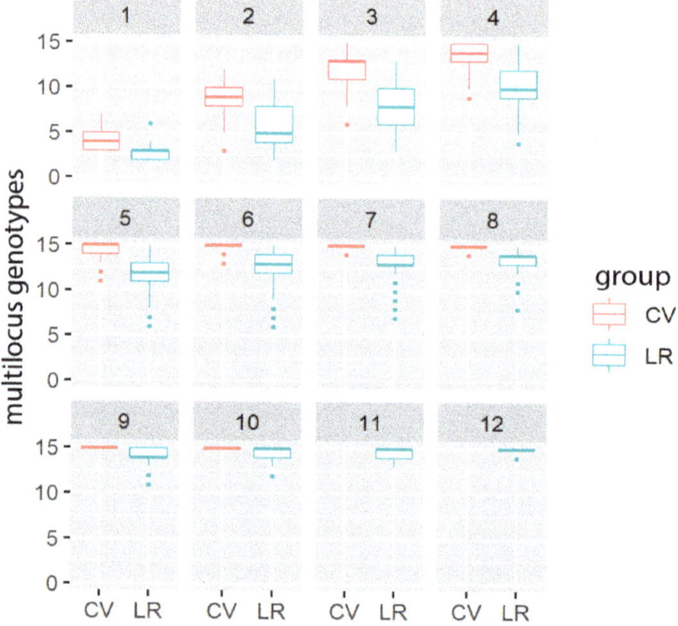

Figure 2. Genotype accumulation analysis in the two pre-defined *S. lycopersium* populations. In each panel, the boxplots summarize the descriptive statistics relative to the quartiles of the number of multilocus genotypes obtained by randomly sampling loci without replacement ($n = 10,000$). The number of sampled SSR loci (from 1 to 12) is indicated in the top dark gray bar. Dots represent outliers (i.e., values outside 1.5 times the interquartile range above the upper and below the lower quartile). Contemporary varieties (respectively, landraces) boxplots are in deep salmon (resp., cyan) color.

3.2. Genetic Structure Analysis of the S. lycopersium Genotypes (Landraces and Contemporary Varieties)

To infer a population subdivision between contemporary varieties and landraces, we performed a Principal Coordinate Analysis. To achieve this goal, we excluded the wild species because of their distant relatedness and limited number of samples. The first two components explained 39.7% of the total variance, which implies that the different SSRs were useful in sampling a mostly uncorrelated, locus-specific genetic variation. The scatter plot of PC1 and PC2 values shows that the samples belonging to the two pre-defined populations are well separated along both PC1 and PC2 (Figure 3). Moreover, the samples of each population were similarly spread on the two-dimensional plane, an indication of a comparable level of diversity.

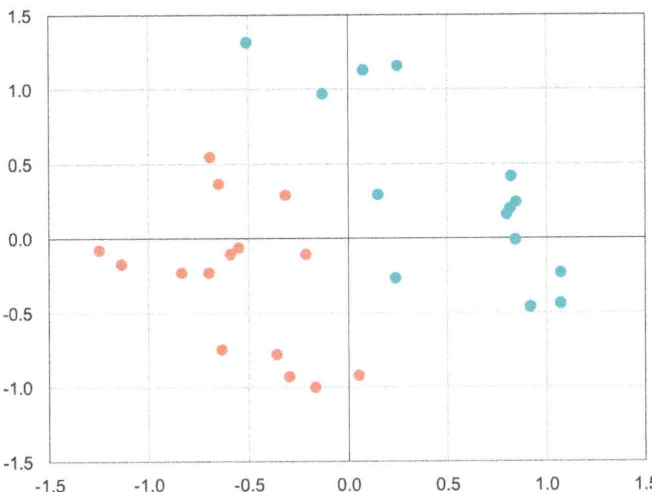

Figure 3. Principal Coordinate Analysis scatter plot of the tomato landraces and contemporary varieties under investigation. Each genotype is represented by a colored circle (deep salmon: contemporary cultivars; cyan: landraces). PC1 (respectively, PC2) explained 22.7 % (resp. 17.0%) of the variance.

Consequently, we performed an Analysis of Molecular Variance to quantify the level of genetic differentiation between the predefined groups (Table 2). A main contribution to the genetic variance in our germplasm originated from the difference between samples in each subpopulation and not predominantly from all samples, as in the case of a panmictic population. Moreover, the differentiation between landraces and contemporary cultivars was high, with a 22% percent of the molecular variation present between the two subpopulations (Table 2). For these reasons, the AMOVA analysis provided evidence to support a population structure.

Table 2. Analysis of the Molecular Variance (AMOVA) table statistics.

Source of Variation	df	SS	MS	Var	%
Between Pops	1	31.583	31.583	0.884	22
Between samples within Pop	28	141.667	5.060	1.938	30
Within samples	30	35.500	1.183	1.183	48
Total	59	208.750		4.006	100

Legend: df, degree of freedom; SS: sum of squares; MS: Mean squares; Var: estimated variance component; %: percentage of genetic variation.

Finally, to validate the AMOVA results, we calculated two widely used measures of population differentiation, Fst and Gst (according to Nei), per locus and globally [45], using a distribution obtained by allele permutations for the statistical significance (Table 3). The data indicated that amount of the genetic differentiation between populations is significant. Interestingly, there were clear differences among loci, with five loci with statistically significant high Fst values. In the presence of a genetic structure, this suggests the possible presence of locus-specific adaptive molecular variation.

Table 3. Statistical analysis of the level of genetic differentiation between landraces and contemporary cultivars according to the *S. lycopersicum* polymorphic SSR loci.

	Fst	*p* Value	Gst	*p* Value
All loci	0.22	0.000	0.15	0.000
LE20592	0.34	0.000	0.23	0.000
LE21085	0.08	0.102	0.15	0.002
LEaat007	0.05	0.237	0.05	0.239
LEat002	0.04	0.284	0.05	0.284
LEctt001	0.20	0.001	0.60	0.000
LEEF1Aa	0.15	0.002	0.11	0.002
LEga003	0.12	0.031	0.08	0.031
LELE25	0.00	1.000	0.03	1.000
LEta003	0.08	0.075	0.07	0.074
LEta003	0.08	0.075	0.07	0.074
LEta015	0.15	0.008	0.11	0.006
LEtat002	0.05	0.237	0.05	0.239

4. Discussion

The tomato is characterized by limited genetic diversity, due to its evolutionary history [23,46]. The breeding sector is dominated by hybrids that have progressively replaced local accessions. Previous molecular characterizations focused on specific landraces, fruit classes or national collections of tomato landraces [29–34]. In this work, we tested the hypothesis that a set of regional tomato landraces of Southern Italy represents a genetically differentiated group compared to contemporary varieties because of their origin and possible non-engagement of the breeding sector. The cladistic classification of the genotypes indicated that every accession had a unique profile. The polymorphic SSRs confirmed to be suitable in distinguish closely related tomato landraces [18]. The most informative SSR was LEEF1Aa, probably because of a composite core motif. This locus was also previously described as highly informative in tomato [47]. As expected, wild relatives had also SSR alleles other from *S. lycopersicum* [33]. The total number of alleles was similar or higher than in other published works [36,37,48,49], indicating a relevant allelic diversity for both landraces and contemporary varieties. It should be added that for landraces, the molecular variation is expected to include both human-driven selection and adaptation to local conditions [50]. Moreover, the selected varieties were chosen to represent different tomato market classes, which associate to different patterns of genetic variation [51]. Also in another paper based on landraces and contemporary tomato varieties [52], the average number of SSR alleles was judged higher compared to the tomato literature, even if the loci employed are not the same of our work. While the presence of distinctive profile for cultivars is not surprising, synonymous cases and intra-varietal variability has been reported for landraces of tomato and other species [53–55]. The explorative hierarchical cluster indicated also biologically coherent and highly balanced clusters of cultivars and landraces at $k = 4$, although the optimal number of clusters was not statistically evaluated. However, the cophenetic correlation was high (≥ 0.80), a reliable signal of a population structure [56]. The relatively low AC value of the UPGMA tree can be justified considering that even for highly structured populations, the "chaining" effect, unique genotypes, and small groups of accessions (together with two or more large groups) can strongly lower (less than 0.6) this index [57]. Finally, also considering the output group, the dendrogram analysis implied a similar level of diversity between the two pre-defined populations in terms of pairwise distances, which should be further explored.

To scan for a genetic structure, we performed a PCoA mainly because it does not require assumptions that may not be hold true for a selected germplasm collection (e.g., HW or linkage equilibrium) [51], and the intrinsic issues of model-based testing for $k = 2$ [58]. The analysis indicated that the *S. lycopersicum* genotypes are well separated according to their pre-defined population [51]. A complete distinction between landraces and contemporary varieties and a confirmation of the UPGMA clustering have not been always found [52]. It is necessary to add that work analyzed a nation-wide group of landraces from geographically diverse collections. Landraces of different secondary centers of diversity (i.e., Spain and Italy) were genetically differentiated [59], suggesting that the complete separation observed in our study could be also due to a more restricted geographic area of origin of our germplasm. The level of population differentiation was high [60], and in line with previous analysis of tomato germplasm that underwent conventional breeding [51]. An important implication is that despite the wide diffusion of contemporary varieties, the admixture between our local landraces and cultivars is very low, considering that we analyzed cultivars that were among the most cultivated in the collection sites. Farmers tend to save seeds from one year to another by choosing the most representative fruits, favoring therefore a maintenance breeding for the fruit morphology (e.g., size and shape), phenological characteristics (e.g., time of flowering and maturity), and quality traits (e.g., color and flavor) they attribute to their accession (Zeven, 2000). We also detected some significant differences among loci in highlighting a genetic differentiation. High Fst values (compared to neutral loci) are typically suggestive of genomic regions that are under divergent selection. Nonetheless, at each locus, Fst values are also influenced by heterozygosity and the mutation rate, two features that are more variable among multiallelic SSRs than in bi-allelic markers. For these reasons, the data motivate further genomic scans to pinpoint the effects of the adaptive and/or breeding processes that led to a differentiation between tomato landraces and contemporary varieties [61].

5. Conclusions

Our study provides a first comparative assessment of the diversity and population structure in a geographically specific collection of tomato landraces from a secondary center of diversity. The data indicated that the landraces constitute a genetically distinct population from common commercial varieties, in addition to having a historical origin and a locally recognized gastronomic identity. Therefore, our work provides a robust justification for implementing measures for in situ and ex situ conservation actions, as well as for creating premium regional brands that can even go beyond the worldwide known, individual landrace names of the Campania region (e.g., San Marzano, piennolo/Pomodorino Vesuviano). Moreover, multivariate analyses indicated that the landraces are characterized by a good, and to a relevant extent, specific diversity. Finally, our work encourages further multidisciplinary studies to unravel the genetic factors responsible for the adaptation to specific environments and potential quality traits of our landraces.

Supplementary Materials: The following are available online at https://www.mdpi.com/2073-4395/11/3/564/s1, Figure S1: an example of the electropherograms at the locus LE20592 showing the allele peaks. For each genotype (indicated on the left; see Table S1 for the code), the height of the peak refers to the scale on the right-hand side (RFU). The top bar indicates the size reference range provided by the genotyping software, used to calibrate the peak data points to their DNA size according to the internal size standard. The rounded dimension of each fragment (bp) is indicated in the rectangle below. Figure S2: heatmap of the pairwise genetic distance (Prevosti) between genotypes (codes are reported in Table S1). The bar on the right side shows the color scale adopted to represent distances, from deep blue (0) to white (1). Table S1: List of the genotypes under investigation and main features of their fruit shape. Table S2: SSR loci employed in this study and their main features.

Author Contributions: Conceptualization, G.C.; formal analysis, Y.R. and G.C.; investigation, M.C. and G.C.; writing—original draft preparation, G.C.; writing—review and editing, Y.R. and G.C. All authors have read and agreed to the published version of the manuscript.

Funding: This research was partially funded by the project "Laboratorio di genomica per l'innovazione e la valorizzazione della filiera pomodoro" (MIUR-art 12 del DM 593/00, Cod. DM17732).

Institutional Review Board Statement: Not applicable.

Informed Consent Statement: Not applicable.

Data Availability Statement: Data is contained within the article or Supplementary Materials.

Conflicts of Interest: The authors declare no conflict of interest.

References

1. Gentilcore, D. Taste and the tomato in Italy: A transatlantic history. *Food Hist.* **2009**, *7*, 125–139. [CrossRef]
2. Bergougnoux, V. The history of tomato: From domestication to biopharming. *Biotechnol. Adv.* **2014**, *32*, 170–189. [CrossRef]
3. Gentilcore, D. *Pomodoro! A History of the Tomato in Italy*; Columbia University Press: New York, NY, USA, 2010.
4. Astarita, T. *The Italian Baroque Table: Cooking and Entertaining from the Golden Age of Naples. Medieval and Renaissance Texts and Studies*; ACMRS Press: Tempe, AZ, USA, 2014; Volume 459.
5. Latini, A. *Lo Scalco Alla Moderna*, 1993 ed.; Biblioteca Culinaria: Lodi, Italy, 1694; Volume 2.
6. Acciarri, N.; Di Candilo, M.; Sanguineti, C.; Soressi, G.P. Tomato. In *Italian Contribution to Plant Genetics and Breeding*; Scarascia Mugnozza, G.T., Pagnotta, M.A., Eds.; Università della Tuscia: Viterbo, Italy, 1998; pp. 501–509.
7. Sumalan, R.M.; Ciulca, S.I.; Poiana, M.A.; Moigradean, D.; Radulov, I.; Negrea, M.; Crisan, M.E.; Copolovici, L.; Sumalan, R.L. The Antioxidant Profile Evaluation of Some Tomato Landraces with Soil Salinity Tolerance Correlated with High Nutraceuticaland Functional Value. *Agronomy* **2020**, *10*, 500. [CrossRef]
8. Landi, S.; De Lillo, A.; Nurcato, R.; Grillo, S.; Esposito, S. In-field study on traditional Italian tomato landraces: The constitutive activation of the ROS scavenging machinery reduces effects of drought stress. *Plant Physiol. Biochem.* **2017**, *118*, 150–160. [CrossRef]
9. Digilio, M.C.; Corrado, G.; Sasso, R.; Coppola, V.; Iodice, L.; Pasquariello, M.; Bossi, S.; Maffei, M.E.; Coppola, M.; Pennacchio, F. Molecular and chemical mechanisms involved in aphid resistance in cultivated tomato. *New Phytol.* **2010**, *187*, 1089–1101. [CrossRef]
10. Corrado, G.; Rao, R. Towards the genomic basis of local adaptation in landraces. *Diversity* **2017**, *9*, 51. [CrossRef]
11. Fullana-Pericàs, M.; Conesa, M.À.; Douthe, C.; El Aou-ouad, H.; Ribas-Carbó, M.; Galmés, J. Tomato landraces as a source to minimize yield losses and improve fruit quality under water deficit conditions. *Agric. Water Manag.* **2019**, *223*, 105722. [CrossRef]
12. Carillo, P.; Kyriacou, M.C.; El-Nakhel, C.; Pannico, A.; dell'Aversana, E.; D'Amelia, L.; Colla, G.; Caruso, G.; De Pascale, S.; Rouphael, Y. Sensory and functional quality characterization of protected designation of origin 'Piennolo del Vesuvio'cherry tomato landraces from Campania-Italy. *Food Chem.* **2019**, *292*, 166–175. [CrossRef] [PubMed]
13. Casals, J.; Rull, A.; Segarra, J.; Schober, P.; Simó, J. Participatory Plant Breeding and the Evolution of Landraces: A Case Study in the Organic Farms of the Collserola Natural Park. *Agronomy* **2019**, *9*, 486. [CrossRef]
14. Grandillo, S.; Mustilli, A.C.; Parisi, M.; Morelli, G.; Giordano, I.; Bowler, C. Tecniche avanzate per la valutazione qualitativa del pomodoro: Il caso Campania. *Agroindustria* **2004**, *3*, 151–159.
15. Villa, T.C.C.; Maxted, N.; Scholten, M.; Ford-Lloyd, B. Defining and identifying crop landraces. *Plant Genet. Resour.* **2005**, *3*, 373–384. [CrossRef]
16. Cleveland, D.A.; Daniela, S.; Smith, S.E. A biological framework for understanding farmers' plant breeding. *Econ. Bot.* **2000**, *54*, 377–394. [CrossRef]
17. Rao, R.; Corrado, G.; Bianchi, M.; Di Mauro, A. (GATA) 4 DNA fingerprinting identifies morphologically characterized 'San Marzano'tomato plants. *Plant Breed.* **2006**, *125*, 173–176. [CrossRef]
18. Caramante, M.; Rao, R.; Monti, L.M.; Corrado, G. Discrimination of 'San Marzano'accessions: A comparison of minisatellite, CAPS and SSR markers in relation to morphological traits. *Sci. Hortic.* **2009**, *120*, 560–564. [CrossRef]
19. Muñoz-Falcón, J.E.; Vilanova, S.; Plazas, M.; Prohens, J. Diversity, relationships, and genetic fingerprinting of the Listada de Gandía eggplant landrace using genomic SSRs and EST-SSRs. *Sci. Hortic.* **2011**, *129*, 238–246. [CrossRef]
20. Muthoni, J.; Nyamongo, D. Seed systems in Kenya and their relationship to on-farm conservation of food crops. *J. New Seeds* **2008**, *9*, 330–342. [CrossRef]
21. Louwaars, N.P. Plant breeding and diversity: A troubled relationship? *Euphytica* **2018**, *214*, 1–9. [CrossRef]
22. Zeven, A.C. Landraces: A review of definitions and classifications. *Euphytica* **1998**, *104*, 127–139. [CrossRef]
23. Bauchet, G.; Causse, M. Genetic diversity in tomato (*Solanum lycopersicum*) and its wild relatives. *Genet. Divers. Plants* **2012**, *8*, 134–162.
24. García-Martínez, S.; Andreani, L.; Garcia-Gusano, M.; Geuna, F.; Ruiz, J.J. Evaluation of amplified fragment length polymorphism and simple sequence repeats for tomato germplasm fingerprinting: Utility for grouping closely related traditional cultivars. *Genome* **2006**, *49*, 648–656. [CrossRef]
25. Hammer, K.; Laghetti, G.; Perrino, P. Collection of plant genetic resources in South Italy, 1988. *Die Kult.* **1989**, *37*, 401–414. [CrossRef]
26. Causse, M.; Giovannoni, J.; Bouzayen, M.; Zouine, M. *The Tomato Genome*; Springer: Berlin/Heidelberg, Germany, 2016.
27. Zhivotovsky, L.A.; Rosenberg, N.A.; Feldman, M.W. Features of evolution and expansion of modern humans, inferred from genomewide microsatellite markers. *Am. J. Hum. Genet.* **2003**, *72*, 1171–1186. [CrossRef] [PubMed]
28. Cornuet, J.-M.; Ravigné, V.; Estoup, A. Inference on population history and model checking using DNA sequence and microsatellite data with the software DIYABC (v1. 0). *BMC Bioinform.* **2010**, *11*, 1–11. [CrossRef]
29. Terzopoulos, P.; Bebeli, P. DNA and morphological diversity of selected Greek tomato (*Solanum lycopersicum* L.) landraces. *Sci. Hortic.* **2008**, *116*, 354–361. [CrossRef]

30. Henareh, M.; Dursun, A.; Abdollahi-Mandoulakani, B.; Haliloğlu, K. Assessment of genetic diversity in tomato landraces using ISSR markers. *Genetika* **2016**, *48*, 25–35. [CrossRef]
31. Parisi, M.; Aversano, R.; Graziani, G.; Ruggieri, V.; Senape, V.; Sigillo, L.; Barone, A. Phenotypic and molecular diversity in a collection of 'Pomodoro di Sorrento'Italian tomato landrace. *Sci. Hortic.* **2016**, *203*, 143–151. [CrossRef]
32. Castellana, S.; Ranzino, L.; Beritognolo, I.; Cherubini, M.; Luneia, R.; Villani, F.; Mattioni, C. Genetic characterization and molecular fingerprint of traditional Umbrian tomato (*Solanum lycopersicum* L.) landraces through SSR markers and application for varietal identification. *Genet. Resour. Crop Evol.* **2020**, *67*, 1807–1820. [CrossRef]
33. Mazzucato, A.; Papa, R.; Bitocchi, E.; Mosconi, P.; Nanni, L.; Negri, V.; Picarella, M.E.; Siligato, F.; Soressi, G.P.; Tiranti, B. Genetic diversity, structure and marker-trait associations in a collection of Italian tomato (*Solanum lycopersicum* L.) landraces. *Theor. Appl. Genet.* **2008**, *116*, 657–669. [CrossRef]
34. Cattáneo, R.A.; McCarthy, A.N.; Feingold, S.E. Evidence of genetic diversity within Solanum Lycopersicum L.'Platense'landrace and identification of various subpopulations. *Genet. Resour. Crop Evol.* **2020**, *67*, 2057–2069. [CrossRef]
35. Scarano, D.; Rubio, F.; Ruiz, J.J.; Rao, R.; Corrado, G. Morphological and genetic diversity among and within common bean (*Phaseolus vulgaris* L.) landraces from the Campania region (Southern Italy). *Sci. Hortic.* **2014**, *180*, 72–78. [CrossRef]
36. Smulders, M.; Bredemeijer, G.; Rus-Kortekaas, W.; Arens, P.; Vosman, B. Use of short microsatellites from database sequences to generate polymorphisms among *Lycopersicon esculentum* cultivars and accessions of other Lycopersicon species. *Theor. Appl. Genet.* **1997**, *94*, 264–272. [CrossRef]
37. He, C.; Poysa, V.; Yu, K. Development and characterization of simple sequence repeat (SSR) markers and their use in determining relationships among *Lycopersicon esculentum* cultivars. *Theor. Appl. Genet.* **2003**, *106*, 363–373. [CrossRef]
38. Verdone, M.; Rao, R.; Coppola, M.; Corrado, G. Identification of zucchini varieties in commercial food products by DNA typing. *Food Control* **2018**, *84*, 197–204. [CrossRef]
39. Peakall, R.; Smouse, P.E. GENALEX 6: Genetic analysis in Excel. Population genetic software for teaching and research. *Mol. Ecol. Notes* **2006**, *6*, 288–295. [CrossRef]
40. Kamvar, Z.N.; Tabima, J.F.; Grünwald, N.J. Poppr: An R package for genetic analysis of populations with clonal, partially clonal, and/or sexual reproduction. *PeerJ* **2014**, *2*, e281. [CrossRef] [PubMed]
41. Prevosti, A.; Ocana, J.; Alonso, G. Distances between populations of *Drosophila subobscura*, based on chromosome arrangement frequencies. *Theor. Appl. Genet.* **1975**, *45*, 231–241. [CrossRef]
42. Kassambara, A. *Practical Guide to Cluster Analysis in R: Unsupervised Machine Learning*; CreateSpace Independent Publishing Platform: Scotts Valley, CA, USA, 2017.
43. Kaufman, L.; Rousseeuw, P.J. *Finding Groups in Data: An Introduction to Cluster Analysis*; John Wiley & Sons: Hoboken, NJ, USA, 2009; Volume 344.
44. Dieringer, D.; Schlötterer, C. Microsatellite analyser (MSA): A platform independent analysis tool for large microsatellite data sets. *Mol. Ecol. Notes* **2003**, *3*, 167–169. [CrossRef]
45. Verity, R.; Nichols, R.A. What is genetic differentiation, and how should we measure it—GST, D, neither or both? *Mol. Ecol.* **2014**, *23*, 4216–4225. [CrossRef]
46. Mariette, S.; Tavaud, M.; Arunyawat, U.; Capdeville, G.; Millan, M.; Salin, F. Population structure and genetic bottleneck in sweet cherry estimated with SSRs and the gametophytic self-incompatibility locus. *BMC Genet.* **2010**, *11*, 1–13. [CrossRef]
47. Miskoska-Milevska, E.; Popovski, Z.T.; Nestorovski, T. Usefulness of a locus LeEF1a in the genetic differentiation of tomato varieties. *J. Agric. Food Environ. Sci.* **2018**, *72*, 56–61.
48. Scarano, D.; Rao, R.; Masi, P.; Corrado, G. SSR fingerprint reveals mislabeling in commercial processed tomato products. *Food Control* **2015**, *51*, 397–401. [CrossRef]
49. Ruiz, J.J.; García-Martínez, S.; Picó, B.; Gao, M.; Quiros, C.F. Genetic variability and relationship of closely related Spanish traditional cultivars of tomato as detected by SRAP and SSR markers. *J. Am. Soc. Hortic. Sci.* **2005**, *130*, 88–94. [CrossRef]
50. Corrado, G.; Piffanelli, P.; Caramante, M.; Coppola, M.; Rao, R. SNP genotyping reveals genetic diversity between cultivated landraces and contemporary varieties of tomato. *BMC Genom.* **2013**, *14*, 1–14. [CrossRef]
51. Sim, S.-C.; Van Deynze, A.; Stoffel, K.; Douches, D.S.; Zarka, D.; Ganal, M.W.; Chetelat, R.T.; Hutton, S.F.; Scott, J.W.; Gardner, R.G. High-density SNP genotyping of tomato (*Solanum lycopersicum* L.) reveals patterns of genetic variation due to breeding. *PLoS ONE* **2012**, *7*, e45520. [CrossRef] [PubMed]
52. Gonias, E.D.; Ganopoulos, I.; Mellidou, I.; Bibi, A.C.; Kalivas, A.; Mylona, P.V.; Osanthanunkul, M.; Tsaftaris, A.; Madesis, P.; Doulis, A.G. Exploring genetic diversity of tomato (*Solanum lycopersicum* L.) germplasm of genebank collection employing SSR and SCAR markers. *Genet. Resour. Crop Evol.* **2019**, *66*, 1295–1309. [CrossRef]
53. Boccacci, P.; Aramini, M.; Valentini, N.; Bacchetta, L.; Rovira, M.; Drogoudi, P.; Silva, A.; Solar, A.; Calizzano, F.; Erdoğan, V. Molecular and morphological diversity of on-farm hazelnut (*Corylus avellana* L.) landraces from southern Europe and their role in the origin and diffusion of cultivated germplasm. *Tree Genet. Genomes* **2013**, *9*, 1465–1480. [CrossRef]
54. Mercati, F.; Longo, C.; Poma, D.; Araniti, F.; Lupini, A.; Mammano, M.M.; Fiore, M.C.; Abenavoli, M.R.; Sunseri, F. Genetic variation of an Italian long shelf-life tomato (*Solanum lycopersicon* L.) collection by using SSR and morphological fruit traits. *Genet. Resour. Crop. Evol.* **2015**, *62*, 721–732. [CrossRef]

55. Foroni, I.; Baptista, C.; Monteiro, L.; Lopes, M.S.; Mendonça, D.; Melo, M.; Carvalho, C.; Monjardino, P.; Lopes, D.J.; da Câmara Machado, A. The use of microsatellites to analyze relationships and to decipher homonyms and synonyms in Azorean apples (*Malus* × *domestica* Borkh.). *Plant Syst. Evol.* **2012**, *298*, 1297–1313. [CrossRef]
56. Odong, T.; Van Heerwaarden, J.; Jansen, J.; van Hintum, T.J.; Van Eeuwijk, F. Determination of genetic structure of germplasm collections: Are traditional hierarchical clustering methods appropriate for molecular marker data? *Theor. Appl. Genet.* **2011**, *123*, 195–205. [CrossRef]
57. Johnson, R.A.; Wichern, D.W. *Applied Multivariate Statistical Analysis*; Prentice Hall: Upper Saddle River, NJ, USA, 2002; Volume 5.
58. Janes, J.K.; Miller, J.M.; Dupuis, J.R.; Malenfant, R.M.; Gorrell, J.C.; Cullingham, C.I.; Andrew, R.L. The K = 2 conundrum. *Mol. Ecol.* **2017**, *26*, 3594–3602. [CrossRef]
59. García-Martínez, S.; Corrado, G.; Ruiz, J.J.; Rao, R. Diversity and structure of a sample of traditional Italian and Spanish tomato accessions. *Genet. Resour. Crop Evol.* **2013**, *60*, 789–798. [CrossRef]
60. Frankham, R.; Ballou, S.E.J.D.; Briscoe, D.A.; Ballou, J.D. *Introduction to Conservation Genetics*; Cambridge University Press: Cambridge, UK, 2002.
61. Kim, M.; Jung, J.-K.; Shim, E.-J.; Chung, S.-M.; Park, Y.; Lee, G.P.; Sim, S.-C. Genome-wide SNP discovery and core marker sets for DNA barcoding and variety identification in commercial tomato cultivars. *Sci. Hortic.* **2021**, *276*, 109734. [CrossRef]

Article

Molecular Diversity within a Mediterranean and European Panel of Tetraploid Wheat (*T. turgidum* subsp.) Landraces and Modern Germplasm Inferred Using a High-Density SNP Array

Paola Ganugi [1], Enrico Palchetti [1,*], Massimo Gori [1], Alessandro Calamai [1], Amanda Burridge [2], Stefano Biricolti [1], Stefano Benedettelli [1] and Alberto Masoni [1,3]

1   Department of Agriculture, Food, Environment and Forestry (DAGRI), University of Florence, Piazzale delle Cascine, 18-50144 Firenze, Italy; paola.ganugi@unifi.it (P.G.); massimo.gori@unifi.it (M.G.); alessandro.calamai@unifi.it (A.C.); stefano.biricolti@unifi.it (S.B.); stefano.benedettelli@unifi.it (S.B.); alberto.masoni@unifi.it (A.M.)

2   School of Biological Science, Life Sciences Building, 24 Tyndall Avenue, Bristol BS8 1TQ, UK; amanda.burridge@bristol.ac.uk

3   Department of Biology, Via Madonna del Piano, 6-50019 Sesto Fiorentino, Italy

*   Correspondence: enrico.palchetti@unifi.it

Abstract: High-density single-nucleotide polymorphism (SNP) molecular markers are widely used to assess the genetic variability of plant varieties and cultivars, which is nowadays recognized as an important source of well-adapted alleles for environmental stresses. In our study, the genetic diversity and population genetic structure of a collection of 265 accessions of eight tetraploid *Triticum turgidum* L. subspecies were investigated using 35,143 SNPs screened with a 35K Axiom®array. The neighbor-joining algorithm, discriminant analysis of principal components (DAPC), and the Bayesian model-based clustering algorithm implemented in STRUCTURE software revealed clusters in accordance with the taxonomic classification, reflecting the evolutionary history of the *Triticum turgidum* L. subspecies and the phylogenetic relationships among them. Based on these results, a clear picture of the population structure within a collection of tetraploid wheats is given herein. Moreover, the genetic potential of landraces and wild relatives for the research of specific traits of interest is highlighted. This research provides a great contribution to future phenotyping and crossing activities. In particular, the recombination efficiency and gene selection programs aimed at developing durum wheat composite cross populations that are adapted to Mediterranean conditions could be improved.

Keywords: Axiom 35K Wheat Breeders array; genetic diversity; population structure; wheat genotyping

Citation: Ganugi, P.; Palchetti, E.; Gori, M.; Calamai, A.; Burridge, A.; Biricolti, S.; Benedettelli, S.; Masoni, A. Molecular Diversity within a Mediterranean and European Panel of Tetraploid Wheat (*T. turgidum* subsp.) Landraces and Modern Germplasm Inferred Using a High-Density SNP Array. *Agronomy* **2021**, *11*, 414. https://doi.org/10.3390/agronomy11030414

Academic Editors: Gregorio Barba-Espín and Jose Ramon Acosta-Motos

Received: 3 December 2020
Accepted: 21 February 2021
Published: 24 February 2021

Publisher's Note: MDPI stays neutral with regard to jurisdictional claims in published maps and institutional affiliations.

1. Introduction

Wheat represents the third most important cereal grain and the most widely grown crop in the world [1]. Bread wheat (*Triticum aestivum* L.) and durum wheat (*T. turgidum* L. ssp. *durum*) are the two subspecies predominantly cultivated, used for bread-making or leavened products (cookies, cakes, and pizza) and for semolina products and pasta, respectively. In addition, both wheat species' byproducts are used for animal feed production.

While bread wheat (*T. aestivum*) is hexaploid ($2n = 6x = 42$ chromosomes, AABBDD genomes), durum wheat belongs to the *T. turgidum* tetraploid subspecies group ($2n = 4x = 28$ chromosomes, AABB genomes) which includes six other subspecies (*Triticum carthlicum*, *Triticum dicoccum*, *Triticum dicoccoides*, *Triticum paleocolchicum*, *Triticum polonicum*, and *Triticum turgidum*) rarely grown commercially [2,3]. Many studies based on cytological and molecular analysis ascribe tetraploid wheat's origin to two different evolutionary steps, which started around 10,000 years ago in the Fertile Crescent [4,5]. The first divergent evolution, of which the original progenitor is unknown, gave rise to diploid species including *Triticum urartu* (A genome), *Aegilops tauschii* (D genome), *Hordeum vulgare* (barley), and *Secale cereale* (rye) [6]. The second evolutionary process was a natural hybridization between *T. urartu* (the A genome donor) and an unknown *Triticum* species, often identified as *Aegilops speltoides* (the B genome donor); this created the wild emmer *T. dicoccoides* ($2n = 4x = 28$, BBAA genomes), the progenitor of durum wheat [7]. The history of durum evolution is the result of domestication starting from wild emmer genotypes and of a transition process from a naked emmer type to durum type [8]. Around 7000 years Before Present (BP), durum genotypes reached the Iberian Peninsula, followed by a rapid spread from the East to the West of the Mediterranean Basin [9,10]. Natural and human selection through thousands of years led to the establishment of wheat landraces characterized by strong adaption to the environmental conditions and cultivation practices of different geographic areas [11]. Local traditional farming communities contributed to the maintenance of these landraces that were characterized by different qualitative and quantitative traits until the first decades of the twentieth century [12].

At the beginning of the 20th century, breeders imposed a strong selection based on commercial purposes: local landrace cultivation was progressively abandoned and replaced with improved, widely adapted, and more productive semi-dwarf varieties, resulting in a reduced level of genetic diversity, especially compared to the wild ancestors [13–15]. Today, this lack of diversity is widely recognized as a limiting factor in the breeding of high-yielding and stress-resistant varieties [16]. Moreover, under the current climate change events (irregular rainfall, high temperatures during the growing season, rainstorms, and drought) that negatively affect wheat cultivation, the development of new resilient varieties or composite cross populations (CCPs) adapted to different cultivation environments and low-input agriculture has become necessary [17–19]. Novel genetic diversity selected by breeders may be introduced into modern genotypes by the introgression of useful alleles from landraces, ancestors, or wild relatives through specific breeding programs [20–22]. Durum wheat landraces and other *Turgidum* subspecies usually show a lower yield when compared to modern varieties [11]; nevertheless, they exhibit reduced productive performance compared to elite germplasm (modern varieties), but their higher genetic variability could be useful, allowing them to cope with environmental stress conditions, and to increase resilience to climate change. They are thus a potential source of favorable alleles to improve grain yield or pest resistance and to give other favorable agronomic traits to new varieties [23,24].

Recent breeding programs have studied and assessed genetic variability or different germplasm panels using different research approaches [25–30,30–32]. Morphological and agronomical markers have been considerably used [25,26], with variable reproducibility depending on environmental conditions. Nevertheless, this has been overtaken with the use of molecular markers that guarantee the opportunity of studying wheat phenotypes, providing reproducible and environment-independent results [27]. Several DNA markers have been developed and largely used to assess genetic diversity in tetraploid wheats [28–31], but the high-density genome coverage provided in recent years by single-nucleotide polymorphism (SNP) markers has made them the best choice for wheat genetic analysis [32].

A few years ago, a novel plant breeding approach—evolutionary plant breeding (EB)—relying on human selection acting on a heterogeneous population (i.e., CCPs) started to represent a valuable method for developing populations adaptable to different agricultural contexts [33,34]. Cultivation conditions can drive the selection of more adaptable genotypes that present increased fitness [35,36]. After several years of cultivation and multiplication in the same area under isolated conditions, these populations may reach equilibrium with stable yields, and the genetic diversity among such populations represents a trait resilient to climate and environmental stress [37].

In this study, we investigated the genetic diversity and population structure of a panel of 265 accessions from seven tetraploid *T. turgidum* subspecies originating from different Mediterranean and European areas using the 35K Wheat Breeders' Axiom®SNP array. This work will prove to be a groundwork for phenotypic analysis, both in the field and in the lab, aimed at identifying the best lines that could be used in a cross-breeding program for the selection of resilient and nutritionally improved wheat CCPs.

2. Materials and Methods

2.1. Plant Material

A large tetraploid wheat germplasm panel of 265 accessions was assembled at the Department of Agriculture (DAGRI) of the University of Florence (Supplementary Table S1). The core collection was represented by seeds of 8 *Turgidum* subspecies—ssp. *carthlicum* (5), *dicoccoides* (3), *dicoccon* (28), *durum* (172), *paleocolchicum* (3), *polonicum* (13), *turanicum* (33), and *turgidum* (7)—collected from the USDA bank (U.S. Department of Agriculture; https://npgsweb.ars-grin.gov/gringlobal/search (accessed on 3 December 2020)), Wageningen CGN Germplasm bank (Centre for Genetic Resources, The Netherlands; https://www.wur.nl/en/Research-Results/Statutory-research-tasks/Centre-for-Genetic-Resources-the-Netherlands-1.htm (accessed on 3 December 2020)), and Istituto di Granicoltura di Caltagirone (www.granicoltura.it (accessed on 3 December 2020)). One *T. aestivum* variety—Bologna—was added to the panel as outgroup genotypes.

Seeds were sown in peat-based soil in single pots and maintained in a climatic chamber at 15 °C during the night and 25 °C during the day, with a cycle of 16 h light and 8 h dark. Six weeks after germination, leaf tissue (5–6 cm section of a true leaf) was harvested from plants, immediately frozen on liquid nitrogen, and then stored at −80 °C prior to nucleic acid extraction. All plants were then transplanted in the field and grown until maturity in order to collect seeds for single-seed line constitution to be used in future field studies.

2.2. DNA Extraction and Genotyping

Frozen leaf tissues were ground in a TissueLyzer II bead mill (Qiagen, Hilden, Germany), with the tissue and plastic adapter having previously been dipped into liquid nitrogen to avoid sample warming. Genomic DNA was extracted from the leaf powder using a standard cetyltrimethylammonium bromide (CTAB) protocol [38] and then treated with RNase-A (New England Biolabs UK Ltd., Hitchin, UK) according to the manufacturer's instructions. DNA was checked for quality and quantity by electrophoresis on 1% agarose gel and Qubit™ fluorimetric assay (Thermofisher), respectively. The 35K Axiom®Wheat breed Genotyping Array (Affymetrix, Santa Clara, US) was used to genotype 265 samples for 35,143 SNPs using the Affymetrix GeneTitan®system at Bristol Genomics Facility (Bristol, UK) according to the procedure described in *Axiom®2.0 Assay Manual Workflow User Guide Rev3* (https://assets.thermofisher.com/TFS-Assets/LSG/manuals/70 2991_6-Axiom-2.0-96F-Man-WrkFlw-SPG.pdf (accessed on 3 December 2020)). This array contains a range of probes that are located on chromosomes belonging to the A, B, and D genomes [39]. Since in tetraploid wheat the D genome is lacking, the effective number of markers that can be investigated is lower, corresponding to 24,240 SNPs. Allele calling was carried out using the Axiom Analysis suite software [40], and a variant call rate threshold of 92% was used instead of the default value (97%) to account for the great heterogeneity of the set analyzed [41]. The number of monomorphic and polymorphic SNP markers, the heterozygosity level, and the types of nucleotide substitution for each accession were evaluated using the same software. Monomorphic SNP markers and those with missing data points were excluded from analysis. SNP markers were then filtered for minimum allele frequency (MAF) greater than 1% and failure rate lower than 20%.

2.3. Statistical Analysis

The levels and patterns of genetic diversity among accessions were investigated starting from the data obtained from SNP genotyping. The Tamura–Nei method [42] for genetic distance evaluation was applied to obtain a matrix of pairwise distances among accessions. An unrooted Bayesian tree was computed by applying the neighbor-joining algorithm [43], implemented in the ape 3.1 package of R software [44].

To obtain a clear picture of the genetic structure of the tetraploid wheat genotypes, we applied the Bayesian model-based clustering algorithm implemented in STRUCTURE software version 2.3.4 [45]. An admixtured and shared allele frequency model was used to determine the number of clusters (K), assumed to be in the range between 2 and 15, with five replicate runs for each assumed group. For each run, the initial burn-in period was set to 10,000 with 10,000 MCMC (Markov chain Monte Carlo) iterations, with no prior information on the origin of individuals. The best fit for the number of clusters, K, was determined using the Evanno method [46] as implemented in the program STRUCTURE HARVESTER [47]. Structure results were then elaborated using the R package pophelper to align cluster assignments across replicate analyses and produce visual representations of the cluster assignments. Discriminant analysis of principal components (DAPC) was used to infer the number of clusters of genetically related individuals [48] using the adegenet package in R-project [49]. The first step of DAPC was data transformation using principal component analysis (PCA), while the second step was discriminant analysis performed on the retained principal components (PCs). Groups were identified using k-means, a clustering algorithm that finds a given number (k) of groups maximizing the variation between them. k-means was run sequentially with increasing values of k to identify the optimal number of clusters, and different clustering solutions were compared using the Bayesian Information Criterion (BIC). The optimal clustering solution should present the lowest BIC [50].

3. Results

After SNP dataset filtering, 21,051 SNP markers were identified and used in the statistical analysis to evaluate the genetic diversity of the 265 tetraploid wheat accessions. The genetic relationships in the panel were assessed through three different approaches—neighbor-joining tree, discriminant analysis of principal components (DAPC), and STRUCTURE software—in order to better detail and define the genetic relationship variability among the tetraploid accessions.

The Bayesian tree obtained by applying the neighbor-joining algorithm revealed groups in the population that highly agreed with the subspecies classification and origin (Figure 1A). Most of the *T. turgidum* ssp. *durum* (shown in yellow in Figure 1A) were placed in a large clade together, with modern varieties that appeared separated from the other accessions. Landraces and old varieties were distributed in branches close together, mostly according to their geographical origin, such as the Syrian and Sicilian accessions. Two other clusters were identified, consisting, respectively, of *T. turgidum* spp. *dicoccon* (shown in orange) and *T. turgidum* ssp. *turgidum* (blue), while *T. turgidum* spp. *turanicum* (brown) was clustered into two groups separated by the set of *T. turgidum* ssp. *polonicum* accessions (grey). The two *T. turgidum* ssp. *paleocolchicum* accessions (light blue) and their cross seemed to be close, while the few accessions belonging to *T. turgidum carthlicum* and *dicoccoides* ssp. appeared to be spread amongst the tree branches.

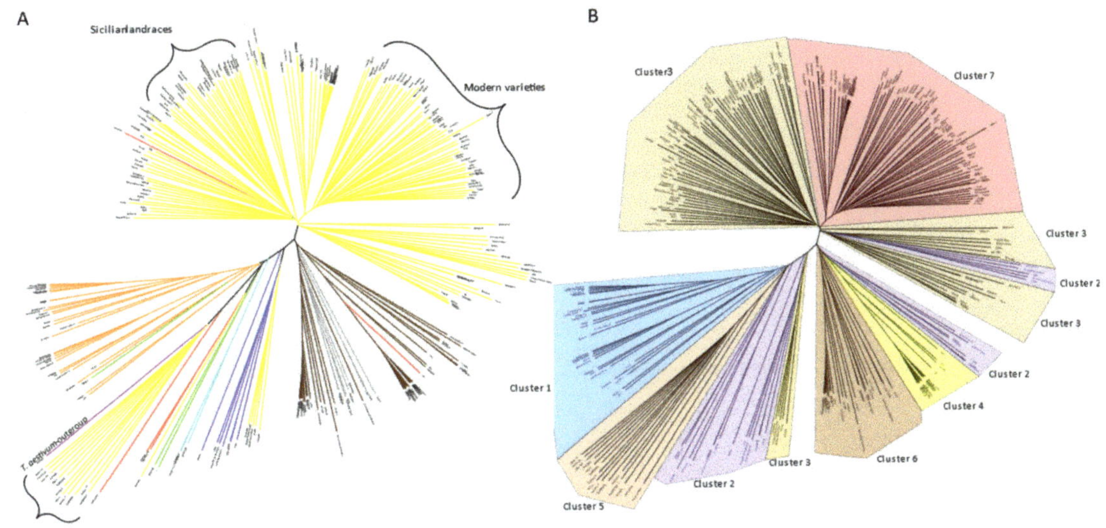

Figure 1. (**A**) Bayesian tree of 265 tetraploid wheat genotypes based on single-nucleotide polymorphism (SNP) genetic markers and colored according to subspecies classification. Branch colors: yellow for *T. turgidum* ssp. *durum*, orange for *T. turgidum* ssp. *dicoccon*, brown for *T. turgidum* ssp. *turanicum*, grey for *T. turgidum* ssp. *polonicum*, blue for *T. turgidum* ssp. *turgidum*, pale blue for *T. turgidum* ssp. *paleocolchicum*, red for *T. turgidum* ssp. *carthlicum*, green for *T.turgidum dicoccoides*, and violet for the *T. aestivum* outgroup accession. (**B**) Phylogenetic tree of 265 tetraploid wheat genotypes based on SNP genetic markers and colored according to discriminant analysis of principal components (DAPC) clusterization.

The wheat genotype arrangement obtained with the Bayesian tree was subsequently confirmed by the DAPC results (Figure 1B, Table S2). Seven clusters (Figure 2) were detected in coincidence with the lowest Bayesian information criterion (BIC) value (Figure S1), and 100 PCs (80% of variance conserved) from PCA were retained. As reported in Figure 1B, the Syrian *T. turgidum* spp. *durum* wheats were pooled in Group 5 and clustered separately in the genetic tree. Most of the old varieties and landraces of the same subspecies were collected in Group 3, while Group 4 was formed by approximately half of the *T. turgidum* spp. *turanicum* accessions, which belonged to the same genetic cluster in the tree. The remaining genotypes of this last subspecies were grouped together with *T. turgidum* spp. *polonicum* wheats which were also clustered in Group 2. Group 1 was entirely composed of *T. turgidum* ssp. *diccocon* accessions, while Group 7 identified the modern varieties of *T. turgidum* ssp. *durum*.

Moreover, the Bayesian tree and the DAPC analysis largely agreed with the accessions' geographic origins. In particular, Syrian (Cluster 5), French (part of the Cluster 7), Moroccan (Cluster 6), and Italian and Algerian (Cluster 3) wheats were almost entirely pooled within the same cluster. Iranian (Clusters 3 and 4) and Portuguese and American (Clusters 2 and 6) accessions were equally divided into two clusters.

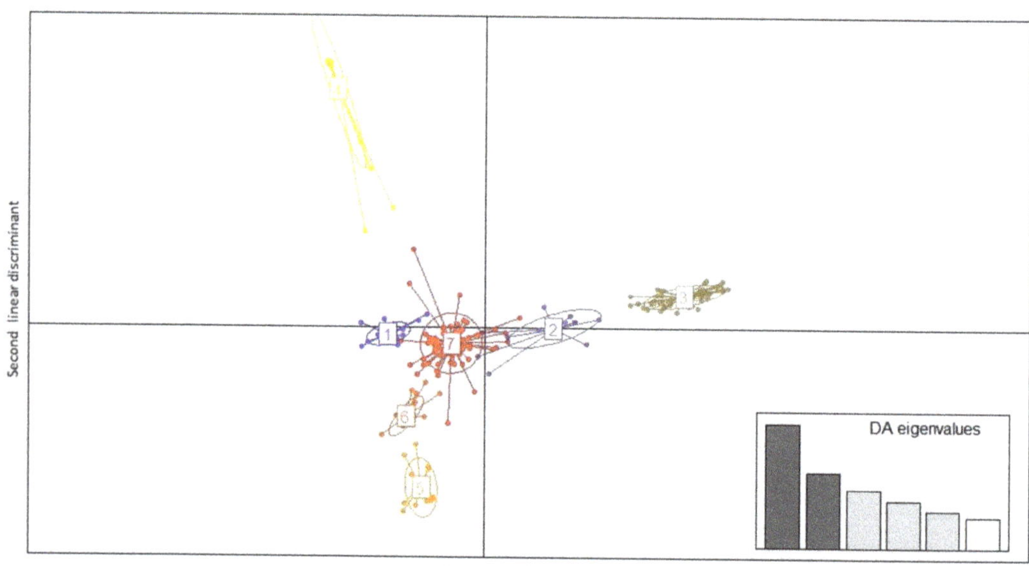

Figure 2. DAPC results for the 265 accessions of *Triticum turgidum* L. used in the analysis. The first two linear discriminants (LDs) are represented by the axes. Each circle with the relative number and color represents one identified cluster, and each dot represents one accession.

The optimum number of subpopulations, *K*, estimated using STRUCTURE software (Figure 3, Table S2) and according to the Evanno method results was 7 (*K* = 7). This indicated the presence of seven subpopulations, as previously found by the Bayesian tree and DAPC analysis, although characterized mostly by different accessions.

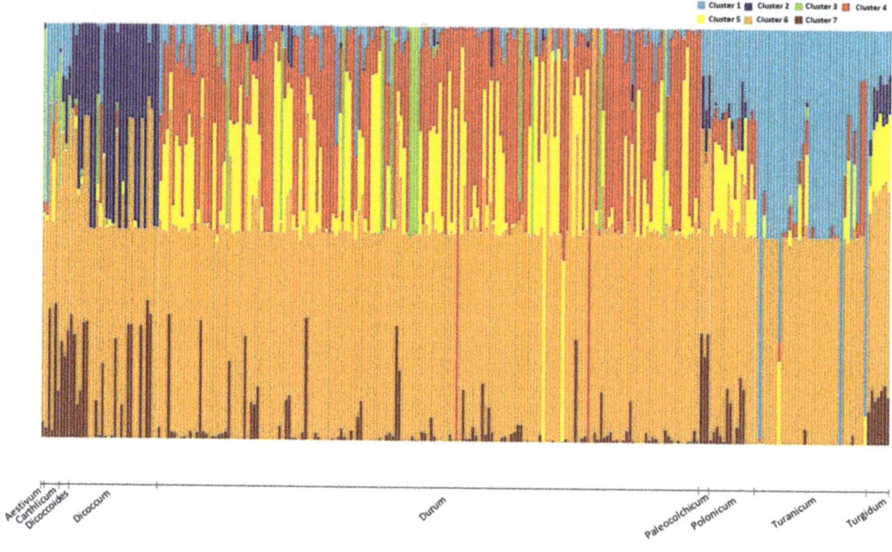

Figure 3. Diversity in admixture analysis by STRUCTURE among the 265 tetraploid wheat accessions. Each individual is represented by a horizontal line. Color codes follow the number of clusters, while the bar line under the graph represents the subspecies groups plus the outgroup genotypes (*T. aestivum*).

4. Discussion

Genetic diversity represents the basis for crop improvement, providing plant breeders with the germplasm necessary to develop cultivars with adaptive traits and good quality characteristics [51]. To better target their crossing schemes, the genetic structure and variability of 265 tetraploid wheats accessions were assessed.

Clustering done via a Bayesian tree and clusters obtained via DAPC revealed a clear classification of genotypes in accordance with their geographical origin, strengthening the results of previous studies of phylogenetic relationships between cultivated wheats and their wild relatives [52,53].

Concerning *T. turgidum* ssp. *durum* accessions, which represented the largest number of genotypes in the panel, their first and second geographical origin centers—Syria and Ethiopia [54]—appeared to be clearly identified in Clusters 5 and 3, respectively. This result agreed with the molecular assessment by Kabbaj et al. [55] regarding a durum wheat collection of cultivars. More interestingly, the Bayesian tree highlighted the proximity between North African (Morocco, Algeria, and Tunisia) and Italian germplasm; this could be linked to the geographical expansion of Romans during the Imperial Period and consequent wheat genotype introduction and cultivation on the African continent, as suggested by Rickman [56].

In addition, the positions of the accessions "Ciceredda", "Bufala rossa lunga", "Bufala nera corta", and "Paola" on the Bayesian tree deserve attention: although they belong to Cluster 3, which grouped almost all the other *T. turgidum* ssp. durum genotypes, they were gathered in a distant cluster between *T. turgidum turgidum* and *polonicum* ssp. The proximity of these accessions could be due to a taxonomic problem, traceable thanks to work by De Cillis [57], which classified these accessions under *T. turgidum turgidum* ssp. *turgidum*.

Finally, another relevant observation on the *T. turgidum* ssp. *durum* accession arrangement concerns the low genetic variability detected in the modern Italian varieties, different from landraces and old varieties. Through the second half of the 20th century, national breeding programs aimed at increasing wheat yield started to establish new varieties characterized by small size, limited sprouting, reduced leaf area, and shorter crop cycle [58]. Due to genetic improvement only, De Vita et al. [59] confirmed in their work a 44% increase in productivity for the main varieties of durum wheat grown in Italy during the 20th century; however, this resulted in pure line selection and the development of varieties with low genetic variability [60]. Our study reflects this strong selection activity: Italian modern varieties were gathered in the same cluster (Figure 1B) and along neighbor branches, highlighting genetic homogeneity.

On the contrary, the subspecies *dicoccon* showed the highest genetic variability, as Laidò [61] et al. verified in their research, confirming this wild germplasm as a powerful source of genes.

Today, the unpredictable climate, characterized by irregular rainfall and long dry periods, results in a rather unstable crop production. Under marginal environments, landraces and old varieties show higher stability in low-input agriculture [62,63]; thus, they could represent valuable genetic resources for breeders in order to develop new cultivars or CCP populations with specific qualitative traits such as resistance to biotic and abiotic stress, ability to efficiently use organic nitrogen and better nutritional qualities [64]. With this aim, our results showed the genetic diversity among accessions belonging to eight tetraploid wheat subspecies and identified the correct numbers of genotypes that explain the screened genetic variability well.

5. Conclusions

The genetic diversity of domesticated wheat accessions has been significantly reduced from that of their wild progenitors through a prolonged selection process for those phenotypic traits that better satisfy human needs. On the contrary, landraces' genetic variability represents a precious source of valuable agronomic traits that could be used for interspecific hybridization and for the introgression of genes and/or alleles into cultivated species. In our work, the genetic diversity and the population structure of 265 tetraploid wheats were investigated in order to understand the genetic relationships between domesticated wheats and their close wild relatives. The results obtained from this research could be used in future phenotyping studies in both field and laboratory tests to select the best lines to be intercrossed for the creation of improved and more resilient durum wheat CCP populations adapted to Mediterranean areas.

Supplementary Materials: The following are available online at https://www.mdpi.com/2073-4395/11/3/414/s1. Figure S1. Statistical determination of the optimum number of clusters by discriminant analysis of principal components (DAPC). The elbow in the curve matches the smallest BIC and clearly indicates that seven cluster should be retained. Table S1. List of wheat accessions used in the experiment. Table S2. Clusters Membership for each accession defined with DAPC and Structure analyses. For Structure, cluster membership probability was reported for each accession.

Author Contributions: Data curation, P.G. and A.C.; Investigation, P.G., M.G., A.B. and A.M.; Methodology, M.G. and A.B.; Project administration, S.B. (Stefano Biricolti) and A.M.; Supervision, E.P., S.B. (Stefano Biricolti) and S.B. (Stefano Benedettelli); Writing—original draft, P.G. All authors have read and agreed to the published version of the manuscript

Funding: This research received no external funding.

Institutional Review Board Statement: Not applicable.

Informed Consent Statement: Not applicable.

Data Availability Statement: Part of the data presented in this study are available on request from the corresponding author.

Acknowledgments: We thank the members of Bristol Genomics Facility Center (http://www.bristol.ac.uk/biology/genomics-facility (accessed on 3 December 2020)) for assistance in the genotyping lab work.

Conflicts of Interest: The authors declare no conflict of interest.

References

1. FAOSTAT. Available online: http://www.fao.org/faostat (accessed on 3 December 2020).
2. Bennici, A. Durum Wheat (*Triticum durum* Desf.). In *Crops I Biotechnology in Agriculture and Forestry*; Bajaj, Y.P.S., Ed.; Springer: Berlin/Heidelberg, Germany, 1986; pp. 89–104.
3. Brenchley, R.; Spannagl, M.; Pfeifer, M.; Barker, G.L.A.; D'Amore, R.; Allen, A.M.; McKenzie, N.; Kramer, M.; Kerhornou, A.; Bolser, D.; et al. Analysis of the bread wheat genome using whole-genome shotgun sequencing. *Nature* **2012**, *491*, 705–710. [CrossRef] [PubMed]
4. Hakan, O.; Willcox, G.; Graner, A.; Salamini, F.; Kilian, B. Geographic distribution and domestication of wild emmer wheat (*Triticum dicoccoides*). *Genet. Resour. Crop Evol.* **2010**, *58*, 11–53.
5. Dubcovsky, J.; Dvorak, J. Genome plasticity a key factor in the success of polyploid wheat under domestication. *Science* **2007**, *316*, 1862–1866. [CrossRef] [PubMed]
6. Ling, H.-Q.; Ma, B.; Shi, X.; Liu, H.; Dong, L.; Sun, H.; Cao, Y.; Gao, Q.; Zheng, S.; Li, Y.; et al. Genome sequence of the progenitor of wheat A subgenome *Triticum urartu*. *Nature* **2018**, *557*, 424–428. [CrossRef] [PubMed]
7. Marcussen, T.; Sandve, S.R.; Heier, L.; Spannagl, M.; Pfeifer, M.; Jakobsen, K.S.; Wulff, B.B.H.; Steuernagel, B.; Mayer, K.F.X.; Olsen, O.-A.; et al. Ancient hybridizations among the ancestral genomes of bread wheat. *Science* **2014**, *345*, 1–4. [CrossRef] [PubMed]
8. Charmet, G. Wheat domestication: Lessons for the future. *Comptes Rendus Biol.* **2011**, *334*, 212–220. [CrossRef]
9. Feldman, M.; Kislev, M.E. Domestication of emmer wheat and evolution of free-threshing tetraploid wheat. *Israel J. Plant Sci.* **2007**, *55*, 207–221. [CrossRef]
10. Feldman, M. Origin of cultivated wheat. In *The World Wheat Book: A History of Wheat Breeding*; Bonjean, A.P., Angus, W.J., Eds.; Lavoisier Publishing: Paris, France, 2001; pp. 3–56.
11. Peng, J.H.; Sun, D.; Nevo, E. Domestication evolution, genetics and genomics in wheat. *Mol. Breed.* **2011**, *28*, 281–301. [CrossRef]
12. Jaradat, A.A. *Wheat Landraces: Genetic Resources for Sustenance and Sustainability*; USDA-ARS: Morris, MN, USA, 2012; pp. 1–20.
13. Kronstad, W.E. Agricultural development and wheat breeding in the 20th Century. In *Wheat: Prospects for Global Improvement. Developments in Plant Breeding*; Braun, H.J., Altay, F., Kronstad, W.E., Beniwal, S.P.S., McNab, A., Eds.; Springer: Dordrecht, The Netherlands, 1997.
14. Waines, J.G.; Ehdaie, B. Domestication and crop physiology: Roots of green-revolution wheat. *Ann. Bot.* **2007**, *100*, 991–998. [CrossRef]

15. Tanksley, S.D.; McCouch, S.R. Seed banks and molecular maps: Unlocking genetic potential from the wild. *Science* **1997**, *277*, 1063–1066. [CrossRef] [PubMed]
16. Haudry, A.; Cenci, A.; Ravel, C.; Bataillon, T.; Brunel, D.; Poncet, C.; Hochu, I.; Poirier, S.; Santoni, S.; Glémin, S.; et al. Grinding up wheat: A massive loss of nucleotide diversity since domestication. *Mol. Biol. Evol.* **2007**, *24*, 1506–1517. [CrossRef]
17. Wheeler, T.; von Braun, J. Climate change impacts on global food security. *Science* **2013**, *341*, 508–513. [CrossRef] [PubMed]
18. Ceccarelli, S.; Grando, S.; Maatougui, M.; Michael, M.; Slash, M.; Haghparast, R.; Rahmanian, M.; Taheri, A.; Al-Yassin, A.; Benbelkacem, A.; et al. Plant breeding and climate changes. *J. Agric. Sci.* **2010**, *148*, 627–637. [CrossRef]
19. Matthews, R.B.; Rivington, M.; Muhammed, S.; Newton, A.C.; Hallett, P.D. Adapting crops and cropping systems to future climates to ensure food security: The role of crop modelling. *Glob. Food Secur.* **2013**, *2*, 24–28. [CrossRef]
20. Tuberosa, R.; Graner, A.; Varshney, R.K. Genomics of plant genetic resources: An introduction. *Plant Genet. Resour.* **2011**, *9*, 151–154. [CrossRef]
21. Kyratzis, A.C.; Nikoloudakis, N.; Katsiotis, A. Genetic variability in landraces populations and the risk to lose genetic variation. The example of landrace 'Kyperounda' and its implications for ex situ conservation. *PLoS ONE* **2019**, *14*, e0224255. [CrossRef] [PubMed]
22. Döring, T.F.; Knapp, S.; Kovacs, G.; Murphy, K.; Wolfe, M.S. Evolutionary plant breeding in cereals into a new era. *Sustainability* **2011**, *3*, 1944–1971. [CrossRef]
23. Mondal, S.; Rutkoski, J.E.; Velu, G.; Singh, P.K.; Crespo-Herrera, L.A.; Guzman, C.G.; Bhavani, S.; Lan, C.; He, X.; Singh, R.P. Harnessing diversity in wheat to enhance grain yield, climate resilience, disease and insect pest resistance and nutrition through conventional and modern breeding approaches. *Front. Plant Sci.* **2016**, *7*, 991. [CrossRef] [PubMed]
24. Dwivedi, S.L.; Ceccarelli, S.; Blair, M.W.; Upadhyaya, H.D.; Are, A.K.; Ortiz, R. Landrace germplasm for improving yield and abiotic stress adaptation. *Trends Plant Sci.* **2016**, *21*, 31–42. [CrossRef]
25. Jain, S.K.; Qualset, C.O.; Bhatt, G.M.; Wu, K.K. Geographical patterns of phenotypic diversity in a world collection of durum wheats. *Crop Sci.* **1975**, *15*, 700–704. [CrossRef]
26. Peccetti, L.; Annicchiarico, P. Grain yield and quality of durum wheat landraces in a dry Mediterranean region of Northern Syria. *Plant Breed.* **1993**, *110*, 243–249. [CrossRef]
27. Moragues, M.; Zarco-Hernández, J.; Moralejo, M.A.; Royo, C. Genetic diversity of glutenin protein subunits composition in durum wheat landraces [*Triticum turgidum* ssp. *turgidum* convar. *durum* (Desf.) MacKey] from the Mediterranean Basin. *Gen. Res. Crop Evol.* **2006**, *53*, 993–1002. [CrossRef]
28. Fahima, T.; Sun, G.L.; Beharav, A.; Krugman, T.; Beiles, A.; Nevo, E. RAPD polymorphism of wild emmer wheat populations, *Triticum dicoccoides*, in Israel. *Theor. Appl. Genet.* **1999**, *98*, 434–447. [CrossRef]
29. Altıntas, S.; Toklu, F.; Kafkas, S.; Kilian, B.; Brandolini, A.; Ozkan, H. Estimating genetic diversity in durum and bread wheat cultivars from Turkey using AFLP and SAMPL markers. *Plant Breed.* **2008**, *127*, 9–14. [CrossRef]
30. Zhuang, P.; Ren, Q.; Li, W.; Chen, G. Genetic diversity of Persian wheat (*Triticum turgidum* ssp. *carthlicum*) accessions by EST-SSR markers. *Am. J. Biochem. Mol. Biol.* **2011**, *1*, 223–230. [CrossRef]
31. Peleg, Z.; Saranga, Y.; Suprunova, T.; Ronin, Y.; Röder, M.S.; Kilian, A.; Korol, A.B.; Fahima, T. High-density genetic map of durum wheat x wildemmer wheat based on SSR and DArTmarkers. *Theor. Appl. Genet.* **2008**, *117*, 103–115. [CrossRef] [PubMed]
32. Rufo, R.; Alvaro, F.; Royo, C.; Soriano, J.M. From landraces to improvedcultivars: Assessment of genetic diversity and population structure of Mediterranean wheat using SNP markers. *PLoS ONE* **2019**, *14*, e0219867. [CrossRef] [PubMed]
33. Phillips, S.; Wolfe, M. Evolutionary plant breeding for low input systems. *J. Agric. Sci.* **2005**, *143*, 245–254. [CrossRef]
34. Masoni, A.; Calamai, A.; Marini, L.; Benedettelli, S.; Palchetti, E. Constitution of Composite Cross Maize (*Zea mays* L.) Populations Selected for the Semi-Arid Environment of South Madagascar. *Agronomy* **2020**, *10*, 54. [CrossRef]
35. Ceccarelli, S.; Grando, S.; Baum, M. Participatory plant breeding in water-limited environments. *Exp. Agric.* **2007**, *43*, 411–435. [CrossRef]
36. Sanchez-Garcia, M.; Álvaro, F.; Martín-Sánchez, J.A.; Sillero, J.C.; Escribano, J.; Royo, C. Breeding effects on the genotype x environment interaction for yield of bread wheat grown in Spain during the 20th century. *Field Crops Res.* **2012**, *126*, 79–86. [CrossRef]
37. Ceccarelli, S.; Guimarães, E.P.; Weltzien, E. *Plant Breeding and Farmer Participation*; Food and Agriculture Organization of the United Nations: Rome, Italy, 2009.
38. Doyle, J.J.; Doyle, J.L. A rapid DNA isolation procedure for small quantities of fresh leaf tissue. *Phyt. Bull.* **1987**, *19*, 11–15.
39. Allen, A.M.; Winfield, M.O.; Burridge, A.J.; Downie, R.C.; Benbow, H.R.; Barker, G.L.; Wilkinson, P.A.; Coghil, J.; Waterfall, C.; Davassi, A.; et al. Characterization of a Wheat Breeders' Array suitable for high-throughput SNP genotyping of global accessions of hexaploid bread wheat (*Triticum aestivum*). *Plant Biotechnol. J.* **2017**, *15*, 390–401. [CrossRef]
40. Ceccarelli, S.; Grando, S. Plant breeding with farmers requires testing the assumptions of conventional plant breeding: Lessons from the ICARDA barley program. In *Farmers, Scientists and Plant Breeding: Integrating Knowledge and Practice*; David, D.A.C., Soleri, D., Eds.; CAB I Publishing International: Wallingford, UK, 2002; pp. 297–332.
41. Winfield, M.O.; Allen, A.M.; Burridge, A.J.; Barker, G.L.A.; Benbow, H.R.; Wilkinson, P.A.; Coghill, J.; Waterfall, C.; Davassi, A.; Scopes, G.; et al. High-density SNP genotyping array for hexaploid wheat and its secondary and tertiary gene pool. *Plant Biotechnol. J.* **2016**, *14*, 1195–1206. [CrossRef]

42. Tamura, K.; Nei, M. Estimation of the number of nucleotide substitutions in the control region of mitochondrial DNA in humans and chimpanzees. *Mol. Biol. Evol.* **1993**, *10*, 512–526.

43. Tamura, K.; Stecher, G.; Peterson, D.; Filipski, A.; Kumar, S. MEGA6: Molecular Evolutionary Genetics Analysis Version 6.0. *Mol. Bio. Evol.* **2013**, *2013*, 2725–2729. [CrossRef]

44. Paradis, C.J.; Strimmer, K. APE: Analyses of phylogenetics and evolution in R language. *Bioinformatics* **2004**, *20*, 289–290. [CrossRef]

45. Pritchard, J.K.; Stephens, M.; Donnelly, P. Inference of population structure using multilocus genotype data. *Genetics* **2000**, *155*, 945–959.

46. Evanno, G.; Regnaut, S.; Goudet, J. Detecting the number of clusters of individuals using the software STRUCTURE: A simulation study. *Mol. Ecol.* **2005**, *14*, 2611–2620. [CrossRef] [PubMed]

47. Ramasamy, R.K.; Ramasamy, S.; Bindroo, B.B.; Naik, V.G. STRUCTURE PLOT: A program for drawing elegant STRUCTURE bar plots in user friendly interface. *SpringerPlus* **2014**, *3*, 431. [CrossRef] [PubMed]

48. Jombart, T.; Devillard, S.; Balloux, F. Discriminant analysis of principal components: A new method for the analysis of genetically structured populations. *BMC Genet.* **2010**, *11*, 94. [CrossRef] [PubMed]

49. Jombart, T.; Collins, C. *A Tutorial for Discriminant Analysis of Principal Components (DAPC) Using Adegenet 2.0.0*; Imperial College London: London, UK, 2015.

50. Chen, S.; Gopalakrishnan, P. Speaker, environment and channel change detection and clustering via the Bayesian Information Criterion. In Proceedings of the DARPA BroadcastNews Transcription and Understanding Workshop, Landsdowne, VA, USA, 8–11 February 1998.

51. Yoshihiro, M. Evolution of Polyploid Triticum Wheats under Cultivation: The Role of Domestication, Natural Hybridization and Allopolyploid Speciation in their Diversification. *Plant Cell Physiol.* **2011**, *52*, 750–764.

52. Feldman, M.; Sears, E.R. The wild gene resources of wheat. *Sci. Am.* **1981**, *244*, 102–113. [CrossRef]

53. Mac Key, J. Wheat: Its concept, evolution, and taxonomy. In *Durum Wheat Breeding: Current Approaches and Future Strategies*, 1st ed.; Royo, C., Nachit, M., Difonzo, N., Araus, J., Pfeiffer, W., Slafer, G., Eds.; The Haworth Press: New York, NY, USA, 2005; pp. 3–61.

54. Harlan, J.R. Ethiopia: A center of diversity. *Econ. Bot.* **1969**, *23*, 309–314. [CrossRef]

55. Kabbaj, H.; Sall, A.T.; Al-Abdallat, A.; Geleta, M.; Amri, A.; Filali-Maltouf, A.; Belkadi, B.; Ortiz, R.; Bassi, F.M. Genetic Diversity within a Global Panel of Durum Wheat (*Triticum durum*) Landraces and Modern Germplasm Reveals the History of Alleles Exchange. *Front. Plant Sci.* **2017**, *8*, 1277. [CrossRef] [PubMed]

56. Rickman, G. *The corn supply of ancient Rome*; Oxford University Press: Oxford, UK, 1980; p. 304.

57. De Cillis, U. *I frumenti Siciliani*, 9th ed.; Stazione sperimentale di granicoltura per la Sicilia: Caltagirone, Italy, 1942; pp. 1–323.

58. De Vita, P.; Matteu, L.; Mastrangelo, A.M.; Di Fonzo, N.; Cattivelli, L. Effects of breeding activity on durum wheat traits breed in Italy during the 20th century. *Ital. J. Agron.* **2007**, 451–462. [CrossRef]

59. De Vita, P.; Nicosia, O.L.D.; Nigro, F.; Platani, C.; Riefolo, C.; Di Fonzo, N.; Cattivelli, L. Breeding progress in morpho-physiological, agronomical and qualitative traits of durum wheat cultivars released in Italy during the 20th century. *Eur. J. Agric.* **2007**, *26*, 39–53. [CrossRef]

60. Fu, Y.B. Understanding crop genetic diversity under modern plant breeding. *Theor. Appl. Genet.* **2015**, *128*, 2131–2142. [CrossRef] [PubMed]

61. Laidò, G.; Mangini, G.; Taranto, F.; Gadaleta, A.; Blanco, A.; Cattivelli, L.; Marone, D.; Matrangelo, A.; Papa, R.; de Vita, P. Genetic Diversity and Population Structure of Tetraploid Wheats (*Triticum turgidum* L.) estimated by SSR, DArT and Pedigree Data. *PLoS ONE* **2013**, *8*, e67280.

62. Murphy, K.; Lammer, D.; Lyon, S.; Carter, B.; Jones, S.S. Breeding for organic and low-input farming systems: An evolutionary-participatory breeding method for inbred cereal grains. *Renew. Agric. Food. Syst.* **2005**, *20*, 48–55. [CrossRef]

63. Ruiz, M.; Aguiriano, E.; Carrillo, J. Effects of N fertilization on yield for low-input production in Spanish wheat landraces (*Triticum turgidum* L. and *Triticum monococcum* L.). *Plant Breed.* **2008**, *127*, 20–23. [CrossRef]

64. Stagnari, F.; Onofri, A.; Codianni, P.; Pisante, M. Durum wheat varieties in N-deficient environments and Organic farming: A comparison of yield, quality and stability performances. *Plant Breed.* **2013**, *132*, 266–275. [CrossRef]

 *agronomy*

Article

Genetic Purity of Cacao Criollo from Honduras Is Revealed by SSR Molecular Markers

Marlon López [1], Massimo Gori [2], Lorenzo Bini [2], Erick Ordoñez [3], Erick Durán [1], Osman Gutierrez [4], Alberto Masoni [5], Edgardo Giordani [2], Stefano Biricolti [2,*] and Enrico Palchetti [5]

1   Cacao and Agroforestry Program, Honduran Foundation for Agricultural Research (FHIA), La Lima Cortés 2067, Honduras; marlonenriquelopez@gmail.com (M.L.); duranerick17@gmail.com (E.D.)

2   Dipartimento di Scienze e Tecnologie Agrarie, Alimentari Ambientali e Forestali (DAGRI), Sezione Colture Arboree, Università degli Studi di Firenze, Viale delle Idee, 30, 50019 Sesto Fiorentino, Italy; massimo.gori@unifi.it (M.G.); lorenzo.bini@unifi.it (L.B.); edgardo.giordani@unifi.it (E.G.)

3   Rural Development Department, National Autonomous University of Honduras, Regional University Center of the Atlantic Littoral (UNAH-CURLA), Carretera CA-13, Desvio Frente a Maxi-Despensa Aeropuerto, La Ceiba 31101, Atlántida, Honduras; erick.trejo@unah.edu.hn

4   USDA-ARS Subtropical Horticulture Research Station, Old Cutler Road, Miami, FL 13601, USA; osman.gutierrez@usda.gov

5   Dipartimento di Scienze e Tecnologie Agrarie, Alimentari Ambientali e Forestali (DAGRI), Sezione Scienze Agronomiche, Genetiche e Gestione del Territorio, Università degli Studi di Firenze, Piazzale delle Cascine 18, 50144 Firenze, Italy; alberto.masoni@unifi.it (A.M.); enrico.palchetti@unifi.it (E.P.)

*   Correspondence: stefano.biricolti@unifi.it

Abstract: The cultivation of cacao represents an income option and a source of employment for thousands of small producers in Central America. In Honduras, due to the demand for fine flavor cacao to produce high-quality chocolate, the number of hectares planted is increasing. In addition, cacao clones belonging to the genetic group named Criollo are in great demand since their white beans lack of bitterness and excellent aroma are used in the manufacturing of premium chocolate. Unfortunately, the low resistance to pests and diseases and less productive potential of Criollo cacao leads to the replacement with vigorous new cultivars belonging to the other genetic groups or admixture of them. In this study, 89 samples showing phenotypic traits of Criollo cacao from four regions of Honduras (Copán, Santa Bárbara, Intibucá, and Olancho) were selected to study their genetic purity using 16 SSR molecular markers. The results showed that some samples belong to the Criollo group while other accessions have genetic characteristics of "Trinitario" or other admixtures cacao types. These results confirm the genetic purity of Criollo cacao in Honduras, reaffirming the theory that Mesoamerica is a cacao domestication center and also serves to generate interest in the conservation of this genetic wealth both in-situ and ex-situ.

Keywords: population structure; Criollo cacao; microsatellites; genetic purity; Central America

Citation: López, M.; Gori, M.; Bini, L.; Ordoñez, E.; Durán, E.; Gutierrez, O.; Masoni, A.; Giordani, E.; Biricolti, S.; Palchetti, E. Genetic Purity of Cacao Criollo from Honduras Is Revealed by SSR Molecular Markers. *Agronomy* **2021**, *11*, 225. https://doi.org/10.3390/agronomy11020225

Academic Editors: Gregorio Barba-Espín and Jose Ramon Acosta-Motos

Received: 15 December 2020

Accepted: 22 January 2021

Published: 26 January 2021

1. Introduction

Cacao (*Theobroma cacao* L.) is an economically important crop because it produces the raw material for making chocolate. It is native to South America, with its center of origin in the upper Amazonian region spanning Peru, Colombia, Brazil, and Ecuador [1–3]. Cacao was initially thought to be domesticated in Southern Mexico and Central America since vessels used by pre-Columbian cultures in Honduras and Mexico contained trace remains of theobromine, confirming the use of cacao products 1800–1000 years BCE [4–9]. However, recent archaeological finds reveal that the upper Amazon region was also a center for the domestication of cacao [10].

Spanish settlers in Honduras were introduced to cacao and realized the importance of this crop [11,12]. Recent studies showed that during this period of domestication there was a high selection for genes related to anthocyanins and theobromine, which caused the cacao Criollo to maintain a high frequency of deleterious mutations [13].

Cacao has high genetic and morphological diversity, with a diploid genome consisting of 10 chromosomes ($2n = 20$) [14] with a genome size between 411 and 494 Mbp [15]. Initially based on phenotype, cacao was classified into three groups Criollo, Forastero, and Trinitario, which is a hybrid between Criollo and Forastero. Forastero cacao is characterized by seeds with purple cotyledons and most of the chocolate consumed in the world is produced from these seeds known as "bulk or ordinary" cacao [3]. Criollo cacao is characterized by having white or pale pink seeds, used for the formulation of premium chocolate that sells for higher prices in the market [16]. Trinitario beans are variable in cotyledon color and produce a liquor with a strong cacao flavor. Criollo and Trinitario are known collectively as "fine of flavor" cacao, along with Amelonado, Nacional, and Refractario (Nacional hybrids). There are different meanings of the term "Criollo", depending on whether it is used by breeders, botanists, or growers. "True Criollo", "Ancient Criollo", and "Criollo" are the ways it is called respectively depending upon who is using the term [16]. Currently, Criollo is defined as a unique genetic group based on the classification proposed by Motamayor et al. [17]. In fact, with the use of molecular tools, a new classification of 10 genetic groups has been proposed, which has furthered the understanding of the genetic and phenotypic diversity of the crop, and "Criollo" cacao is one of the identified groups [17].

Knowing the genetic diversity of cacao has been of great interest to breeders and growers. The first classifications of the *Theobroma* species were made concerning their phenotypic characteristics, evidencing that one of the marked differences between Criollo and the others, was the content of anthocyanins in its seeds [2,4,18]. The first genetic diversity studies of cacao were made using RAPD and RFLP markers [19–23], followed by SSR markers [24–32], and SNP markers [33–38]. Some of these studies made it possible to identify Criollo as a different genetic group. A high level of homozygosity and significantly low genetic diversity within the Criollo population was also found [24] in a great part attributable to Criollo self-compatibility.

The results of the genetic diversity of cacao from Honduras and Nicaragua with SNP markers confirmed the existence of a population of ancient cacao Criollo. Furthermore, the Trinitario varieties of cacao used by growers are more influenced by the Amelonado genetic group than the Criollo genetic group [7]. The study also showed that although the varieties of cacao type Trinitario used by growers in both countries are different, the samples of ancient cacao Criollo analyzed were grouped in the same cluster [7]. Several studies on the genetic diversity of cacao have used samples of cacao Criollo from Honduras as a reference of the purity of the Criollo genetic group [2,7,16,29,35,39]. A phenotypic characterization study of cacao from Honduras determined that 72% of the production comes from cacao Trinitario type varieties and 28% comes from Amelonado type varieties [40]. The cacao growers in Honduras have the technical and scientific support provided by the Honduran Foundation for Agricultural Research (FHIA) for over 36 years and the foundation's agroforestry program has been collecting genetic material and maintaining a germplasm bank in order to preserve elite material for the improvement of cacao cultivation. However, the genetic purity of Criollo in Honduras and its diversity has not been fully investigated, even if it would be highly beneficial to the cacao genetic improvement and the conservation of Criollo in-situ and ex-situ. Therefore, the goals of this study were to assess the genetic purity and diversity of 89 cacao accessions, which have been selected based on the Criollo phenotype from four different departments of Honduras, using single sequence repeat (SSR) markers. The results of the analyses have also been compared with a robust SSR dataset of 116 samples obtained from the original classification developed by Motamayor et al. [17] representing the ten cacao genetic groups.

2. Materials and Methods

2.1. Plant Material

In this study, eighty-nine cacao plants were selected from four different departments in Honduras: Copán, Santa Bárbara, Olancho, and Intibucá (Figure 1). Cacao trees were previously identified by the farmers, who had preserved the genetic material for many years, and only trees showing phenotypical characteristics of Criollos were sampled (Figure 2). Generally, they were located inside small and medium-sized family farms, isolated, and not properly maintained. Two young fully expanded leaves were selected from each tree and conserved and transported inside 20 mL plastic tubes filled with 96% ethanol [41,42] for DNA extraction. The GPS coordinates were recorded for each tree for subsequent mapping (Table S1).

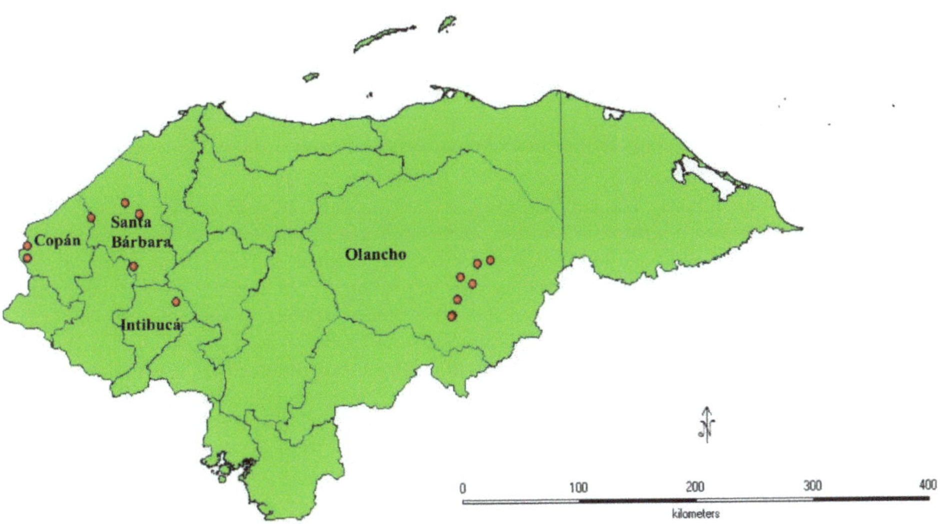

Figure 1. Geographical sites (red dots) where cacao's leaves were collected in Honduras.

Figure 2. Pod samples of cacao Criollo from (**A**). Copán, (**B**). Intibucá, (**C**). Olancho, and (**D**). Santa Bárbara.

2.2. DNA Extraction and Quantification

For DNA extraction, 40–50 mg of dry leaf sample was placed in a 2 mL tube with tungsten carbide beads, frozen in liquid nitrogen, and finely ground in a tissue homogenizer (Tissue Lyser, QIAGEN, Hilden, Germany). DNA was extracted using an Invisorb Spin Plant Mini Kit (Stratec Molecular, Berlin, Germany). DNA has been electrophoresed on an agarose gel to validate quality and quantity.

2.3. SSR Markers

Sixteen previously reported SSR markers were used in this study [43,44]. The list of the SSR markers' names, the EMBL accession, the chromosome where the locus is mapped, the specific primers, the observed size range of the amplicon, the motifs, and the references are reported in Table 1.

2.4. PCR Amplification and Electrophoresis

PCR amplifications were carried out in a final volume of 25 µL, containing 20 ng of genomic DNA, 2 mM $MgCl_2$, 0.2 mM dNTPs, 0.2 µM each of the forward and reverse primer, 3% pure formamide, 1X GoTaq® colorless Reaction Buffer (Promega, Madison, WI, USA), and 1 Unit of GoTaq® Flexi DNA Polymerase (Promega, Madison, WI, USA). The PCR reactions were based on the following amplification protocol: a denaturation step of 4 min at 96 °C; 35 cycles of 40-s denaturation at 96 °C, 40-s primer annealing at appropriate temperature for each primer pairs, 40-s extension at 72 °C; and 5 min final extension at 72 °C. This protocol was performed on a Primus 96 advanced (PEQLAB Biotechnologie Gmbh, Erlanger, Germany) thermocycler. Each marker had a fluorescent tag on the forward primer (6-FAM, VIC, NED or PET) (Applied Biosystems, Inc., Waltham, MA, USA) for scoring fragment length. Fragments analysis of the PCR products was performed on two capillary electrophoresis instruments: an ABI-3130 Genetic Analyzer (Applied Biosystems, Inc., Waltham, MA, USA) using performance optimized polymer (POP7) at the University of Florence and an ABI-3730 Genetic Analyzer (Applied Biosystems, Inc. Waltham, MA, USA) using performance optimized polymer (POP7) at USDA-ARS in Miami. The accuracy and comparability of the sizing performance of the two sequencers were tested by genotyping nine shared reference accessions. The data generated were analyzed with GeneMapper 3.0 (Applied Biosystems, Inc. Waltham, MA, USA). An SSR dataset of 116 samples from Motamayor et al. [17] was included as reference accessions (Table S2) representing the ten cacao genetic groups identified by Motamayor et al. [17]: Amelonado, Contamana, Criollo, Curaray, Guiana, Iquitos, Marañon, Nacional, Nanay, and Purús. Two SSR data sets were generated, one including only the samples from Honduras, named "Honduras" and the other data set including both Honduras and reference accessions, named "Complete". For the "Complete" dataset a representative number of reference accessions was analyzed along with our samples to assess the comparability of the two datasets.

2.5. Data Analysis

Despite gathering samples in a vast area, Honduran accessions sharing identical SSR profile were retrieved. Therefore, to reduce noise and to avoid any alteration of the results due to the presence of identical genotypes, a preliminary analysis was performed with the software Cervus 3.0.9 [45]. In this case, only the samples that shared 100% of their allele's identity were considered. From this procedure, 32 identical samples were detected and then not taken into consideration in the subsequent evaluation unless they do not belong to the same department, reducing the number of examined genotypes to 57 (Table S3).

Concerning the "Honduras" dataset, for each SSR locus, the number of alleles (Na), the effective number of alleles (Ne), frequency of the predominant allele (Fa), the observed (Ho) and expected heterozygosity (He), and the fixation index (F) were calculated using Genealex 6.5 [46]. The polymorphic information content (PIC) was calculated with PowerMarker 3.25 [47].

Table 1. List of the 16 SSR loci used for the analysis of the cacao samples: SSR names, EMBL accessions, chromosome number, forward and reverse primer sequences, observed size, repeat motifs, and references.

SSR Name	EMBL Accession	Chromosome No.	Forward Primer	Reverse Primer	Size (bp)	Motif	Ref.
MTcCIR6	Y16980	6	5′-TTCCCTCTAAACTACCCT AAAT-3′	5′-TAAAGCAAAGCAAT CTAACATA-3′	224–246	$(TG)_7(GA)_{13}$	[43]
MTcCIR7	Y16981	7	5′-ATGCGAATGACAACTGG T-3′	5′-GCTTTCAGTCCTTTG CTT-3′	147–159	$(GA)_{11}$	[43]
MTcCIR8	Y16982	9	5′-CTAGTTTCCCATTTACCA-3′	5′-TCCTCAGCATTTTCTT TC-3′	286–302	$(TC)_5TT(TC)_{17}TTT(CT)_4$	[43]
MTcCIR11	Y16985	2	5′-TTTGGTGATTATTAGCA G-3′	5′-GATTCGATTTGATG TGAG-3′	290–316	$(TC)_{13}$	[43]
MTcCIR12	Y16986	4	5′-TCTGACCCCAAACCTGT A-3′	5′-ATTCCAGTTAAAGC ACAT-3′	187–219	$(CATA)_4N_{18}(TG)_6$	[43]
MTcCIR18	Y16991	4	5′-GATAGCTAAGGGGATTG AGGA-3′	5′-GGTAATTCAATCATT TGAGGATA-3′	331–345	$(GA)_{12}$	[43]
MTcCIR22	Y16995	1	5′-ATTCTCGCAAAAACTTA G-3′	5′-GATGGAAGGAGTGT AAATAG-3′	277–289	$(TC)_{12}N_{146}(CT)_{10}$	[43]
MTcCIR24	Y16996	9	5′-TTTGGGGTGATTTCTTC TGA-3′	5′-TCTGTCTCGTCTTTTG GTGA-3′	184–200	$(AG)_{13}$	[43]
MTcCIR26	Y16998	8	5′-GCATTCATCAATACAT TC-3′	5′-GCACTCAAAGTTCA TACTAC-3′	289–307	$(TC)_9C(CT)_4TT(CT)_{11}$	[43]
MTcCIR33	AJ271826	4	5′-TGGGTTGAAGATTTG GT-3′	5′-CAACAATGAAAATA GGCA-3′	270–344	$(TG)_{11}$	[43]
MTcCIR37	AJ271942	10	5′-CTGGGTGCTGATAGAT AA-3′	5′-AATACCCTCCACAC AAAT-3′	144–184	$(GT)_{15}$	[43]
MTcCIR40	AJ271943	3	5′-AATCCGACAGTCTTTAA TC-3′	5′-CCTAGGCCAGAGAA TTGA-3′	271–285	$(AC)_{15}$	[43]
MTcCIR60	AJ271958	2	5′-CGCTACTAACAAACATCA AA-3′	5′-AGAGCAACCATCAC TAATCA-3′	188–212	$(CT)_7(CA)_{20}$	[43]
MTcCIR115	AJ566457	4	5′-GTGATTCAAATTCAAATA TG-3′	5′-AATAGCAAGAGAGT GATGAG-3′	189–195	$(TC)_{11}$	[44]
MTcCIR158	AJ566489	4	5′-TGTAGGTTATGCAGCGTG TTC-3′	5′-GATGAGGGGTGTAG CTGTTTG-3′	211–217	$(CT)_8$	[44]
MTcCIR168	AJ566497	4	5′-GGTAGTATTGAGGTGCGT AT-3′	5′-GTGAATGAATGGAT GTGAAA-3′	174–186	$(TC)_9$	[44]

2.6. Cluster Analysis

The Bayesian approach implemented in the software STRUCTURE v.2.3.4 [48] was used to define the population structure in both datasets, "Honduras" and "Complete", as previously described. An admixtured and shared allele frequencies was adopted to determine the number of populations (K), assumed in the range from 1 to 10 in "Honduras" and from 1 to 20 in "Complete". For each K value, 10 runs with 100,000 burn-in periods and 200,000 MCMC (Markov Chain Monte Carlo) iterations were performed. The ΔK of the Evanno method [49] was calculated to estimate the best value of K that fits the data, by using STRUCTURE HARVESTER [50]. The structure plot was obtained with the package pophelper v2.3.0 [51] in the R-project. The Criollo reference (Criollo 13) was included in the "Honduras" dataset to establish which Honduran samples were pure Criollo.

A further STRUCTURE analysis was performed with the 116 reference accessions from Motamayor et al. [17] along with the eight Honduran pure Criollo cacaos, which had the highest coefficient of membership (higher than 0.99) to the Criollo cluster in the STRUCTURE analysis of "Complete" dataset (Cop_22, Cop_23, Cop_24, Cop_28, Cop_29, San_106, San_107, and San_109) to sustain the fitness with the 10 genetic groups established in Motamayor et al. [17].

Moreover, concerning the "Complete" dataset, a matrix of Manhattan distances was obtained to produce a principal coordinates analysis (PCoA) with the packages adegenet 2.1.3 [52] and Ade4 [53] in the R-project. The result was plotted to show graphically if individuals tended to be grouped according to the 10 cacao genetic groups identified by Motamayor et al. [17].

3. Results

3.1. Descriptive Genetic Parameters of the SSR Loci

In total, the 16 SSR markers were able to amplify 88 alleles across all cacao samples of the "Honduras" dataset. As shown in Table 2, Na varied from 3 (MTcCIR22, MTcCIR115, and MTcCIR158) to 11 (MTcCIR33) with an average of 5.5 alleles for each locus. The effective number of alleles (Ne) ranged from 1.664 (MTcCIR60) to 3.926 (MTcCIR33) with a mean of 2.270. The frequency of a predominant allele (Fa) in polymorphic loci varied between 0.327 (MTcCIR33) and 0.746 (MTcCIR60) with an average of 0.610. The expected heterozygosity (He) ranged from 0.399 (MTcCIR60) to 0.745 (MTcCIR33) with a mean of 0.540, while observed heterozygosity (Ho) had values between 0.140 (MTcCIR158) and 0.386 (MTcCIR7) with an average of 0.293. For all loci, He was higher the Ho. The fixation index (F) was higher than zero for all loci and ranged between 0.077 (MTcCIR60) to 0.692 (MTcCIR158). The PIC varied from 0.360 (MTcCIR158) to 0.703 (MTcCIR33) with an average of 0.481.

3.2. Cluster Analysis

The STRUCTURE analysis results indicated that the 57 Honduran samples were grouped in two clusters (Figure 3). Only the accessions with a coefficient of membership higher than 0.70 were considered belonging to one of the two clusters. Trees sampled in Copán and Santa Bárbara (except for San_5, San_6, San_17, San_87, San_88, and San_89) were clustered in the same population along with Criollo 13, which is the reference carrying the genetic characteristics of pure Criollo cacao. Only two accessions sampled in Intibucá (Int_54 and Int_55) were grouped along with Criollo 13. Additionally, some accessions collected in Olancho (Ola_11, Ola_14, Ola_15, Ola_57, Ola_60, Ola_62, Ola_63, Ola_69, Ola_79, and Ola_85) were found to be pure Criollo, while ten samples were grouped in the second population, being represented by the blue color of the bar (Ola_8, Ola_9, Ola_12, Ola_16, Ola_59, Ola_61, Ola_65, Ola_68, Ola_75, and Ola_84). The remaining genotypes from Intibucá and Ola_10, Ola_86, San_3, San_4, and San_18 showed an admixture of the genetic characteristics of both two clusters (being the bar roughly half blue and half green).

Table 2. Genetic parameters from 16 SSR loci. For each locus, the number of alleles (Na), the effective number of alleles (Ne), frequency of the predominant allele (Fa), the observed (Ho) and expected heterozygosity (He), the fixation index (F), and the polymorphic information content (PIC) of the "Honduras" dataset are shown.

Locus	Na	Ne	Fa	Ho	He	F	PIC
MTcCIR6	6.0	2.10	0.63	0.26	0.52	0.50	0.46
MTcCIR7	5.0	2.24	0.60	0.39	0.55	0.30	0.49
MTcCIR8	5.0	2.01	0.66	0.33	0.50	0.34	0.45
MTcCIR11	5.0	2.04	0.62	0.28	0.51	0.45	0.43
MTcCIR12	7.0	2.18	0.61	0.35	0.54	0.35	0.48
MTcCIR18	5.0	1.97	0.67	0.30	0.49	0.39	0.44
MTcCIR22	3.0	1.80	0.69	0.25	0.44	0.44	0.37
MTcCIR24	5.0	2.79	0.52	0.28	0.64	0.56	0.59
MTcCIR26	6.0	2.71	0.54	0.26	0.63	0.58	0.58
MTcCIR33	11.0	3.93	0.33	0.38	0.74	0.49	0.70
MTcCIR37	8.0	2.67	0.54	0.30	0.62	0.52	0.58
MTcCIR40	6.0	2.50	0.60	0.30	0.60	0.50	0.57
MTcCIR60	5.0	1.67	0.75	0.37	0.40	0.08	0.35
MTcCIR115	3.0	2.06	0.63	0.28	0.51	0.45	0.44
MTcCIR158	3.0	1.84	0.66	0.14	0.46	0.69	0.36
MTcCIR168	5.0	1.82	0.71	0.21	0.45	0.53	0.41
Mean	5.5	2.27	0.61	0.29	0.54	0.45	0.48

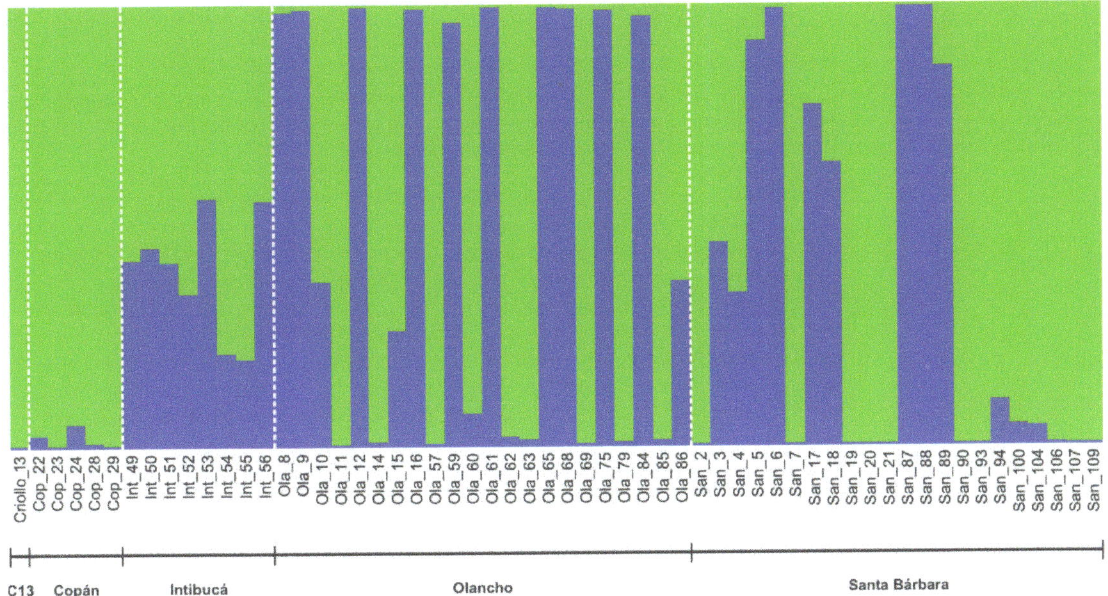

Figure 3. Population structure of *Theobroma cacao* L. of the "Honduras" dataset based on 16 SSR markers, using the best number of clusters according to the Evanno method (K = 2) [49]. Each individual sample is represented by a vertical line divided into K colored segments that represent the membership fraction for each cluster. The bar line under the plot represents the 4 Honduran departments. Criollo 13 has been used as Criollo reference and to identify the pure Honduran Criollo cacaos.

The ΔK of the Evanno method separated the "Complete" dataset into two main clusters (Figure S1). No other K was identified as a probable population number. The first cluster (in blue) consisted of all pure reference accessions of Criollo and most samples from

Honduras with a different degree of Criollo purity (Figure 4). The second cluster included the other cacaos accessions presenting admixture population characteristics. This group contained the remaining genetic groups identified by Motamayor et al. [17]. The divergence between Criollo and the other groups was much higher than the genetic divergence among the remaining nine clusters, and the presence of many Criollo samples in the dataset determined the clustering of the other cacaos into the same group.

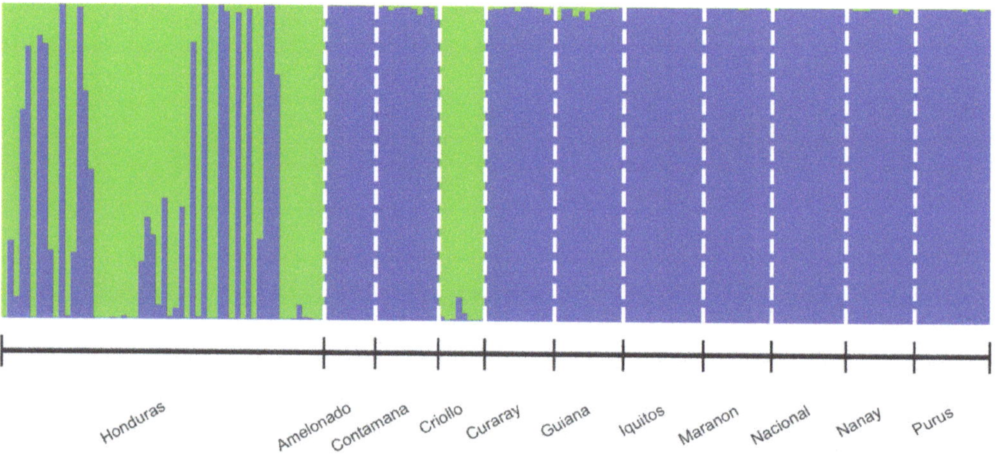

Figure 4. Population structure of *Theobroma cacao* L. of the "Complete" dataset based on 16 SSR markers, using the best assignment result retrieved according to the Evanno method (K = 2) [49]. Each individual sample is represented by a vertical line divided into K colored segments that represent the membership fraction for each cluster. The bar line under the plot represents the Honduras samples and the ten genetic groups identified by Motamayor et al. [17].

Honduran accessions included also Trinitario samples that showed genetic characteristics of Criollo and other genetic groups [54].

According to the Manhattan distances, a principal coordinates analysis was produced from the first two dimensions to show the genetic differences in the "Complete" dataset. The first dimension accounted for 24.42% of the total variation, while the second dimension for 9.69%. PCoA revealed the same result of the STRUCTURE analysis: two main clusters were identified. In the group located on the left of the plot all Criollo samples clustered together either those collected in Honduras either the reference Criollo accessions from Motamayor et al. [17], while the other nine genetics groups were clustered on the right side of the plot (Figure 5) along with some Honduran samples mainly distributed among the Amelonado, Guiana, and Marañon groups. Moreover, other Honduran samples, showing features of both of the two clusters and covering the genetic distances between Criollo and other genetic groups (Amelonado, Guiana, and Marañon), should be ascribed to as Trinitario.

A further STRUCTURE analysis was performed with the 116 references accessions from Motamayor et al. [17] along with the eight Honduran pure Criollo cacaos: Cop_22, Cop_23, Cop_24, Cop_28, Cop_29, San_106, San_107, and San_109. This reduction of the samples representative of the Honduran population was necessary to diminish the statistical weight of the Criollo accessions, which, due to their genetic distance with the other groups and to their low variability, determined a peak of the ΔK corresponding to K = 2 and not congruent with that of [17]. The ΔK of the Evanno method showed that the most probable population number is K = 2, highlighting a great genetic distance between Criollo and all the other genetic groups; however, peaks of K= 5 and K = 10 were also identified as the possible number of clusters (Figure S2). The value K = 10 showed the same results obtained by Motamayor et al. [17] (Figure 6). All Criollo accessions collected in Honduras correctly clustered with the Criollo references, while, all the other remaining genetic groups were identified.

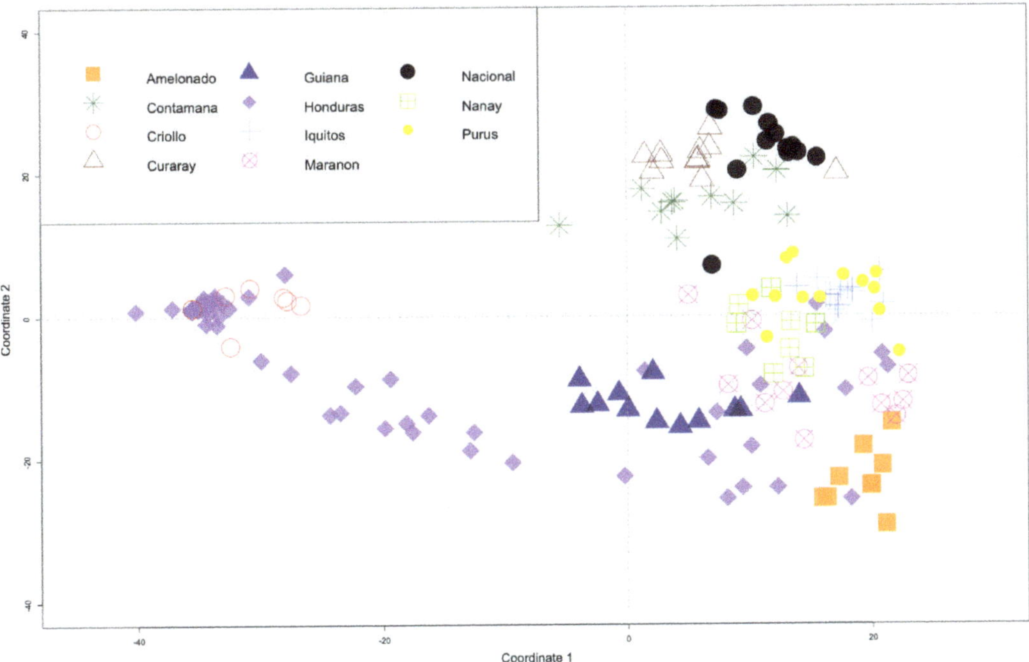

Figure 5. Principal coordinates analysis of the "Complete" dataset. The references accessions are divided based on the ten genetic groups [17], while all Honduran samples are identified by the same symbol to better visualize their wide distribution.

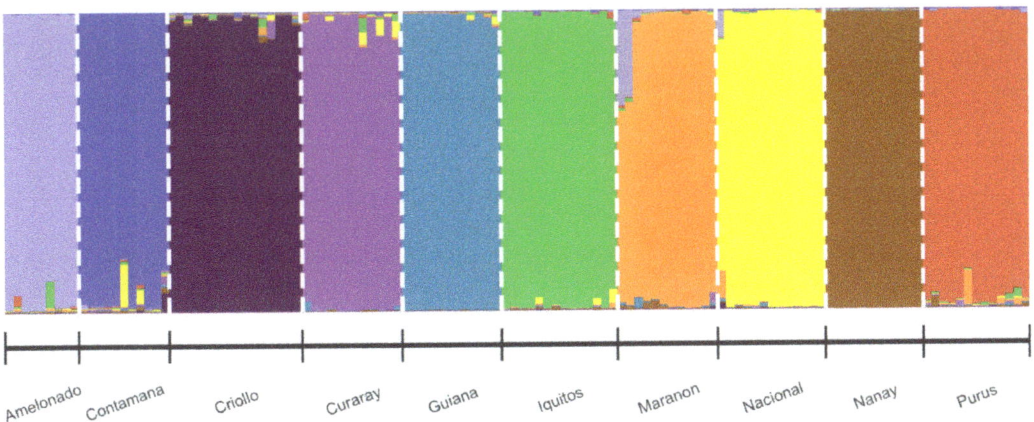

Figure 6. Population structure of *Theobroma cacao* L. of the 116 reference accessions along with 8 representative Honduran pure Criollo samples based on 16 SSR markers, using K = 10 according to the second best ΔK. Each individual sample is represented by a vertical line divided into K colored segments that represent the membership fraction for each cluster. The bar line under the plot represents the ten genetic groups identified by [17]. The 8 representative Honduras pure Criollo samples cluster with the Amazon Criollo reference accessions.

4. Discussion

The cultivation of cacao is of great economic importance for small producers in Honduras. In recent years, the number of hectares planted was increased, and therefore the cacao export volumes have also grown, especially to Europe. The increase in the demand for Honduran cacao is due to the high quality chocolate produced by the fine flavor cacao genetically influenced by the mixture of Criollo cacao. However most of the cacao cultivated fields are of the Trinitario type, which is a mixture between Criollo and Amelonado [7,55], because of the advanced lack of interest of Criollo cultivation in Honduras where no commercial plantations are available. Since about 10 years ago, ex-situ conservation began, taking some trees of Criollo cacao from the farmers to the FHIA germplasm bank. The cacao cultivation in Honduras has a historical and cultural significance, it is known from anthropological discoveries that ancient populations such as the Mayas, Lencas, Chortis, Tolupanes, and others used cacao as a form of trade and even came to have importance in their ceremonial rites, and in the form of commodities exchange (used as money) [10]. It is believed that the Criollo cacao was the only cultivated in Mesoamerica before the arrival of the Europeans [2,24]. Some investigations indicate that cacao was domesticated in Mesoamerica, from Southern Mexico to Panama and that is why it is common that in these countries Criollo cacao is found in a dispersed way. Possibly, the Mesoamerican cacao producers stopped cultivating Criollo cacao due to the arrival of new cacao cultivars (Forastero type) that are more resistant to pests and diseases and overall more productive [7,16]. Although Criollo cacao did not continue to be cultivated, it is found in small plots or isolated areas that are about to disappear. It is observed that some Criollo cacao trees are not vigorous, possibly the genetic isolation has caused an increase in the level of homozygosity (as confirmed by the observed heterozygosity and F index detected in the present study) hindering their growth, development, and adaptability to other environmental conditions, but Criollo cacao is superior in quality [40].

There are few studies on the characterization of the genetic diversity of Criollo cacao of Honduras using genomic tools. The most important study was carried out by Ji et al. [7] using SNP markers, where eleven out of the fourteen samples collected as Criollo from Honduras turned out to be pure. Motilal et al. [39], investigating the genetic diversity of Belizean Criollo cacao found that eleven different genotypes belonged to the Criollo group out of 77 samples analyzed. In this study, five accessions of Criollo cacao from Honduras (Honduras 6, 10, 11, 13, and 18), which belong to the international cacao genebank in Trinidad and Tobago (ICG, T), were used as references. In both studies [7,39] the authors used Criollo 13 as a reference as was done in the present research. Criollo 13 belongs to the CATIE cacao germplasm bank in Costa Rica [16]. The results of the present study are similar to those found by [7,39]: in fact, thirty out of the fifty-seven samples considered were shown to be pure Criollo, using Criollo 13 as a reference in the analysis, and they are distributed in Copán, Santa Bárbara, and Olancho. However, although samples were taken from trees that phenotypically have characteristics of Criollo cacao, the results show that not all the collected samples have genetic purity: in fact, eleven cacaos (from Intibucá, Olancho, and Santa Bárbara) have a mixture of Criollo and other genetic groups and therefore they could be considered as Trinitario type (Figures 3 and 4). Moreover, the remaining Honduras samples show a genetic profile clustering mainly with the Amelonado, Guiana, and Marañon genetic groups (Figure 5) and they are predominantly located in the Olancho departments, where the cultivation of cacao has been recently introduced. The geographical areas where the samples of Criollo cacao were collected in Honduras (Copán, Santa Bárbara, Intibucá, and Olancho) represent areas currently inhabited by indigenous groups that preserve the tradition of cacao cultivation, especially in Copán, which was in past times the most important ceremonial and astronomical center of the Mayan culture [24,56].

The observed heterozygosity level detected in the "Honduras" dataset is low and comparable with that observed within the cacao populations analyzed by [7,39]. This low level of heterozygosity is probably due to the fact that samples has been collected either in small-medium family farms either from wild sources where they live in isolation or in limited groups without any farming care. This condition might be conducive to inbreeding or possible crossover between siblings that could account for the low variability within the Criollo population. The low variability is also supported by the PCoA where the pure Criollos are located in a restricted area on the left side of the graphic (Figure 5). Moreover, the distribution of the groups in the PCoA shows a great affinity with those of the PcoA in Ji et al. [7] and of the multivariate analysis of the genetic distances in Motilal et al. [39], indicating that a fairly relative number of SSR marker is sufficient to identify the main cacao groupings.

The present study evidenced that notwithstanding the collection of samples has been carried on according to the morphological characteristics of Criollo, some samples showed a different genetic clustering. Therefore, the morphological characterization is not sufficient to identify Criollo cacao and genetic analysis is requested for a correct determination of the samples.

5. Conclusions

The SSR markers used in the current study were able to separate the samples of pure Criollo cacao from the other remaining genetic groups and admixture populations. Results highlighted the great genetic distance between Criollo and other genetic groups. Criollo 13 cacao continues to be a good reference for studies of cacao genetic diversity where Criollo samples are used. Finally, it was possible to determine the genetic purity of cacao Criollos from Honduras and affirm the results of other archaeological investigations that indicate that Mesoamerica is the center of cacao domestication. These results can be used to include these samples of Criollo cacao in a genetic improvement program and especially for the conservation of this genetic wealth that we have inherited from our ancestral peoples.

Supplementary Materials: The following are available online at https://www.mdpi.com/2073-4395/11/2/225/s1, Table S1: List of the 89 cacao accessions collected in Honduras. The ID, the department where samples were collected and the GPS coordinates are shown, Table S2: List of 116 reference from Motamayor et al. [17], Table S3: List of the allelic profiles of the 57 Honduran samples for the 16 SSR loci, Figure S1: The ΔK of the Evanno method for the "Complete" dataset, Figure S2: The ΔK of the Evanno method for the STRUCTURE analysis of the 116 references accessions along with the 8 representative Honduran pure Criollo.

Author Contributions: Conceptualization, M.L. and S.B.; methodology, M.G., L.B., M.L., O.G. and E.O.; formal analysis, L.B. and A.M.; investigation, E.O., E.D. and E.G.; resources, E.P. and E.G.; data curation, M.G. and L.B.; writing—original draft preparation, M.L., S.B. and L.B.; writing—review and editing, O.G., M.L., S.B. and L.B.; supervision, E.P. and M.L.; funding acquisition, E.P. All authors have read and agreed to the published version of the manuscript.

Funding: This research received no external funding.

Informed Consent Statement: Not applicable.

Data Availability Statement: The set of data regarding Honduras collected samples presented in this study are available in supplementary material here. Other data used for supporting the analyses carried out in this article must be requested to the corresponding author of the papers cited in the article.

Acknowledgments: This research was performed in the frame of the Master student grant program AICS-MAECI and the Master Course in Natural Resources Management for Tropical Rural Development—DAGRI—University of Florence—Italy, in cooperation with the National Autonomous University of Honduras (UNAH) La Ceiba, Atlántida—Honduras and the World Cocoa Foundation Fellowship.

Conflicts of Interest: The authors declare no conflict of interest. The funders had no role in the design of the study; in the collection, analyses, or interpretation of data; in the writing of the manuscript, or in the decision to publish the results.

References

1. Cheesman, E. *Notes on the Nomenclature, Classification and Possible Relationships of Cacao Populations*; IPC Science and Technology Press: Guildford, UK, 1944.
2. Bartley, B.G. *The Genetic Diversity of Cacao and Its Utilization*; CABI: Oxfordshire, UK, 2005; ISBN 1-84593-024-X.
3. Flood, J.; Murphy, R. *Cocoa Futures: A Source Book of Some Important Issues Facing the Cocoa Industry*; Federación de Cafetaleros de Colombia: Bogotá, Colombia; CABI: London, UK, 2004; ISBN 958-97441-1-7.
4. Cuatrecasas, J. *Cacao and Its Allies: A Taxonomic Revision of the Genus Theobroma*; Smithsonian Institution: Washington, DC, USA, 1964; Volume 35.
5. Henderson, J.S.; Joyce, R.A.; Hall, G.R.; Hurst, W.J.; McGovern, P.E. Chemical and archaeological evidence for the earliest cacao beverages. *Proc. Natl. Acad. Sci. USA* **2007**, *104*, 18937–18940. [CrossRef] [PubMed]
6. Powis, T.G.; Cyphers, A.; Gaikwad, N.W.; Grivetti, L.; Cheong, K. Cacao use and the San Lorenzo Olmec. *Proc. Natl. Acad. Sci. USA* **2011**, *108*, 8595–8600. [CrossRef] [PubMed]
7. Ji, K.; Zhang, D.; Motilal, L.A.; Boccara, M.; Lachenaud, P.; Meinhardt, L.W. Genetic diversity and parentage in farmer varieties of cacao (*Theobroma cacao* L.) from Honduras and Nicaragua as revealed by single nucleotide polymorphism (SNP) markers. *Genet. Resour. Crop Evol.* **2013**, *60*, 441–453. [CrossRef]
8. Crown, P.L.; Hurst, W.J. Evidence of cacao use in the Prehispanic American Southwest. *Proc. Natl. Acad. Sci. USA* **2009**, *106*, 2110–2113. [CrossRef]
9. Wood, G.; Lass, R. *Cocoa Tropical Agriculture*; Wiley-Blackwell: Hoboken, NJ, USA, 1985.
10. Zarrillo, S.; Gaikwad, N.; Lanaud, C.; Powis, T.; Viot, C.; Lesur, I.; Fouet, O.; Argout, X.; Guichoux, E.; Salin, F. The use and domestication of *Theobroma cacao* during the mid-Holocene in the upper Amazon. *Nat. Ecol. Evol.* **2018**, *2*, 1879–1888. [CrossRef]
11. Coe, S.D.; Coe, M.D. *The True History of Chocolate*; Thames & Hudson: London, UK, 2013; ISBN 0-500-77093-X.
12. Bergmann, J.F. The distribution of cacao cultivation in pre-Columbian America. *Ann. Assoc. Am. Geogr.* **1969**, *59*, 85–96. [CrossRef]
13. Cornejo, O.E.; Yee, M.-C.; Dominguez, V.; Andrews, M.; Sockell, A.; Strandberg, E.; Livingstone, D.; Stack, C.; Romero, A.; Umaharan, P. Population genomic analyses of the chocolate tree, *Theobroma cacao* L., provide insights into its domestication process. *Commun. Biol.* **2018**, *1*, 1–12. [CrossRef]
14. Bekele, F.; Bekele, I. A sampling of the phenetic diversity of cacao in the International Cocoa Gene Bank of Trinidad. *Crop Sci.* **1996**, *36*, 57–64. [CrossRef]
15. Argout, X.; Salse, J.; Aury, J.-M.; Guiltinan, M.J.; Droc, G.; Gouzy, J.; Allegre, M.; Chaparro, C.; Legavre, T.; Maximova, S.N.; et al. The genome of *Theobroma cacao*. *Nat. Genet.* **2011**, *43*, 101–108. [CrossRef]
16. Lachenaud, P.; Motamayor, J.C. The Criollo cacao tree (*Theobroma cacao* L.): A review. *Genet. Resour. Crop Evol.* **2017**, *64*, 1807–1820. [CrossRef]
17. Motamayor, J.C.; Lachenaud, P.; Mota, J.W.d.S.e.; Loor, R.; Kuhn, D.N.; Brown, J.S.; Schnell, R.J. Geographic and genetic population differentiation of the Amazonian chocolate tree (*Theobroma cacao* L.). *PLoS ONE* **2008**, *3*, e3311. [CrossRef]
18. Pound, F. A note on the cocoa population of South America. *Rep. Proc. Cocoa Res. Conf.* **1945**, *1*, 131–133.
19. N'Goran, J.A.; Laurent, V.; Risterucci, A.-M.; Lanaud, C. Comparative genetic diversity studies of *Theobroma cacao* L. using RFLP and RAPD markers. *Heredity* **1994**, *73*, 589–597. [CrossRef]
20. Lerceteau, E.; Robert, T.; Pétiard, V.; Crouzillat, D. Evaluation of the extent of genetic variability among *Theobroma cacao* accessions using RAPD and RFLP markers. *Theor. Appl. Genet.* **1997**, *95*, 10–19. [CrossRef]
21. Whitkus, R.; De la Cruz, M.; Mota-Bravo, L.; Gómez-Pompa, A. Genetic diversity and relationships of cacao (*Theobroma cacao* L.) in southern Mexico. *Theor. Appl. Genet.* **1998**, *96*, 621–627. [CrossRef]
22. Sounigo, O.; Umaharan, R.; Christopher, Y.; Sankar, A.; Ramdahin, S. Assessing the genetic diversity in the International Cocoa Genebank, Trinidad (ICG, T) using isozyme electrophoresis and RAPD. *Genet. Resour. Crop Evol.* **2005**, *52*, 1111–1120. [CrossRef]
23. dos Santos, R.C.; Pires, J.L.; Lopes, U.V.; Gramacho, K.P.G.; Flores, A.B.; Bahia, R.d.C.S.; Ramos, H.C.C.; Corrêa, R.X.; Ahnert, D. Assessment of genetic diversity on a sample of cocoa accessions resistant to witches' broom disease based on RAPD and pedigree data. *Bragantia* **2005**, *64*, 361–368. [CrossRef]
24. Motamayor, J.C.; Risterucci, A.-M.; Lopez, P.A.; Ortiz, C.F.; Moreno, A.; Lanaud, C. Cacao domestication I: The origin of the cacao cultivated by the Mayas. *Heredity* **2002**, *89*, 380–386. [CrossRef]
25. Zhang, D.; Mischke, S.; Johnson, E.S.; Phillips-Mora, W.; Meinhardt, L. Molecular characterization of an international cacao collection using microsatellite markers. *Tree Genet. Genomes* **2009**, *5*, 1–10. [CrossRef]
26. Zhang, D.; Boccara, M.; Motilal, L.; Mischke, S.; Johnson, E.S.; Butler, D.R.; Bailey, B.; Meinhardt, L. Molecular characterization of an earliest cacao (*Theobroma cacao* L.) collection from Upper Amazon using microsatellite DNA markers. *Tree Genet. Genomes* **2009**, *5*, 595–607. [CrossRef]
27. Zhang, D.; Martínez, W.J.; Johnson, E.S.; Somarriba, E.; Phillips-Mora, W.; Astorga, C.; Mischke, S.; Meinhardt, L.W. Genetic diversity and spatial structure in a new distinct *Theobroma cacao* L. population in Bolivia. *Genet. Resour. Crop Evol.* **2012**, *59*, 239–252. [CrossRef]
28. Stack, J.C.; Royaert, S.; Gutiérrez, O.; Nagai, C.; Holanda, I.S.A.; Schnell, R.; Motamayor, J.-C. Assessing microsatellite linkage disequilibrium in wild, cultivated, and mapping populations of *Theobroma cacao* L. and its impact on association mapping. *Tree Genet. Genomes* **2015**, *11*, 19. [CrossRef]

29. Motilal, L.A.; Zhang, D.; Mischke, S.; Meinhardt, L.W.; Umaharan, P. Microsatellite-aided detection of genetic redundancy improves management of the International Cocoa Genebank, Trinidad. *Tree Genet. Genomes* **2013**, *9*, 1395–1411. [CrossRef]
30. Everaert, H.; Rottiers, H.; Pham, P.H.D.; Ha, L.T.V.; Nguyen, T.P.D.; Tran, P.D.; De Wever, J.; Maebe, K.; Smagghe, G.; Dewettinck, K. Molecular characterization of Vietnamese cocoa genotypes (*Theobroma cacao* L.) using microsatellite markers. *Tree Genet. Genomes* **2017**, *13*, 99. [CrossRef]
31. Everaert, H.; De Wever, J.; Tang, T.K.H.; Vu, T.L.A.; Maebe, K.; Rottiers, H.; Lefever, S.; Smagghe, G.; Dewettinck, K.; Messens, K. Genetic classification of Vietnamese cacao cultivars assessed by SNP and SSR markers. *Tree Genet. Genomes* **2020**, *16*, 43. [CrossRef]
32. Guiraud, B.S.H.B.; Tahi, M.G.; Fouet, O.; Trebissou, C.I.; Pokou, D.; Rivallan, R.; Argout, X.; Koffi, K.K.; Koné, B.; Zoro, B.I.A.; et al. Assessment of genetic diversity and structure in cocoa trees (*Theobroma cacao* L.) in Côte d'Ivoire with reference to their susceptibility to Cocoa swollen shoot virus disease (CSSVD). *Tree Genet. Genomes* **2018**, *14*, 52. [CrossRef]
33. Lindo, A.A.; Robinson, D.E.; Tennant, P.F.; Meinhardt, L.W.; Zhang, D. Molecular characterization of cacao (*Theobroma cacao*) germplasm from Jamaica using single nucleotide polymorphism (SNP) markers. *Trop. Plant Biol.* **2018**, *11*, 93–106. [CrossRef]
34. Adenet, S.; Regina, F.; Rogers, D.; Bharath, S.; Argout, X.; Rochefort, K.; Cilas, C. Study of the genetic diversity of cocoa populations (*Theobroma cacao* L.) of Martinique (FWI) and potential for processing and the cocoa industry. *Genet. Resour. Crop Evol.* **2020**, *67*, 1–11. [CrossRef]
35. Arevalo-Gardini, E.; Meinhardt, L.W.; Zuñiga, L.C.; Arévalo-Gardni, J.; Motilal, L.; Zhang, D. Genetic identity and origin of "Piura Porcelana"—A fine-flavored traditional variety of cacao (Theoborma cacao) from the Peruvian Amazon. *Tree Genet. Genomes* **2019**, *15*, 11. [CrossRef]
36. Cosme, S.; Cuevas, H.; Zhang, D.; Oleksyk, T.; Irish, B. Genetic diversity of naturalized cacao (*Theobroma cacao* L.) in Puerto Rico. *Tree Genet. Genomes* **2016**, *12*, 88. [CrossRef]
37. Mata-Quirós, A.; Arciniegas-Leal, A.; Phillips-Mora, W.; Meinhardt, L.W.; Motilal, L.; Mischke, S.; Zhang, D. Assessing hidden parentage and genetic integrity of the "United Fruit Clones" of cacao (*Theobroma cacao*) from Costa Rica using SNP markers. *Breed. Sci.* **2018**, 18057. [CrossRef] [PubMed]
38. Olasupo, F.O.; Adewale, D.B.; Aikpokpodion, P.O.; Muyiwa, A.A.; Bhattacharjee, R.; Gutierrez, O.A.; Motamayor, J.C.; Schnell, R.J.; Ebai, S.; Zhang, D. Genetic identity and diversity of Nigerian cacao genebank collections verified by single nucleotide polymorphisms (SNPs): A guide to field genebank management and utilization. *Tree Genet. Genomes* **2018**, *14*, 32. [CrossRef]
39. Motilal, L.A.; Zhang, D.; Umaharan, P.; Mischke, S.; Mooleedhar, V.; Meinhardt, L.W. The relic Criollo cacao in Belize–genetic diversity and relationship with Trinitario and other cacao clones held in the International Cocoa Genebank, Trinidad. *Plant Genet. Resour. Charact. Util.* **2010**, *8*, 106. [CrossRef]
40. Durán, E.; Dubón, A. *Tipos Genéticos de Cacao y Distribución Geográfica en Honduras*; Centro de Comunicación Agrícola: Lima, Honduras, 2016.
41. Murray, M.G.; Pitas, J.W. Plant DNA from alcohol-preserved samples. *Plant Mol. Biol. Report.* **1996**, *14*, 261–265. [CrossRef]
42. Bressan, E.A.; Rossi, M.L.; Gerald, L.T.; Figueira, A. Extraction of high-quality DNA from ethanol-preserved tropical plant tissues. *BMC Res. Notes* **2014**, *7*, 268. [CrossRef] [PubMed]
43. Saunders, J.A.; Mischke, S.; Leamy, E.A.; Hemeida, A.A. Selection of international molecular standards for DNA fingerprinting of *Theobroma cacao*. *Theor. Appl. Genet.* **2004**, *110*, 41–47. [CrossRef]
44. Pugh, T.; Fouet, O.; Risterucci, A.-M.; Brottier, P.; Abouladze, M.; Deletrez, C.; Courtois, B.; Clément, D.; Larmande, P.; N'Goran, J.A. A new cacao linkage map based on codominant markers: Development and integration of 201 new microsatellite markers. *Theor. Appl. Genet.* **2004**, *108*, 1151–1161. [CrossRef]
45. Kalinowski, S.T.; Taper, M.L.; Marshall, T.C. Revising how the computer program CERVUS accommodates genotyping error increases success in paternity assignment. *Mol. Ecol.* **2007**, *16*, 1099–1106. [CrossRef]
46. Peakall, R.; Smouse, P.E. GENALEX 6: Genetic analysis in Excel. Population genetic software for teaching and research. *Mol. Ecol. Notes* **2006**, *6*, 288–295. [CrossRef]
47. Liu, K.; Muse, S.V. PowerMarker: An Integrated Analysis Environment for Genetic Marker Analysis. *Bioinformatics* **2005**, *21*, 2128–2129. [CrossRef]
48. Pritchard, J.K.; Stephens, M.; Donnelly, P. Inference of population structure using multilocus genotype data. *Genetics* **2000**, *155*, 945–959. [PubMed]
49. Evanno, G.; Regnaut, S.; Goudet, J. Detecting the number of clusters of individuals using the software STRUCTURE: A simulation study. *Mol. Ecol.* **2005**, *14*, 2611–2620. [CrossRef] [PubMed]
50. Earl, D.A. Structure harvester: A website and program for visualizing STRUCTURE output and implementing the Evanno method. *Conserv. Genet. Resour.* **2012**, *4*, 359–361. [CrossRef]
51. Francis, R.M. Pophelper: An R package and web app to analyse and visualize population structure. *Mol. Ecol. Resour.* **2017**, *17*, 27–32. [CrossRef]
52. Jombart, T. Adegenet: A R package for the multivariate analysis of genetic markers. *Bioinformatics* **2008**, *24*, 1403–1405. [CrossRef]
53. Chessel, D.; Dufour, A.B.; Thioulouse, J. The ade4 package-I-One-table methods. *R News* **2004**, *4*, 5–10.
54. Motamayor, J.C.; Risterucci, A.-M.; Heath, M.; Lanaud, C. Cacao domestication II: Progenitor germplasm of the Trinitario cacao cultivar. *Heredity* **2003**, *91*, 322–330. [CrossRef]

55. Boza, E.J.; Motamayor, J.C.; Amores, F.M.; Cedeño-Amador, S.; Tondo, C.L.; Livingstone, D.S.; Schnell, R.J.; Gutiérrez, O.A. Genetic Characterization of the Cacao Cultivar CCN 51: Its Impact and Significance on Global Cacao Improvement and Production. *J. Am. Soc. Hortic. Sci.* **2014**, *139*, 219–229. [CrossRef]
56. De la Cruz, M.; Whitkus, R.; Gómez-Pompa, A.; Mota-Bravo, L. Origins of cacao cultivation. *Nature* **1995**, *375*, 542–543. [CrossRef]

Article

Grain Quality and Allelic Variation of the *Badh2* Gene in Thai Fragrant Rice Landraces

Phukjira Chan-in [1], Sansanee Jamjod [1,2], Narit Yimyam [3], Benjavan Rerkasem [1] and Tonapha Pusadee [1,4,*]

[1] Plant Genetic Resource and Nutrition lab (CMUPNlab), Department of Plant and Soil Science, Faculty of Agriculture, Chiang Mai University, Chiang Mai 50200, Thailand; phukjira_c@cmu.ac.th (P.C.-i.); sansanee.j@cmu.ac.th (S.J.); benrerkasem@gmail.com (B.R.)

[2] Research Center for Development of Local Lanna Rice and Rice Products, Chiang Mai University, Chiang Mai 50200, Thailand

[3] Department of Highland Agriculture Natural Resources and Environment, Faculty of Agriculture, Chiang Mai University, Chiang Mai 50200, Thailand; narit.y@cmu.ac.th

[4] Innovative Agriculture Research Center, Faculty of Agriculture, Chiang Mai University, Chiang Mai 50200, Thailand

* Correspondence: tonapha.p@cmu.ac.th; Tel.: +66-85-375-8160

Received: 17 April 2020; Accepted: 29 May 2020; Published: 30 May 2020

Abstract: Fragrance, which plays an important role in determining the economic value of rice to growers and consumers, is known to be controlled by the *Badh2* gene. This study evaluated the grain quality characteristics and allelic variation of the *Badh2* gene in 22 fragrant rice landraces from Thailand. The rice seed samples from farmers' storage facilities in northern, northeastern and southern Thailand, plus two advanced breeding lines and three check varieties, were evaluated for seed morphology and grain quality, and their *Badh2* genes covering intron 4 to intron 8 were re-sequenced. Almost all of the landraces were classified as large grain types, with medium to high gelatinization temperatures. The variation in the *Badh2* gene by haplotype analysis correlated with grain aroma by sensory evaluation. The *badh2*-E7 was found in haplotype 1 with a strong aroma in KH, NDLP, and PLD, as in KDML105 and the moderately aromatic BNM-CMU, BNM4, and SKH, along with PTT1. Three haplotypes had different positions of SNP on the *Badh2* gene with varying results in the sensory test. The present results suggest that some rice varieties could be potentially introduced as genetic resources for fragrant rice breeding programs or could be developed to highly palatable cultivars with geographical indications to increase the income of highland farmers.

Keywords: aromatic rice; local variety; gelatinization temperature; *badh2*-E7 allele

---

1. Introduction

The eating quality of rice, also known as rice palatability, is a very important factor that determines the economic value of rice for producers and consumers—the aroma of rice plays a significant part of this. Fragrant rice is quality rice with a good taste, softness, and a unique aroma, making it very popular [1]. Aromatic or fragrant rice is the most expensive rice in the global rice market [2–4]. Among the numerous volatile compounds associated with the aroma in rice, the main aromatic compound has been identified as 2-acetyl-1-pyrroline (2AP) [5], produced under the control of the recessive gene *Badh2* [6–8]. In aromatic rice genotypes, there is an eight base-pairs deletion in exon 7 of the *Badh2* gene [9], but with the 2AP concentration and aroma varying greatly depending on the environment [10]. The most widely known aromatic rice are Basmati rice of India and Pakistan [2], and Thailand's Hom Mali [11]. Basmati, with extra-long slender grains of soft and fluffy texture when cooked, belongs to the *japonica* group, have while Hom Mali and other jasmine type varieties, with long slender grains, soft and moist texture when cooked, are more closely with the *indica* group [12]. In spite of their distinctive grain quality features, the *Badh2* gene and 2AP are common to both groups of aromatic rice. Most of India's Basmati rice is grown from the variety Pusa Basmati 1121 [13], while Thai Hom Mali, a worldwide registered trademark [14], is required by law to be grown from the traditional photoperiod sensitive varieties KDML105 and RD15 in the wet season only [15]. However, in addition to these mega-varieties, numerous other fragrant rice varieties are found in rice-growing countries of Asia [2]. The mutations on the *Badh2* gene have led to the introduction of a premature stop codon to produce a protein that disables the *Badh2* enzyme, leading to the accumulation of the 2AP substrate [16,17]. Further studies have revealed that numerous *Badh2* alleles have been detected in diverse rice (*Oryza sativa* L.) germplasms. Fragrant rice accessions have been reported in three different genetic subpopulations of rice, including *Group V* (Basmati and Sadri varieties), *indica* (Jasmine varieties), and *tropical japonica* [18]. There were many kinds of mutations in the *Badh2* gene among different rice varieties, namely the common 8 bp deletion and three single nucleotide polymorphism sites (SNPs) in the 7th exon of the *Badh2* gene (*badh2.1* or *badh2-E7*) [9,19]; a 7-bp deletion in exon 2 (*badh2-E2.1*) [16]; a 7-bp insertion in exon 8 (*badh2-E8*) [20]; an additional eight alleles in exons 1 (*badh2-E1.1*), 10 (*badh2-E10.1*, *badh2-E10.2*, *badh2-E10.3*) 13 (*badh2-E13.1* and *badh2-E13.2*) and 14 (*badh2-E14.1* and *badh2-E14.2*) [12]; a 803-bp deletion between exons 4 and 5 (*badh2-E4-5.2*) [21]; a SNP at exon 10 (*badh2-E10.4*); a 75-bp deletion in exon 2 (*badh2-E2.2*); a 806-bp deletion between exons 4 and 5 (*badh2-E4-5.1*) [22]; a 3-bp deletion in the 5′ UTR (*badh2-p-5′ UTR*); an 8-bp insertion in the promoter [23]; an SNP at the exon1–intron1 junction (*badh2-E1.2*) [24]; and a 3-bp deletion in exon 12 (*badh2-E12*) [25].

Detailed studies have been conducted on the *Badh2* gene in the aromatic rice germplasm of Myanmar [26,27], while outstanding commercial potential has been seen with Cambodian Phka varieties [28,29]. Several aromatic rice genotypes have been identified in Thailand, including those with special quality characteristics, e.g., those with pigmented pericarp, Leum Pua [30,31] and Hom-nil [32,33]. Accessions of a fragrant rice landrace, commonly known in some parts of northern Thailand's highlands as Bue Nur Moo (fragrant non-glutinous rice in the Karen language, here referred to as BNM), collected from farmers have been found to range widely both in the content of the aromatic compound 2AP and in the allelic variation of the aroma gene *Badh2* [34]. Two accessions of the landrace were found, with the 2AP content approaching that of KDML105 and the same key deletion of the *Badh2* gene, designated BNM-CMU and BNM4, while others did not have the important deletion, containing little to no fragrance. The BNM accessions also exhibited strong interaction effects of genotype × location on their grain quality characteristics, the head rice yield, gelatinization temperature and 2AP concentration [35]. The dominance of KDML105 in fragrant rice production is a relatively recent development in Thailand, where fragrance was once a common feature among the diversity of grain quality and adaptation traits of local rice landraces [36]. The present study set out to examine the grain quality of fragrant rice landraces from different regions of the country and to explore the allelic variation of the *Badh2* gene.

In Thailand, Hom Mali is grown mostly from the variety Khao Dawk Mali 105 (KDML105), which was judged the world's best rice for two consecutive years at the 9th The Rice Trader (TRT) World Rice Conference 2017 [37]. However, in the following years the title was lost to the Cambodian Malys Angkor, and Vietnam's ST24 [38]. This causes some concern regarding the competitiveness of Thai Hom Mali. With an economic advantage of much higher price than non-aromatic rice, studies of aromatic rice germplasm should contribute towards maintenance and improvement in the rice quality standards.

Thailand is one of the most significant and unique countries for plant genetic resources and crop diversity, especially for rice (*Oryza sativa* L.) [39]. As Thailand lies partly in the center of rice diversity and in the region where rice was originally domesticated [40,41], the characterization of local varieties and landrace collections are critical for the utilization of these resources. The farmers of this region still use their traditional or local cultivars, which not only suit their taste, but also provide crop security. Rice landraces have unique characteristics, including special quality traits such as their aroma, adaptation to the local environment and resistance to biotic and abiotic stress [42]. In addition, local rice contains high levels of genetic diversity, which provides an opportunity for plant breeders to select and improve new cultivars, which include both farmers and breeder's preferred traits. Hence, the development of new modern rice varieties has depended on the availability of genetic diversity [43–46].

Local fragrant rice germplasm of northern Thailand are high-value genetic resources as they contain special qualities. Many fragrant rice varieties of northern Thailand have been collected and selected by Chiang Mai University's breeding team for their high 2AP content in brown rice compared to the famous jasmine rice of Thailand. Initial studies focused on nine populations of one Thai fragrant rice landrace, Buer Ner Moo (BNM), from nine farmers and the two advanced breeding lines of JPD and, BNM-CMU, a cross between BNM and PTT1. The different local fragrant rice varieties showed different levels of 2AP content in brown rice, even within one variety name, due to the variation within landraces. Moreover, *Badh2* allelic variation was found, which illustrated the *badh2-E7* deletion on the *Badh2* gene in BNM4, and BNM-CMU. They had high 2AP contents in brown rice with *badh2-E7* deletion similar to that in KDML105 and PTT, the popular fragrant rice in Thailand [34]. In addition, a significant correlation was also found between 2AP content in brown rice and elevation. The 2AP content in brown rice of most of the Bue Ner Moo populations significantly increased when grown at high altitude [35]. Therefore, the local rice varieties that profess to be fragrant rice could have the potential to be developed into a new fragrant rice variety with geographical indications to increase income for the highland farmers. However, the studies of the fragrant gene in landrace rice are not well understood. To investigate germplasm resources for crop utilization and improvement programs, it is essential to describe and evaluate the morphological characteristics of existing germplasm resources to effectively identify and differentiate each cultivar, including identification of the interested genes' functional alleles.

Therefore, the objectives of this study were to evaluate the grain qualities, to investigate the allelic variation of the *Badh2* gene in fragrant rice landraces, and to introduce a genetic resource for fragrant rice breeding programs by pure-line selection method or improved variety. This would also provide a basis for the development of other varieties to be consistent with the increasing demand of the market and consumers.

2. Materials and Methods

2.1. Plant Materials

Seed samples of fragrant rice landraces were collected from farmers' seed storages, including nine from the north, eight from the northeast and five from the south of Thailand. Two promising landraces previously identified by a rice breeding program at Chiang Mai University, BNM-CMU and Jow Pluak Dam (JPD), plus two elite fragrant varieties, Khao Dawk Mali 105 (KDML105) and Pathum Thani 1 (PTT1), and one high yielding non-fragrant variety Suphan Buri 1 (SPR1) were also included for comparison (Table 1). The seeds of each population were germinated in petri dishes for 14 days and then transplanted to 30 cm diameter undrained pots, with ten plants per pot, two pots per variety and two replications. The plants were grown as wetland rice, with 15 cm of water maintained above the soil surface at Chiang Mai University, Thailand, in the wet season of 2017/2018. At the tillering stage, leaf samples of each individual were collected and kept at −20 °C for DNA extraction. At grain maturity, seed samples were collected from 10 individual plants from each variety.

Table 1. List of the fragrant rice landraces from three regions of Thailand, the elite varieties, and the advanced breeding lines studied for comparison.

Rice Samples	Source	Code	Rice Variety Name	Source (Province)
Rice landrace	North	BNM1	Buer Ner Moo	Chiang Mai
		BNM2	Buer Ner Moo	Chiang Mai
		BNM3	Buer Ner Moo	Chiang Mai
		BNM4	Buer Ner Moo	Chiang Mai
		BNM5	Buer Ner Moo	Chiang Mai
		BNM6	Buer Ner Moo	Chiang Mai
		BNM7	Buer Ner Moo Pho Phi	Chiang Mai
		BNM8	Buer Ner Moo Pho Phi	Chiang Mai
		BNM9	Buer Ner Moo Phardo	Chiang Mai
	Northeast	EL	E-Leuang	Loei
		KH	Kaow Hawm	Loei
		HS	Hawm Sa-ngium	Loei
		NDLP	Niaw Dam Luem Pua	Loei
		NU	Niaw Ubon	Loei
		PS	Pla Sew	Loei
		PLD	Phayaa Luem Dang	Loei
		SKH	Sew Kliang Hawm	Loei
	South	BH62	Baow Hawm 62	Songkhla
		BH96	Baow Hawm96	Songkhla
		HB	Hawm Baow	Songkhla
		HBK	Hawm Bang Kaew	Songkhla
		HNK	Hawm Na Kaow	Songkhla
Breeding Line		BNM-CMU	Buer Ner Moo-CMU	Department of Agriculture
		JPD	Jow Pluak Dam	Department of Agriculture
Elite rice		KDML105	Khao Dawk Mali 105	Department of Agriculture
(as check varieties)		PTT1	Pathum Thani 1	Department of Agriculture
		SPR1	Suphan Buri 1	Department of Agriculture

2.2. Seed Morphological Characterization

The rice seed morphological characterization was conducted in accordance with the rice descriptor [47]. One hundred seeds of each sample were recorded individually for husk color, pericarp color, and awning. Seed sizes (grain length, width, and thickness) of unhusked seeds were measured by a Digital Vernier caliper (Draper Tools. Ltd., Chandler's Ford, United Kingdom) and classified into shape by the scheme of Matsuo graph [48]. Diversity in seed morphological characteristics was determined by the Shannon–Weaver index (H') [49], defined as:

$$H' = -\sum_{i=1}^{k} pi \ In \ pi \tag{1}$$

where k is the number of phenotypic classes for a trait and *ln pi* is natural log of the proportion of individuals in the *i*th class of the trait.

2.3. Grain Qualities

Alkali spreading assay was performed to determine the gelatinization temperatures [47,50]. One hundred whole grains of milled rice from each experimental unit were placed in a petri dish, with 20 grains per dish. Subsequently, 10 mL of 1.7% KOH was added to the petri dishes and kept at room temperature for 23 h. After the 23 h incubation, the seeds were evaluated visually and given a score in accordance with the following seven point scale: (1) grain not affected; (2) grain swollen; (3) grain swollen, collar incomplete and narrow; (4) grain swollen, collar complete and wide; (5) grain split or segmented, collar complete and wide; (6) grain dispersed, merging with collar; and (7) grain completely dispersed and intermingled [51]. The alkali spreading value corresponded to the gelatinization temperature as follows: 1–2, high temperature (74–80 °C) and hard when left to cool after cooking; 3–5, intermediate temperature (70–73 °C) and medium hardness when left to cool after cooking; and 6–7, low temperature (<70 °C) and soft when left to cool after cooking [52]. KDML105 (low gelatinization temperature) and RD4 (high gelatinization temperature) were used as checks.

The fragrance status of each cultivars was identified by tasting the milled grain. One milliliter of 1.7% potassium hydroxide (KOH) solution was applied to ten de-husked grains of each sample for 10 min at room temperature. The presence or absence of aroma was scored from 0–3: 0 for non-aromatic, (1) for slightly aromatic, (2) for moderately aromatic, and (3) for strongly aromatic. Each individual sample was inspected multiple times by 10 trained persons to confirm the phenotype (modified from the method of [53]).

2.4. DNA Extraction and Badh2 Sequences

In order to understand the genetic basis of the fragrance in Thai rice landraces, genomic DNA was extracted from leaf samples of each individual landraces using the CTAB method [54]. The DNA of each sample was examined and sequenced for the allele *Badh2/badh2* in exon 7 of *Badh2* gene by using primers Badh2P5 (F: 5′-CCTCCGTGTTAATGCAGCTC-3′, R: 5′-CATAGCAAGTGGCATGTACC-3′) and Badh2P6 (F: 5′-GGTTGGTCTTCCTTCAGGTG-3′, R: 5′-GTCCTTCCTAACTGCCTTCC-3′) [21]. Each 50 µL reaction contained 5 µL 10X PCR buffer, 2.5 µL 2.5mM MgCl2, 0.4 µL 0.2mM Deoxyribonucleotides (dNTP), 5 µL of each primer (10 ng/ µL), 0.4 µL 0.5U Taq DNA polymerase (Thermo scientific), 40 µL dH2O and 5 µL genomic DNA (50 ng/ µL).

The amplification consisted of 94 °C/2 min, followed by 40 cycles of 94 °C/45 s, 50 °C/45 s, and 72 °C/1 min, ending with 72 °C for 5 min as the final extension. Amplified products were genotyped using 1.5% agarose gel electrophoresis. Then, staining was carried out with MaestroSafe[TM] Nucleic Acid Stains (MAESTROGEN, Xiangshan, Hsinchu, Taiwan) and visualized under a UV transilluminator (BioDoc-It[2] imaging systems, Analytik Jena, Upland, CA, USA) before samples were sent for sequencing at Macrogen, Inc. (Seoul, South Korea). The sequence size was 1323 bp, from intron 4 to intron 8 of the *Badh2* gene.

The sequences were assembled using the DNA baser assembler v5.15.0 trial version (Heracle BioSoft S.R.L., Arges, Romania). The resulting contigs were used as BLAST queries using the data of *Badh2/badh2*, which were reported in the GenBank database of National Center for Biotechnology Information (NCBI, Bethesda, Maryland, USA) including the fragrant rice variety, SuYuNuo, and the non-fragrant rice variety, Nanjing11 (accession numbers EU7703020.1 and EU710319.1, respectively). The aligned sequences were imported into MEGA7 software [55] to compare sample sequences. Haplotype data were generated in DNAsp 5.10.01 [56]. The relationship between haplotypes was investigated by constructing median networks using Network version 5.0 [57]

2.5. Statistical Analysis

Analysis of variance was used to determine the significant difference between the different morphological characteristics at $p < 0.05$ by statistical analysis STATISTIX 8.0 (Tallahassee, FL, USA). Least significant difference (LSD) was used to indicate the mean differences in the grain size, weight, and shape between the varieties in regards to grain morphology. *T*-tests were performed between the mean sensory test score of the varieties in each haplotype pool versus the score of each check line (high and low separately).

3. Results

3.1. Seed Morphological Variation of Thai Fragrant Rice Landraces

The seed morphological traits of 22 Thai fragrant rice landrace populations are moderately heterogenous (Table 2 and Figure 1). The variation within populations was found in husk color and seed awning. A variation of husk color was found in BNM1, BNM2, BNM6, BNM7, HB, HBK, and HNK with straw colored husks mixed with brown furrows on straw. The Shannon–Weaver index (H′) varied from 0.991 in HB to 0.098 in BNM2 and BNM6 while the rest of the landraces including BNM4 and BNM-CMU and three check varieties had uniform straw-colored husks. Ten Thai fragrant rice landrace populations, BNM-CMU and PTT1 had varying seed awning, from short awn and partly awn, while the remaining ten rice landrace populations, JPD, KDML105 and SPR1 were awnless. No variation was found within populations in the pericarp color; 26 rice populations had colorless pericarp except NDLP which had a purple pericarp (Table 3 and Table S2).

Table 2. Grain size and shape of paddy rice of 22 fragrant rice landrace accessions, 2 breeding lines, and 3 elite variety checks.

Varieties	Grain Length (mm)	Grain Width (mm)	Grain Thickness (mm)	Grain Shape [#]
BNM1	9.87	3.11	2.18	Large type
BNM2	10.20	3.03	2.10	Large type
BNM3	10.34	3.25	2.21	Large type
BNM4	10.41	3.12	2.15	Large type
BNM5	10.41	3.10	2.19	Large type
BNM6	10.29	3.29	2.21	Large type
BNM7	10.40	3.23	2.19	Large type
BNM8	11.08	3.28	2.18	Large type
BNM9	10.38	3.24	2.14	Large type
EL	11.08	3.50	2.22	Slender type
HS	9.98	2.51	1.79	Large type
KH	11.25	3.91	2.45	Slender type
NDLP	11.07	3.53	2.15	Slender type
NU	10.65	2.62	1.94	Large type
PLD	11.22	4.04	2.51	Slender type
PS	10.72	3.09	2.16	Slender type
SKH	10.79	3.13	1.89	Slender type
BH62	9.23	2.50	1.77	Large type
BH96	9.06	2.40	1.90	Large type
HB	9.08	2.12	1.70	Slender type
HBK	9.91	2.48	1.75	Large type
HNK	9.86	2.64	1.94	Large type
BNM-CMU	10.83	2.96	2.12	Large type
JPD	10.05	3.05	2.09	Large type
KDML105	10.52	2.49	1.88	Slender type
PTT1	10.56	2.49	1.94	Slender type
SPR1	9.60	2.38	2.06	Slender type
Mean	10.33	2.98	2.07	
SD	0.62	0.48	0.20	
CV (%)	6.16	5.77	6.46	
F-test	***	***	***	
LSD $_{(0.05)}$	0.32	0.09	0.07	

CV (%), Coefficient of variation; LSD, least significant difference; ***, Significant at the 1% level; [#] classified with the scheme of Matsuo (1952).

Table 3. Seed morphology of 22 landrace fragrant rice varieties, 2 breeding line varieties, and 3 check varieties.

| Varieties | Husk Color | | Pericarp Traits | Awning | |
	Phenotype	*H'*		Phenotype	*H'*
BNM1	Straw and brown furrows on straw	0.135	Colorless	Short and partly awned	0.135
BNM2	Straw and brown furrows on straw	0.098	Colorless	Short and partly awned	0.227
BNM3	Straw	0	Colorless	Short and partly awned	0.135
BNM4	Straw	0	Colorless	Short and partly awned	0.168
BNM5	Straw	0	Colorless	Short and partly awned	0.456
BNM6	Straw and brown furrows on straw	0.098	Colorless	Absent	0
BNM7	Straw and brown furrows on straw	0.168	Colorless	Short and partly awned	0.325
BNM8	Straw	0	Colorless	Short and partly awned	0.423
BNM9	Straw	0	Colorless	Short and partly awned	0.254
EL	Straw	0	Colorless	Absent	0
HS	Brown furrows on straw	0	Colorless	Absent	0
KH	Straw	0	Colorless	Absent	0
NDLP	Brown furrows on straw	0	Purple	Absent	0
NU	Straw	0	Colorless	Absent	0
PLD	Straw	0	Colorless	Absent	0
PS	Straw	0	Colorless	Short awned	0
SKH	Straw	0	Colorless	Absent	0
BH62	Straw	0	Colorless	Absent	0
BH96	Straw	0	Colorless	Short awned	0
HB	Straw and brown furrows on straw	0.991	Colorless	Short and partly awned	0.683
HBK	Straw and brown furrows on straw	0.637	Colorless	Absent	0
HNK	Straw and brown furrows on straw	0.673	Colorless	Short and partly awned	0.598
BNM-CMU	Straw	0	Colorless	Short and partly awned	0.135
JPD	Black	0	Colorless	Absent	0
KDML105	Straw	0	Colorless	Absent	0
PTT1	Straw	0	Colorless	Short and partly awned	0.199
SPR1	Straw	0	Colorless	Absent	0

Sample = 100 seeds (*n* = 100).

The grain size of the paddy rice was found to vary between the different rice samples collected ($p < 0.05$) (Table 2 and Table S1). The grain length, width, and thickness were found to vary at 10.33 ± 0.62 mm, 2.98 ± 0.48 mm, and 2.07 ± 0.20 mm, respectively. The range of the grain length, width, and thickness of landrace rice were found to be 9.06–11.25 mm, 2.12–4.04 mm, and 1.70–2.51 mm, respectively. The two advanced breeding lines and two elite checks had ranges of 9.06–10.83 mm grain length, 2.38–2.96 mm width, and 1.88–2.12 mm thickness, and classification by the Matsuo (1952) scheme found fifteen of the rice landrace varieties along with the advanced breeding lines BNM-CMU and JPD belong to the large grain type, and the remaining seven as slender type, similar to the elite varieties, KDML105, PTT1 and SPR1 (Figure 1).

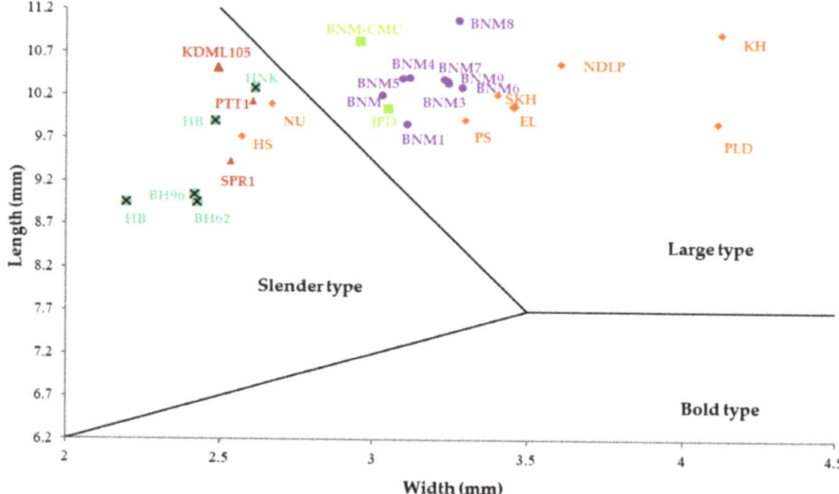

Figure 1. Shape of unhusked grains of 22 fragrant rice landrace varieties, with 2 breeding line varieties, and 3 check varieties, classed with the scheme of Matsuo (1952). Different colors and symbols identify different regions and rice groups: The nine purple circles, northern; the eight orange diamonds, the northeastern; the five blue squares, the southern; the two light green squares, promising landraces; and the three red triangles, elite fragrant varieties.

3.2. Grain Quality

Of the landraces locally recognized as fragrant when grown at Chiang Mai University and rated by a sensory test, KH, NDLP and PLD were found to be strongly aromatic in the same range as KDML105. Those found to be moderately aromatic similar to the elite variety PTT1 were BNM4, BNM-CMU and SKH. BH62, BH96, HS, JPD and PS were slightly aromatic, with no aroma detected in the remaining 43% of the germplasm (Table 4 and Table S4).

Gelatinization temperature assessed by alkali spreading assay was found to be low in four accessions of the landraces (BNM1, BNM2, EL and HB), comparable to KDML105 and PTT1, which remained soft when the cooked rice was allowed to cool to room temperature. Nine accessions (BNM3, BNM4, BNM5, BNM6, BNM7, BNM8, BNM9, JPD, and NU) had intermediate gelatinization temperatures similar to the advanced breeding line BNM-CMU, and the remaining 43% of the germplasm had high gelatinization temperatures similar to SPR1, a standard firm textured rice (Table S3).

Table 4. Nucleotide sequences of *Badh2/badh2* allele on exon 7, sensory test and alkaline test of 22 landrace fragrant rice varieties, 2 breeding line varieties, and 3 check varieties.

Haplotype	Variety	Nucleotide Position								Sensory Test [#]	Gelatinization Temperature
		2415	2901	2903	2905-2912	2913	3233	3386	3482		
Haplotype 1	BNM4	Insertion G	T	T	8 deletion	T	G	C	T	Moderately aromatic	Medium
	KH	Insertion G	T	T	8 deletion	T	G	C	T	Strongly aromatic	High
	NDLP	Insertion G	T	T	8 deletion	T	G	C	T	Strongly aromatic	High
	PLD	Insertion G	T	T	8 deletion	T	G	C	T	Strongly aromatic	High
	BNM-CMU	Insertion G	T	T	8 deletion	T	G	C	T	Moderately aromatic	Medium
	KDML105	Insertion G	T	T	8 deletion	T	G	C	T	Strongly aromatic	Low
	PTT1	Insertion G	T	T	8 deletion	T	G	C	T	Moderately aromatic	Low
	SuYuNuo[δ]	insertion G	T	T	8 deletion	T	G	C	T	*ND*	*ND*
Haplotype 2	BH62	Insertion G	A	A	No deletion	C	G	C	T	Slightly aromatic	High
	BH96	Insertion G	A	A	No deletion	C	G	C	T	Slightly aromatic	High
	HS	Insertion G	A	A	No deletion	C	G	C	T	Slightly aromatic	High
	JPD	Insertion G	A	A	No deletion	C	G	C	T	Slightly aromatic	Medium
	PS	Insertion G	A	A	No deletion	C	G	C	T	Slightly aromatic	High
	SKH	Insertion G	A	A	No deletion	C	G	C	T	Moderately aromatic	High
Haplotype 3	EL	Insertion G	A	A	No deletion	C	A	G	C	Non-aromatic	Low
	HBK	Insertion G	A	A	No deletion	C	A	G	C	Non-aromatic	High
	HNK	Insertion G	A	A	No deletion	C	A	G	C	Non-aromatic	High
	NU	Insertion G	A	A	No deletion	C	A	G	C	Non-aromatic	Medium
Haplotype 4	BNM1	-	A	A	No deletion	C	A	G	C	Non-aromatic	Low
	BNM2	-	A	A	No deletion	C	A	G	C	Non-aromatic	Low
	BNM3	-	A	A	No deletion	C	A	G	C	Non-aromatic	Medium
	BNM5	-	A	A	No deletion	C	A	G	C	Non-aromatic	Medium
	BNM6	-	A	A	No deletion	C	A	G	C	Non-aromatic	Medium
	BNM7	-	A	A	No deletion	C	A	G	C	Non-aromatic	Medium
	BNM8	-	A	A	No deletion	C	A	G	C	Non-aromatic	Medium
	BNM9	-	A	A	No deletion	C	A	G	C	Non-aromatic	Medium
	HB	-	A	A	No deletion	C	A	G	C	Non-aromatic	Low
	SPR1	-	A	A	No deletion	C	A	G	C	Non-aromatic	High
	Nanjing11 [δδ]	-	A	A	No deletion	C	A	G	C	*ND*	*ND*

[#] Sensory test using 10 testers, [δ] fragrant and [δδ] non-fragrant varieties obtained from NCBI accession numbers EU770320.1 and EU710319.1 respectively, *ND* = not determined.

3.3. Badh2 Sequence Analysis

Sequences of the *Badh2* gene in exon 7 of 27 rice varieties were compared with the sequences of the two rice varieties, SuYuNuo and Nanjing1, obtained from the NCBI database, accession numbers EU7703020.1 and EU710319.1, respectively. The 1323 base pair segments of *Badh2* gene, covering from intron 4 to intron 8, revealed 8 base pair deletion in exon 7 and three single nucleotide polymorphisms (SNPs), similar to those found in the *badh2*-E7 allele of the check genotype SuYuNuo. Deletion of *badh2*-E7 was found in the strongly aromatic KH, NDLP, PLD and KDML105 and moderately aromatic BNM4, BNM-CMU, and PTT1.

Four haplotypes were identified in the analysis of the *Badh2* gene in exon 7 sequence (Figure 2). The strongly aromatic KH, NDLP, PLD and moderately aromatic BNM4, together with the advanced breeding line BNM-CMU and the elite varieties KDML105 and PTT1 were identified as haplotype H1. Haplotype H2 comprised the moderately aromatic SKH, and the slightly aromatic BH62, BH96, HS, PS, and JPD, containing non-8 bp deletion and 3 SNP at 2901, 2903 and 2913. Haplotype H3 comprised EL, HBK, HNK and NU, displaying similar SNP to haplotype H2, except at the 3233 sites, which showed the G/A transition and the 3482 sites showing the T/C transition. The last haplotype, H4, contained BNM1, BNM2, BNM3, BNM5, BNM6, BNM7, BNM8, BNM9 and HB, which had no aroma in the sensory test, along with the non-aromatic check, SPR1, containing similar sequences to non-aromatic Nanjing1 (Figure 3).

Figure 2. The haplotype network of the *Badh2* gene. Letters inside the circle correspond to the haplotype name. The numbers on the lines indicate the position of nucleotide substitution polymorphisms. The size of each circle is proportional to the number of individuals varieties possessing that haplotype.

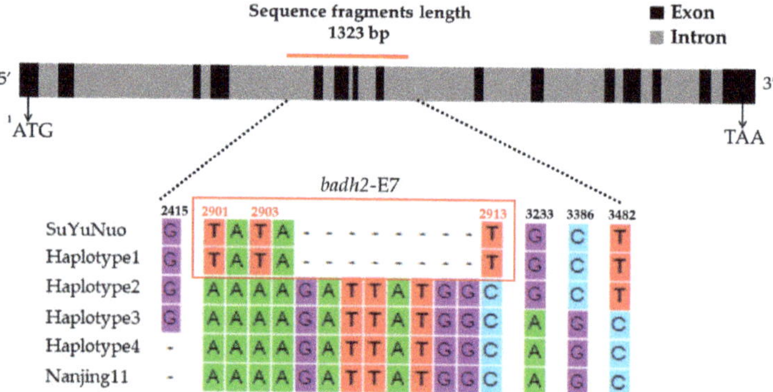

Figure 3. Extended haplotypes. Haplotype analysis of the sequence covering in the vicinity of *badh2*-E7 regions. 2415 site was in intron 4; 3233, 3386 and 3482 sites were in intron 8 of the *Badh2* gene.

Statistical T-test analysis identified significant differences between the sensory test score of the varieties in each haplotype pool versus the score of the high check line (Figure 4a) and low check line (Figure 4b).

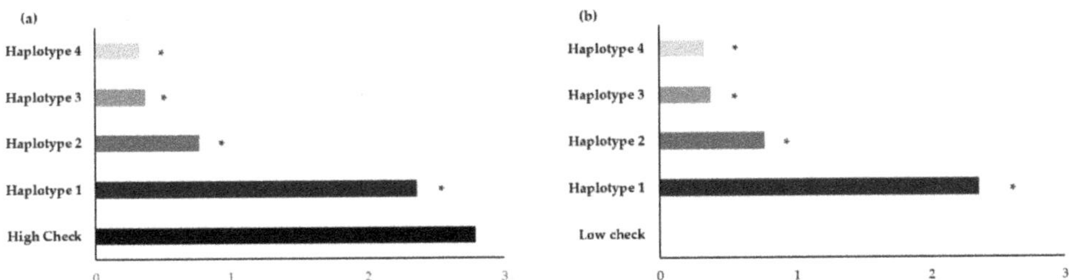

Figure 4. Statistical T-test analysis between the mean sensory test score of the varieties in each haplotype pool versus the score of high check line (**a**) and low check line (**b**); *, significant at the 5% level. Aroma was scored as 0 for non-aromatic, 1 for slightly aromatic, 2 for moderately aromatic, and 3 for strongly aromatic.

4. Discussion

The 22 rice landraces from northern, northeastern, and southern Thailand and two advanced breeding lines recognized locally as fragrant varieties have been found to vary considerably in their grain morphology, cooking quality, aroma and allelic variation of the fragrant gene *Badh2*. The landraces were classified mainly as large grain type, especially those from the highland of the north and northeast, where large grain type rice is preferred, as in the neighboring Lao PDR [58–60]. Diversity of local taste was also indicated in gelatinization temperature variation, ranging from low, in the same range as KDML105 and PTT1; intermediate, similar to the advanced breeding line BNM-CMU; to high as in the elite high yielding, non-aromatic variety SPR1. The premier Thai rice variety KDML105, with its long slender grain, low amylose content and low gelatinization temperature, has become the international standard for aromatic jasmine rice, just as it has long been in Thailand [61]. The grain quality features of KDML105 provide the benchmark for the selection of jasmine type grain quality in Thai breeding programs [31,36], and other areas [62–64], which commonly utilize KDML105 or other genotypes with similar grain quality characteristics as the parent. The rice landraces in this study clearly differentiate from KDML105 in their grain quality. The majority of the accessions belong to the large grain type. The few slender grain accessions either had high gelatinization temperatures or were non-aromatic when grown in Chiang Mai. Similar to the local Indian aromatic rice varieties of small and medium grain that are classed as non-Basmati [2], the local landraces in this study may be classed as non-jasmine type aromatic rice. It should be recognized as a distinctive and valuable set of rice genetic resources utilized and preserved on-farm. Unique among the landraces studied was NDLP, which is a glutinous rice with pigmented pericarp, sold as a premium priced, special quality rice [30]. It has potential health and pharmaceutical applications due to its anthocyanin content, phenolics and flavonoids, and high anti-oxidative properties [65].

The presence of the aromatic *badh2*-E7 allele encoding the production of the compound 2AP [19] confirmed the strong aroma in KH, NDLP and PLD, as in KDML105. However, the presence of *badh2*-E7 in the moderately aromatic BNM-CMU, BNM4 and SKH, along with PTT1 that was previously shown to have a significantly lower 2AP concentration than KDML105 [34], suggests some yet to be identified genotype-specific attenuating factors. Variation in aroma and 2AP content of KDML105 by environment and management in the lowlands, on the other hand, is well established [66–69]. A strong G x E interaction effect on the concentration of 2AP, as well as the aroma by sensory test was found among accessions of Bue Nur Moo (BNM) grown at 330 and 800 m elevations, having no effect on the 2AP content of KDML105 [35]. The absence of aroma in the BNM accessions except for BMN4 in the present study, therefore, suggest a possible environmental effect on the aroma of these landraces from the highlands, and similarly, in the accessions from other regions. The absence of the *badh2*-E7 allele in many of the accessions, however, suggests that local perception of the aroma in rice may be complex and not determined simply by the compound 2AP.

The fragrance of rice is mainly controlled by the major gene, *Badh2*, but expression of the gene, and thus, the concentration of the aromatic compound 2AP is influenced by climatic conditions and crop management [70]. For instance, the soil conditions, nutrients, plant growth regulators, planting density, irrigation draining, harvesting, post-harvest practices and storage temperature all have an effect [71]. However, these rice landraces were all grown together under the same environment and management. The complete lack of aroma in the genotypes with haplotype 4 may either indicate a different local definition of aroma that does not involve the compound 2AP, or a case of misnaming as the haplotype also includes the non-aromatic elite variety SPR1. An as yet unknown regulatory mechanism controlling the expression of the *Badh2* gene and the strength of the aroma is suggested by the variation in the aroma among the genotypes of haplotype 1, including the well-established difference between KDML105 and PTT1 [34]. The occurrence of deletion/insertion at other locations of the *Badh2* gene or the presence of other genes related to aroma characteristics increasing or decreasing of 2AP formation in aromatic rice [10]. This appeared to have taken place with the difference in the nucleotide sequences between haplotype 1 and haplotype 2. However, it is yet to be verified if the absence of the aroma in the genotypes with haplotype 3 is location-specific or not.

In conclusion, the aromatic rice landraces from different regions of Thailand had grain qualities that differentiated them from the elite varieties of KDML105 and PTT1. The majority were large grain with medium to high gelatinization temperatures in contrast to the slender grain with a low gelatinization temperature of the typical jasmine varieties KDML105 and PTT1. Only four of the landraces shared the *badh2*-E7 haplotype with the elite varieties of KDML105 and PTT1, and the previously selected BNM-CMU. The absence of the *badh2*-E7 allele in most of the accessions suggests that the 2AP compound may not be the only determinant of aroma in rice as perceived by local consumers. These locally recognized aromatic rice landraces make up a distinctive set of rice genetic resources preserved on-farm. One with commercial potential is exemplified by the glutinous genotype with pigmented pericarp, NDLP. Therefore, the landraces found to belong to the H1 group will be considered for further evaluation and introduced as genetic resources for a fragrant rice breeding program to develop highly palatable cultivars by pure-line selection or improving variety.

Factors affecting 2AP biosynthesis in fragrant rice will be identified in further work. Not only the fragrant genes, but also the environmental factors and crop management practices [72,73] should be examined to identify genes and develop functional markers for fragrant rice landrace breeding. These *badh2* alleles and functional markers are important for the development and identification of new fragrant rice varieties using marker-assisted selection.

Supplementary Materials: The following are available online at http://www.mdpi.com/2073-4395/10/6/779/s1, Table S1: Mean value and standard errors (3 replications) of the grain length (mm), grain width (mm), and grain thickness (mm) of paddy rice of 22 fragrant rice landrace accessions, 2 advanced breeding lines, and 3 elite variety checks; Table S2: Seed morphology of 22 landraces fragrant rice varieties, 2 breeding line varieties, and 3 check varieties (n=100); Table S3: Alkali spreading test (%) at six levels of 22 landraces fragrant rice varieties, 2 breeding line varieties, and 3 check varieties (n=100) Level; Table S4: Sensory scores by using 10 testers of 22 landraces fragrant rice varieties, 2 breeding line varieties, and 3 check varieties

Author Contributions: Conceptualization, T.P., S.J. and N.Y.; Funding acquisition, T.P., S.J. and B.R.; Investigation, P.C.-i. and T.P.; Methodology, P.C.-i., T.P., N.Y. and S.J.; Project administration, T.P. and S.J.; Supervision, T.P., B.R. and S.J.; Writing—original draft, P.C.-i. and T.P.; Writing—review and editing, P.C.-i., T.P., and B.R. All authors have read and agreed to the published version of the manuscript.

Funding: This research was funded by Thailand Research Fund (Research and Researcher for Industries) grant number PHD62I0031.

Acknowledgments: We are grateful to the farmers for sharing their knowledge and rice seeds. We thank the members of CMUPN lab, Chiang Mai University, for advice throughout this study and Asst Prof. Supranee Sitthiphrom, Dr. Anupong Wongtamee, and Dr. Nantiya Panomjan for collecting the rice seeds. The first author is the recipient of a Research and Researcher for Industries scholarship (PHD62I0031). This research work was partially supported by Chiang Mai University.

Conflicts of Interest: The authors declare no conflict of interest. The funders had no role in the design of the study; in the collection, analyses, or interpretation of data; in the writing of the manuscript, or in the decision to publish the results.

References

1. Sakthivel, K.; Sundaram, R.M.; Shobha Rani, N.; Balachandran, S.M.; Neeraja, C.N. Genetic and Molecular Basis of Fragrance in Rice. *Biotechnol. Adv.* **2009**, *27*, 468–473. [CrossRef] [PubMed]
2. Singh, R.K.; Singh, U.S.; Khush, G.S. *Aromatic Rices*; Oxford and India Book House Publishing Co. Pvt. Ltd.: New Delhi, India, 2000.
3. FAO Rice Price Update. Available online: http://www.fao.org/economic/est/publications/rice-publications/the-fao-rice-price-update/en (accessed on 13 April 2020).
4. TREA. Thai Rice Exporters Association. Available online: http://www.thairiceexporters.or.th/price.htm (accessed on 13 April 2020).
5. Buttery, R.G.; Ling, L.C.; Juliano, B.O.; Turnbaugh, J.G. Cooked Rice Aroma and 2-Acetyl-1-Pyrroline. *J. Agric. Food Chem.* **1983**, *31*, 823–826. [CrossRef]
6. Ahn, S.N.; Bollich, C.N.; Tanksley, S.D. RFLP Tagging of a Gene for Aroma in Rice. *Appl. Genet.* **1992**, *84*, 825–828. [CrossRef]
7. Yi, M.; Nwe, K.T.; Vanavichit, A.; Chai-arree, W.; Toojinda, T. Marker Assisted Backcross Breeding to Improve Cooking Quality Traits in Myanmar Rice Cultivar Manawthukha. *Field Crop. Res.* **2009**, *113*, 178–186. [CrossRef]
8. Fitzgerald, T.L.; Waters, D.L.E.; Brooks, L.O.; Henry, R.J. Fragrance in Rice (*Oryza Sativa*) Is Associated with Reduced Yield under Salt Treatment. *Environ. Exp. Bot.* **2010**, *68*, 292–300. [CrossRef]
9. Bradbury, L.M.T.; Fitzgerald, T.L.; Henry, R.J.; Jin, Q.; Waters, D.L.E. The Gene for Fragrance in Rice. *Plant Biotechnol. J.* **2005**, *3*, 363–370. [CrossRef] [PubMed]
10. Itani, T.; Tamaki, M.; Hayata, Y.; Fushimi, T.; Hashizume, K. Variation of 2-Acetyl-1-Pyrroline Concentration in Aromatic Rice Grains Collected in the Same Region in Japan and Factors Affecting Its Concentration. *Plant Prod. Sci.* **2004**, *7*, 178–183. [CrossRef]
11. Vanavichit, A.; Kamolsukyeunyong, W.; Siangliw, M.; Siangliw, J.L.; Traprab, S.; Ruengphayak, S.; Chaichoompu, E.; Saensuk, C.; Phuvanartnarubal, E.; Toojinda, T.; et al. Thai Hom Mali Rice: Origin and Breeding for Subsistence Rainfed Lowland Rice System. *Rice* **2018**, *11*, 20. [CrossRef]
12. Kovach, M.J.; Calingacion, M.N.; Fitzgerald, M.A.; McCouch, S.R. The Origin and Evolution of Fragrance in Rice (*Oryza Sativa* L.). *Proc. Natl. Acad. Sci. USA* **2009**, *106*, 14444–14449. [CrossRef]
13. Singh, V.; Singh, A.K.; Mohapatra, T.; Ellur, R.K. Pusa Basmati 1121—A Rice Variety with Exceptional Kernel Elongation and Volume Expansion after Cooking. *Rice* **2018**, *11*, 19. [CrossRef]
14. Promotional and Public Relations Activities on Thai Hom Mali Rice Standard. Available online: http://www.dft.go.th/th-th/Service-DFT/Service-Information/DATA-Group-Product/Detail-DATA-Group-Product/ArticleId/7494/dft-Thai-jasmine-rice-22 (accessed on 1 April 2020).
15. Thai Ministry of Commerce. Thailand Standards for Rice. Available online: http://www.dft.go.th/th-th/ShareDocument1/ArticleId/8860/-Thailand-Standards-for-Rice (accessed on 13 April 2020).
16. Shi, W.; Yang, Y.; Chen, S.; Xu, M. Discovery of a New Fragrance Allele and the Development of Functional Markers for the Breeding of Fragrant Rice Varieties. *Mol. Breed.* **2008**, *22*, 185–192. [CrossRef]
17. Sakthivel, K.; Rani, N.S.; Pandey, M.K.; Sivaranjani, A.K.P.; Neeraja, C.N.; Balachandran, S.M.; Madhav, M.S.; Viraktamath, B.C.; Prasad, G.S.V.; Sundaram, R.M. Development of a Simple Functional Marker for Fragrance in Rice and Its Validation in Indian Basmati and Non-Basmati Fragrant Rice Varieties. *Mol. Breed.* **2009**, *24*, 185–190. [CrossRef]
18. Gaur, A.; Wani, S.; Deepika, P.; Bharti, N.; Malav, A.; Shikari, A.; Bhat, A. Understanding the Fragrance in Rice. *Rice Res. Open Access* **2016**, *4*. [CrossRef]
19. Bradbury, L.M.T.; Henry, R.J.; Jin, Q.; Reinke, R.F.; Waters, D.L.E. A Perfect Marker for Fragrance Genotyping in Rice. *Mol. Breed.* **2005**, *16*, 279–283. [CrossRef]

20. Amarawathi, Y.; Singh, R.; Singh, A.K.; Singh, V.P.; Mohapatra, T.; Sharma, T.R.; Singh, N.K. Mapping of Quantitative Trait Loci for Basmati Quality Traits in Rice (*Oryza Sativa* L.). *Mol. Breed. Mol. Breed. New Strat. Plant Improv.* **2008**, *21*, 49–65. [CrossRef]

21. Shao, G.N.; Tang, A.; Tang, S.Q.; Luo, J.; Jiao, G.A.; Wu, J.L.; Hu, P.S. A New Deletion Mutation of Fragrant Gene and the Development of Three Molecular Markers for Fragrance in Rice. *Plant Breed.* **2011**, *130*, 172–176. [CrossRef]

22. Shao, G.; Tang, S.; Chen, M.; Wei, X.; He, J.; Luo, J.; Jiao, G.; Hu, Y.; Xie, L.; Hu, P. Haplotype Variation at *Badh2*, the Gene Determining Fragrance in Rice. *Genomics* **2013**, *101*, 157–162. [CrossRef]

23. Shi, Y.; Zhao, G.; Xu, X.; Li, J. Discovery of a New Fragrance Allele and Development of Functional Markers for Identifying Diverse Fragrant Genotypes in Rice. *Mol. Breed.* **2014**, *33*, 701–708. [CrossRef]

24. Ootsuka, K. Genetic Polymorphisms in Japanese Fragrant Landraces and Novel Fragrant Allele Domesticated in Northern Japan. *Breed. Sci.* **2014**, *64*, 115–124. [CrossRef]

25. He, Q.; Yu, J.; Kim, T.-S.; Cho, Y.-H.; Lee, Y.-S.; Park, Y.-J. Resequencing Reveals Different Domestication Rate for *BADH1* and *BADH2* in Rice (*Oryza Sativa*). *PLoS ONE* **2015**, *10*, e0134801. [CrossRef]

26. Myint, K.M.; Arikit, S.; Wanchana, S.; Yoshihashi, T.; Choowongkomon, K.; Vanavichit, A. A PCR-Based Marker for a Locus Conferring the Aroma in Myanmar Rice (*Oryza Sativa* L.). *Appl. Genet.* **2012**, *125*, 887–896. [CrossRef] [PubMed]

27. Oo, K.S.; Kongjaimun, A.; Khanthong, S.; Yi, M.; Myint, T.T.; Korinsak, S.; Siangliw, J.L.; Myint, K.M.; Vanavichit, A.; Malumpong, C.; et al. Characterization of Myanmar Paw San Hmwe Accessions Using Functional Genetic Markers. *Rice Sci.* **2015**, *22*, 53–64. [CrossRef]

28. Chan, S. Malys Angkor Crowned World's Best Rice. Available online: https://www.khmertimeskh.com/540271/malys-angkor-crowned-worlds-best-rice/ (accessed on 13 April 2020).

29. Bairagi, S.; Mohanty, S.; Custodio, M.C. Consumers' Preferences for Rice Attributes in Cambodia: A Choice Modeling Approach. *J. Agribus. Dev. Emerg. Econ.* **2019**, *9*, 94–108. [CrossRef]

30. Saetan, J. Seed Development and Maturation and Panicle Position on Seed Quality of Upland Rice Cv. Leum Pua. Master's Thesis, Prince of Songkla University, Songkla, Thailand, 2017.

31. TRKB. Rice Variety. Available online: http://www.ricethailand.go.th/rkb3/Varieties.htm (accessed on 13 April 2020).

32. Sadabpod, K.; Tongyonk, L.; Kangsadalampai, K. Effect of Hom Nil Rice and Black Glutinous Rice Extracts during Treatment of Chicken Extract with Sodium Nitrite Using Ames Test. *J. Health Sci. Med. Res.* **2014**, *32*, 139–149.

33. KURDI. Hom-nil, A Black Non-Glutinous Rice with High Nutritional Quality; Kasetsart University Research and Development Institute: KURDI. Available online: https://www3.rdi.ku.ac.th/?p=27452 (accessed on 13 April 2020).

34. Chan-in, P.; Jamjod, S.; Yimyam, N.; Pusadee, T. Diversity of *BADH2* Alleles and Microsatellite Molecules in Highland Fragrant Rice Landraces. *J. Agric.* **2019**, *35*, 23–35.

35. Tejakum, P.; Khumto, S.; Jamjod, S.; Yimyam, N.; Pusadee, T. Yield, Grain Quality and Fragrance of a Highland Fragrant Rice Landrace Variety, Bue Ner Moo. *Khon Kaen Agric. J.* **2019**, *47*, 317–326. [CrossRef]

36. Sarkarung, S.; Somrith, B.; Chitrakorn, S. Aromatic Rice of Thailand. In *Aromatic Rices*; Singh, R.K., Singh, U.S., Khush, G.S., Eds.; Mohan Primlani for Oxford and India Book House Publishing Co. Pvt. Ltd.: New Delhi, India, 2000; pp. 180–183.

37. The Rice Trader: TRT World's Best Rice News. Available online: https://thericetrader.com/news/ (accessed on 13 April 2020).

38. Arunmas, P. Farmers Urge Action after Top Rice Fails to Win Prize. Available online: https://www.bangkokpost.com/business/1796249/farmers-urge-action-after-top-rice-fails-to-win-prize (accessed on 30 December 2019).

39. Harakotr, B.; Prompoh, K.; Boonyuen, S.; Suriharn, B.; Lertrat, K. Variability in Nutraceutical Lipid Content of Selected Rice (*Oryza Sativa* L. Spp. *Indica*) Germplasms. *Agronomy* **2019**, *9*, 823. [CrossRef]

40. Harlan, J.R. Crops and Man. *Syst. Bot.* **1977**, *2*, 227. [CrossRef]

41. Londo, J.P.; Chiang, Y.-C.; Hung, K.-H.; Chiang, T.-Y.; Schaal, B.A. Phylogeography of Asian Wild Rice, *Oryza Rufipogon*, Reveals Multiple Independent Domestications of Cultivated Rice, *Oryza Sativa*. *Proc. Natl. Acad. Sci. USA* **2006**, *103*, 9578–9583. [CrossRef]

42. Pusadee, T.; Jamjod, S.; Chiang, Y.-C.; Rerkasem, B.; Schaal, B.A. Genetic Structure and Isolation by Distance in a Landrace of Thai Rice. *Proc. Natl. Acad. Sci. USA* **2009**, *106*, 13880–13885. [CrossRef]

43. Khush, G.S.; Toenniessen, G.H. *Rice Biotechnology*; International Rice Research Institute: Los Baños, Philippines, 1991.

44. Dowling, N.G.; Greenfield, S.M.; Fischer, K.S. *Sustainability of Rice in the Global Food System*; International Rice Research Institute: Los Baños, Philippines, 1998.

45. Pusadee, T.; Oupkaew, P.; Rerkasem, B.; Jamjod, S.; Schaal, B.A. Natural and Human-mediated Selection in a Landrace of Thai Rice (*Oryza Sativa*). *Ann. Appl. Biol.* **2014**, *165*, 280–292. [CrossRef]

46. Govindaraj, M.; Vetriventhan, M.; Srinivasan, M. Importance of Genetic Diversity Assessment in Crop Plants and Its Recent Advances: An Overview of Its Analytical Perspectives. *Genet. Res. Int.* **2015**, *2015*, e431487. [CrossRef] [PubMed]

47. IBPGR-IRRI Rice Advisory Committee and International Board for Plant Genetic Resources. *Descriptors for Rice, Oryza Sativa L.*; International Rice Research Institute: Los Baños, Philippines, 1980.

48. Matsuo, T. Genecological Studies on Cultivated Rice. *Bull. Natl. Inst. Agric. Sci. Jpn. D* **1952**, *3*, 1–111.

49. Shannon, C.E. A Mathematical Theory of Communication. *Bell Syst. Tech. J.* **1948**, *27*, 379–423. [CrossRef]

50. Bhattacharya, K.R. Gelatinization Temperature of Rice Starch and Its Determination. In *Chemical Aspects of rice Grain Grain Quality*; International Rice Research Institute: Los Baños, Philippines, 1979; pp. 231–250.

51. Mariotti, M.; Fongaro, L.; Catenacci, F. Alkali Spreading Value and Image Analysis. *J. Cereal Sci.* **2010**, *52*, 227–235. [CrossRef]

52. Perez, C.M.; Juliano, B.O. Modification of the Simplified Amylose Test for Milled Rice. *Starch-Stärke* **1978**, *30*, 424–426. [CrossRef]

53. Sood, B.C.; Siddiq, E.A. A Rapid Technique for Scent Determination in Rice. *Indian J. Genet. Plant Breed.* **1978**, *38*, 268–275.

54. Doyle, J.J.; Doyle, J.L. A Rapid DNA Isolation Procedure for Small Quantities of Fresh Leaf Tissue. *Phytochem. Bull.* **1987**, *19*, 11–15.

55. Kumar, S.; Stecher, G.; Tamura, K. MEGA7: Molecular Evolutionary Genetics Analysis Version 7.0 for Bigger Datasets. *Mol. Biol. Evol.* **2016**, *33*, 1870–1874. [CrossRef]

56. Librado, P.; Rozas, J. DnaSP v5: A Software for Comprehensive Analysis of DNA Polymorphism Data. *Bioinformatics* **2009**, *25*, 1451–1452. [CrossRef]

57. Bandelt, H.J.; Forster, P.; Röhl, A. Median-Joining Networks for Inferring Intraspecific Phylogenies. *Mol. Biol. Evol.* **1999**, *16*, 37–48. [CrossRef] [PubMed]

58. Heidhues, F.; Rerkasem, B. *IRRI's Upland Rice Research Follow-up Review*; Giar Science Council: Rome Italy, 2006.

59. Xiongsiyee, V.; Rerkasem, B.; Veeradittakit, J.; Saenchai, C.; Lordkaew, S.; Prom-u-thai, C.T. Variation in Grain Quality of Upland Rice from Luang Prabang Province, Lao PDR. *Rice Sci.* **2018**, *25*, 94–102. [CrossRef]

60. Jamjod, S.; Yimyam, N.; Lordkaew, S.; Prom-u-thai, C.; Rerkasem, B. Characterization of On-Farm Rice Germplasm in an Area of the Crop's Center of Diversity. *Chiang Mai Univ. J. Nat. Sci.* **2017**, *16*, 85–98. [CrossRef]

61. Juliano, B.O.; Duff, B. Rice Grain Quality as an Emerging Priority in National Rice Breeding Programmes. In *Rice Grain Marketing and Quality Issues*; International Rice Research Institute: Manila, Philippines, 1991; pp. 55–64.

62. Bollich, C.N. Release of New Rice Cultivar Jasmine 85 in USA. *Int. Rice Res. Newsl.* **1989**, *14*, 12.

63. Marchetti, M.A.; Bollich, C.N.; Webb, B.D.; Jackson, B.R.; McClung, A.M.; Scott, J.E.; Hung, H.H. Registration of 'Jasmine 85' Rice. *Crop. Sci.* **1998**, *38*, 896. [CrossRef]

64. Sha, X.Y.; Linscombe, S.D.; Jodari, F.; Chu, Q.R.; Groth, D.E.; Blanche, S.B.; Harrell, D.L.; White, L.M.; Oard, J.H.; Chen, M.H.; et al. Registration of 'Jazzman' Aromatic Long-Grain Rice. *J. Plant Regist.* **2011**, *5*, 304–308. [CrossRef]

65. Seekhaw, P.; Mahatheeranont, S.; Sookwong, P.; Luangkamin, S.; Neonplab, A.N.L.; Puangsombat, P. Phytochemical Constituents of Thai Dark Purple Glutinous Rice Bran Extract [Cultivar Luem Pua (*Oryza Sativa L.*)]. *Chiang Mai J. Sci.* **2018**, *45*, 1383–1395.

66. Suwanarit, A.; Kreetapirom, S.; Buranakarn, S.; Varanyanond, W.; Tungtrakul, P.; Somboonpong, S.; Rattapat, S.; Ratanasupa, S.; Romyen, P.; Wattanapayapkul, S.; et al. Effects of nitrogen fertilizer on grain qualities of Khaw Dauk Mali-105 aromatic rice. *J. The Kasetsart (Nat. Sci.)* **1996**, *30*, 458–474.

67. Leesawatwong, M.; Jamjod, S.; Rerkasem, B.; Pinjai, S. Determinants of a Premium-Priced, Special-Quality Rice. *Int. Rice Res. Notes* **2003**, *28*, 33.

68. Yoshihashi, T.; Huong, N.T.T.; Inatomi, H. Precursors of 2-Acetyl-1-Pyrroline, a Potent Flavor Compound of an Aromatic Rice Variety. *J. Agric. Food Chem.* **2002**, *50*, 2001–2004. [CrossRef]

69. Yoshihashi, T.; Nguyen, T.T.H.; Kabaki, N. Area Dependency of 2-Acetyl-1-Pyrroline Content in an Aromatic Rice Variety, Khao Dawk Mali 105. *Jpn. Agric. Res. Q. Jarq* **2004**, *38*, 105–109. [CrossRef]

70. Golam, F.; Norzulaani, K.; Jennifer, A.H.; Subha, B.; Zulqarnain, M.; Osman, M.; Nazia, A.M.; Zulqarnian, M.; Mohammad, O. Evaluation of Kernel Elongation Ratio and Aroma Association in Global Popular Aromatic Rice Cultivars in Tropical Environment. *Afr. J. Agric. Res.* **2010**, *5*, 1515–1522.

71. Wakte, K.; Zanan, R.; Hinge, V.; Khandagale, K.; Nadaf, A.; Henry, R. Thirty-Three Years of 2-Acetyl-1-Pyrroline, a Principal Basmati Aroma Compound in Scented Rice (*Oryza Sativa* L.): A Status Review. *J. Sci. Food Agric.* **2016**. [CrossRef]

72. Gay, F.; Maraval, I.; Roques, S.; Gunata, Z.; Boulanger, R.; Audebert, A.; Mestres, C. Effect of Salinity on Yield and 2-Acetyl-1-Pyrroline Content in the Grains of Three Fragrant Rice Cultivars (*Oryza Sativa* L.) in Camargue (France). *Field Crop. Res.* **2010**, *117*, 154–160. [CrossRef]

73. Li, M.; Ashraf, U.; Tian, H.; Mo, Z.; Pan, S.; Anjum, S.A.; Duan, M.; Tang, X. Manganese-Induced Regulations in Growth, Yield Formation, Quality Characters, Rice Aroma and Enzyme Involved in 2-Acetyl-1-Pyrroline Biosynthesis in Fragrant Rice. *Plant Physiol. Biochem.* **2016**, *103*, 167–175. [CrossRef]

MDPI
St. Alban-Anlage 66
4052 Basel
Switzerland
Tel. +41 61 683 77 34
Fax +41 61 302 89 18
www.mdpi.com

Agronomy Editorial Office
E-mail: agronomy@mdpi.com
www.mdpi.com/journal/agronomy